RNAi Technology

RNAi
Technology

Editors

R.K. Gaur
Mody Institute of Technology and Science
Lakshmangarh, Sikar, India

Yedidya Gafni
Department of Genetics and Vegetable Research
Volcani Center, Bet Dagan, Israel

P. Sharma
Directorate of Wheat Research, Karnal, India

V.K. Gupta
National University of Ireland, Galway, Ireland

CRC Press
Taylor & Francis Group
Boca Raton London New York

CRC Press is an imprint of the
Taylor & Francis Group, an **informa** business
A SCIENCE PUBLISHERS BOOK

CRC Press
Taylor & Francis Group
6000 Broken Sound Parkway NW, Suite 300
Boca Raton, FL 33487-2742

First issued in paperback 2017

© 2011 by Taylor & Francis Group, LLC
CRC Press is an imprint of Taylor & Francis Group, an Informa business

ISBN 13: 978-1-138-11434-0 (pbk)
ISBN 13: 978-1-57808-716-7 (hbk)

Library of Congress Cataloging-in-Publication Data

RNAi technology / editors, R.K. Gaur ... [et al.]. -- 1st ed.
    p. ; cm.
  Includes bibliographical references and index.
 ISBN 978-1-57808-716-7 (hardback)
1. Small interfering RNA. 2. Gene silencing. I. Gaur, Rajarshi
   Kumar.
   [DNLM: 1. RNA Interference. 2. Biomedical Technology. QU 475]
   QP623.5.S63I76 2011
   572.8′8--dc22
                                                        2011000003

Foreword

RNA interference (RNAi) is a small-RNA-based molecular mechanism which can regulate genes in a sequence-dependent manner. RNAi functions in most multicellular eukaryotes to temporally and spatially control endogenous gene expression during growth and development. RNAi is also an important defense mechanism against invading and parasitic nucleic acids (i.e., viruses, *Agrobacterium* T-DNA and transposons). RNAi functions through the interaction of a small single-stranded (ss) RNA molecule and an RNA-induced silencing complex (RISC). Different RISCs contain one of a set of Argonaute (AGO) proteins which defines their activity in transcriptional or post-transcriptional gene silencing (TGS and PTGS, respectively). TGS or RNA-induced transcriptional silencing (RITS) functions through DNA methylation and chromatin remodeling to regulate the transcription of a sequence complementary to the AGO-bound small RNA. TGS can be maintained by the cell and can be heritable. PTGS, on the other hand, affects existing RNA transcripts and can function via two different mechanisms: specific RNA transcript cleavage or translational inhibition. The former is the main, but not exclusive, mode of action in plants and the latter is predominant in animals.

There are two main types of small RNA molecules which can "program" the RISC, based on their biosynthetic origin: micro RNA (miRNA) and small-interfering RNA (siRNA). miRNA precursors are transcribed by RNA polymerase II from non-coding genomic regions. These ssRNAs fold into a characteristic hairpin which is processed by a specific double-stranded (ds) RNase of the Dicer family into a dsRNA duplex. This duplex contains a single, sequence-defined miRNA and a passenger-strand miRNA* which is subsequently degraded. In contrast, siRNAs can be produced from any perfectly complementary dsRNA, usually *in vivo* by an RNA-dependent RNA polymerase (RdRP). This dsRNA is cleaved by a Dicer into a series of short duplexes, in either a random or synchronized manner. Although siRNAs are produced from the whole length of the dsRNA, the terminal nucleotides and thermodynamic properties of the duplex dictate which molecules, from either strand, are functional, and which are degraded. These miRNAs or siRNAs (which have been first separated into single-stranded molecules) are incorporated into the RISC machinery. In PTGS, the small RNA molecule in RISC hybridizes with a target ssRNA, such as a messenger RNA (mRNA) or a viral ssRNA, leading to its cleavage or translational repression. Interestingly, short RNA molecules can also lead to activation of gene expression, most likely via histone demethylation, in a process that is generally referred to as dsRNA-induced gene activation (RNAa). In some cases, short RNA molecules travel from cell to cell and even between different organs, spreading the silencing and regulation signal. This is utilized by plants, for example, to combat a virus' systemic spread.

The discovery of the importance of RNAi for gene regulation and defense led to the idea that this pathway can be used for functional gene analysis. Indeed, several RNAi-mediated PTGS strategies have been developed and are currently being used for the induction of gene knockdown in various model organisms and cell types. Perhaps the simplest method of silencing a given target gene is to introduce a matching dsRNA molecule into the target cell and rely on the cell's Dicer to initiate the formation of siRNA. siRNA can also be directly delivered into the target cell or be expressed from transfected plasmids or from DNA which has been stably integrated into the chromosome of the target organism. It is interesting to note the unique methods which have been developed for the delivery of dsRNA into *Caenorhabditis elegans*. In this system, worms are fed with *Escherichia coli* bacteria expressing the dsRNA molecules and these molecules are delivered to target cells from the worm's intestinal tract. Also worth noting is the potential of RNAi technology for plant biotechnology and medicine. Effort has been invested in producing crop plants with reduced allergens and toxins by suppressing specific metabolic pathways, and studies in cell cultures have revealed the potential of RNAi technology for treating cancer via the suppression of tumor-formation-related genes. Thus, a better understanding of the RNAi machinery will not only contribute to our understanding of the basic cellular processes governing cell and organism growth and development, but will also yield diagnostic assays for diseases, novel tools for basic cell research, and new agricultural products and bio-pharmaceuticals.

Dr. Yoel Shiboleth
Prof. Tzvi Tzfira
University of Michigan
Ann Arbor, MI, USA

Preface

RNA interference (RNAi) is a system within living cells that helps to regulate which genes need to be active at a point of time, and their level of activity. Two types of small RNA molecules—microRNA (miRNA) and small interfering RNA (siRNA)—are central to RNA interference. The discovery of RNAi was preceded first by observations of transcriptional inhibition by antisense RNA expressed in transgenic plants and more directly by reports of unexpected outcomes in experiments performed by plant scientists in USA and The Netherlands in the early 1990s. RNAi technology are used for large-scale screens that systematically shut down each gene in the cell, which can help identify the components necessary for a particular cellular process or an event such as cell division. Exploitation of the pathway is also a promising tool in biotechnology and medicine. We are including different approaches of gene silencing in a single book. The advent of RNA interference (RNAi) a decade ago has greatly enhanced the ease of studying gene function allowing us to develop hitherto unimaginable tools. As an example, it is today possible to use high-trough put platforms to perform multiparametric analysis of gene function after genome-wide RNAi in invertebrate and mammalian cells, which may soon become a standard approach for basic research and identification of possible therapeutic targets *in vitro*.

This book reviews the selective and robust effect of RNAi on gene expression which makes it a valuable research tool, both in cell culture and in living organisms because synthetic dsRNA introduced into cells can induce suppression of specific genes of interest. The book will introduce new technology in the study of RNA interference going on microorganisms, plant and animal.

Editors

Contents

List of Contributors

A. O'Donovan: School of Natural Sciences, National University of Ireland (NUI), University Road, Galway, Ireland. email: odonovan-anthonia@hotmail.com

A. Manimaran: School of Natural Sciences, National University of Ireland (NUI), University Road, Galway, Ireland. email: manimaran.ayyachamy@nuigalway.ie

Abhijit Datta: Department of Biotechnology (DBT)–Centre for Bioinformatics, Presidency College, 86/1, College Street, Kolkata, West Bengal 700073, India. email: abhijit_datta21@yahoo.com

Allison C. Mallory: Institut Jean-Pierre Bourgin, Institut National de la Recherche Agronomique (INRA), 78026 Versailles Cedex, France. email: Allison.Mallory@versailles.inra.fr

Assaf Eybishtz: The Otto Warburg Minerva Center for Agricultural Biotechnology and the Institute of Plant Science and Genetics in Agriculture, The Hebrew University of Jerusalem, Rehovot 76100, Israel.

Athar Ali: Centre for Transgenic Plant Development, Dept. of Biotechnology, Jamia Hamdard, New Delhi 110062, India.

B.M. Bashyal: Division of Plant Pathology, Indian Agricultural Research Institute, Pusa Campus, New Delhi 110012, India. email: bishnumayabashyal@gmail.com

Cláudia Gasque Schoof: Department of Genetic and Evolutionary Biology, University of Sao Paulo, Rua do Matão, 277, 05508-090, Sao Paulo, Brazil.

Dagan Sade: The Otto Warburg Minerva Center for Agricultural Biotechnology and the Institute of Plant Science and Genetics in Agriculture, The Hebrew University of Jerusalem, Rehovot 76100, Israel.

Duncan Ayers: Center for Medical Genetics Ghent, Ghent University Hospital, 185, 9000 Ghent Belgium, Belgium. email: Duncan.Ayers@ UGent.be

Eder Leite da Silva Botelho: Department of Biochemistry, University Federal of Sao Paulo, Rua Mirassol, 207, Vila Clementino, 04044-020, Sao Paulo, Brazil.

Favi Vidavski: The Otto Warburg Minerva Center for Agricultural Biotechnology and the Institute of Plant Science and Genetics in Agriculture, The Hebrew University of Jerusalem, Rehovot 76100, Israel. email: vidavaski@agri.huji.ac.il.

Federico Calegari: DFG-Research Center and Cluster of Excellence for Regenerative Therapies Dresden, Medical Faculty, Technische Universität Dresden, Tatzberg 47/49, 01307, Dresden, Germany. email: federico.calegari@crt-dresden.de

Henryk Czosnek: The Otto Warburg Minerva Center for Agricultural Biotechnology and the Institute of Plant Science and Genetics in Agriculture, The Hebrew University of Jerusalem, Rehovot 76100, Israel. email: czosnek@agri.huji.ac.il

Hervé Vaucheret: Institut Jean-Pierre Bourgin, Institut National de la Recherche Agronomique (INRA), 78026 Versailles Cedex, France. email: herve.vaucheret@versailles.inra.fr

Hila Beeri: The Otto Warburg Minerva Center for Agricultural Biotechnology and the Institute of Plant Science and Genetics in Agriculture, The Hebrew University of Jerusalem, Rehovot 76100, Israel.

Imran Amin: Agricultural Biotechnology Division, National Institute for Biotechnology and Genetic Engineering, PO Box 577, Jhang Road, Faisalabad, Punjab 38000, Pakistan. email: imran@nibge.org

Indranil Dasgupta: Department of Plant Molecular Biology, University of Delhi, South Campus Benito Juarez Road, New Delhi 110021, India. email: indranil58@yahoo.co.in, indasgup@south.du.ac.in

Iris Sobol: The Otto Warburg Minerva Center for Agricultural Biotechnology and the Institute of Plant Science and Genetics in Agriculture, The Hebrew University of Jerusalem, Rehovot 76100, Israel.

K.M. Turner: School of Natural Sciences, National University of Ireland (NUI), University Road, Galway, Ireland. email: km.turner@ nuigalway.ie

Luciana dos Reis Vasques: Department of Biochemistry, University Federal of Sao Paulo, Rua Mirassol, 207, Vila Clementino, 04044-020, Sao Paulo, Brazil. email: lrvasques@gmail.com

M. Tuohy: School of Natural Sciences, National University of Ireland (NUI), University Road, Galway, Ireland. email: maria.tuohy nuigalway.ie

M.Z. Abdin: Centre for Transgenic Plant Development, Dept. of Biotechnology, Jamia Hamdard, New Delhi 110062, India. email: mzabdin@rediffmail.com

Masato Ikegami: NODAI Research Institute, Tokyo University of Agriculture, 1-1-1 Sakuragaoka, Setagaya-ku, Tokyo 156-8502, Japan. email: m3ikegam@nodai.ac.jp

Mather Ali Khan: Centre for Transgenic Plant Development, Dept. of Biotechnology, Jamia Hamdard, New Delhi 110062, India.

Muhammad Saeed: Agricultural Biotechnology Division, National Institute for Biotechnology and Genetic Engineering, PO Box 577, Jhang Road, Faisalabad, Punjab 38000, Pakistan. email: saeed_hafeez@yahoo.com

Nandini Verma: University School of Biotechnology, Guru Gobind Singh Indraprastha University Room # AFR108, Dwarka, Sector-16 C, New Delhi 110 403, India. email: nandini22dec@gmail.com

P.K. Jain: Dept. of Science, Faculty of Arts, Science and Commerce, Mody Institute of Technology and Science, Lakshmangarh, Sikar 332311, India. email: pankajbiotech2001@yahoo.com

Pradeep Sharma: Division of Crop Improvement, Directorate of Wheat Research, 132001 Karnal, India. email: neprads@gmail.com

Pranjal Yadava: International Centre for Genetic Engineering and Biotechnology, ICGEB Campus, Aruna Asaf Ali Marg, New Delhi, 110067. email: pranjal@icgeb.res.in

Pravej Alam: Centre for Transgenic Plant Development, Dept. of Biotechnology, Jamia Hamdard, New Delhi 110062, India. email: alamprez@gmail.com

R.K. Gaur: Department of Science, Faculty of Arts, Science and Commerce, Mody Institute of Technology and Science, Lakshmangarh, Sikar 332311, India. email: gaurrajarshi@hotmail.com

Rakha H. Das: Guru Gobind Singh Indraprastha University Room # AFR108, Dwarka, Sector-16 C, New Delhi 110 403, India. email: rakhahdas@yahoo.com

Rajender Singh: Division of Crop Improvement, Directorate of Wheat Research, 132001, Karnal, India. email: rajenderkhokhar@yahoo.com

Rashmi Aggarwal: Division of Plant Pathology, Indian Agricultural Research Institute, Pusa Campus, New Delhi 110012, India. email: rashmi.aggarwal2@gmail.com

Rena Gorovits: The Otto Warburg Minerva Center for Agricultural Biotechnology and the Institute of Plant Science and Genetics in Agriculture, The Hebrew University of Jerusalem, Rehovot 76100, Israel.

Rob W. Briddon: Agricultural Biotechnology Division, National Institute for Biotechnology and Genetic Engineering, PO Box 577, Jhang Road, Faisalabad, Punjab 38000, Pakistan. email: rob.briddon@gmail.com

S. Abdolhamid Angaj: Department of Biology, Tarbiat Moallem University, Tehran 31979-37551, Iran. email: ershad110@yahoo.com

Sakshi Issar: Department of Science, Faculty of Arts, Science and Commerce, Mody Institute of Technology and Science, Lakshmangarh, Sikar 332311, India. email: sakshi.issar@gmail.com

Sayak Ganguli: Department of Biotechnology (DBT)–Centre for Bioinformatics, Presidency College, 86/1, College Street, Kolkata, West Bengal 700073, India. email: sayakbif@yahoo.com

Shahid Mansoor: Agricultural Biotechnology Division, National Institute for Biotechnology and Genetic Engineering, PO Box 577, Jhang Road, Faisalabad, Punjab 38000, Pakistan. email: smansoor@nibge.org

Somayeh Darvishani: Department of Biology, Tarbiat Moallem University, Tehran 31979-37551, Iran.

Tatiana V. Komarova: A.N. Belozersky Institute of Physico-Chemical Biology, Moscow State University and N.I. Vavilov Institute of General Genetics, Russian Academy of Science, Moscow 119992, Russia. email: komarova@genebee.msu.su

Tatsuya Kon: Department of Plant Pathology, University of California, One Shields Ave., Davis, CA 95616-8751, USA. email: tkon@ucdavis.edu

V.K. Gupta: Department of Science, Faculty of Arts, Science and Commerce, Mody Institute of Technology and Science, Lakshmangarh, Sikar 332311, India. email: vizaifzd@gmail.com

Vincent Jauvion: Institut Jean-Pierre Bourgin, Institut National de la Recherche Agronomique (INRA), 78026 Versailles Cedex, France.

Vitaly Citovsky: Department of Biochemistry and Cell Biology, State University of New York, Life Science Building, Stony Brook, NY 11794-5215, USA. email: vitaly.citovsky@stonybrook.edu

Xiaobin Luo: Department of Medicine, Montreal Heart Institute, University of Montreal, 5000 Belanger East, Montreal, PQ H1T 1C8 Canada.

Yuri L. Dorokhov: A.N. Belozersky Institute of Physico-Chemical Biology, Moscow State University and N.I. Vavilov Institute of General Genetics, Russian Academy of Science, Moscow 119992, Russia. email: dorokhov@genebee.msu.su

Zhiguo Wang: Department of Medicine, Montreal Heart Institute, University of Montreal, 5000 Belanger East, Montreal, PQ H1T 1C8 Canada. email: wangz.email@gmail

CHAPTER 1

Acute RNA Interference for Basic Research and Therapy

Federico Calegari

ABSTRACT

Manipulation of gene expression in living organisms is of paramount importance for basic research and development of novel therapeutic approaches. In mammals, several systems have been developed that allow the tissue-specific control of gene expression in genetically manipulated animals, most typically mouse. While very powerful for basic research, generation of transgenic organisms is time-consuming and, clearly, cannot be used for therapy. Recently, the advent of RNA interference (RNAi) has provided us with the means to overcome these limitations allowing the silencing of gene expression to be performed acutely in wildtype organisms. However, use of RNAi *in vivo* is still hampered by intrinsic difficulties in delivering small interfering RNAs (siRNA) to the desired tissue, which calls for the development of new technologies for efficient targeting of the relevant cell types. In this chapter, I will discuss the various approaches that have been used to induce acute and tissue-specific RNAi during mammalian embryonic development or adulthood and report on the first attempts to use these promising tools in therapy.

Keywords: Cell targeting, tissue-specific RNAi, therapy.

INTRODUCTION

The advent of RNA interference (RNAi) a decade ago (Fire et al. 1998) has greatly increased the ease of studying gene function allowing us to develop hitherto unimaginable tools. As one example, it is today possible to use high-trough put platforms to perform multiparametric analysis of gene function after genome-wide RNAi in invertebrate (Fraser et al. 2000, Gonczy et al. 2000) and mammalian (Collinet et al. 2010, Kittler et al. 2004)

* *Corresponding author e-mail*: federico.calegari@crt-dresden.de

cells, which may soon become a standard approach for basic research and identification of possible therapeutic targets *in vitro*.

As over the years more sophisticated studies have expanded and diversified the use of RNAi *in vitro*, a major limitation still remains, which is the difficulty to apply these tools *in vivo*. In fact, while targeting cells on a petri dish is relatively easy, delivering siRNAs to the desired tissue and cell type in alive multicellular organisms is not.

In addition to plants (Napoli et al. 1990, Hamilton et al. 1999), certain animal species, such as *C. elegans*, revealed easy to target and it was in fact in this nematode that RNAi was first observed (Guo et al. 1995) and explained (Fire et al. 1998). Yet, superficially similar species, such as planarians (the similarity is superficial indeed because planarians are platyhelminthes and not nematodes) are to date totally resistant to any sort of genetic manipulation.

The ease of performing RNAi in *C. elegans* is based on the possibility to inject double stranded RNAs (dsRNA) of 50 or more bases in length in the parental germ cells, which triggers RNAi for several cell divisions in the resulting embryos (Fire et al. 1998). Additionally, worms can simply be fed with bacteria expressing dsRNAs, which triggers their internalization and RNAi in animal cells by yet undefined mechanisms (Fraser et al. 2000, Timmons et al. 1998). Similarly, germ line injection with dsRNAs was performed in flies (Kennerdell et al. 1998) and the same strategy was then used to target the single-cell mouse zygote (Wianny et al. 2000). Finally, *D. melanogaster* was the first organism in which systemic RNAi has been achieved after generation of transgenic lines expressing a short hairpin RNA (shRNA), i.e., a RNA with a partial, self-complementary sequence that after the formation of a loop and annealing creates a structure functionally similar to a dsRNA (Kennerdell et al. 1998). While efficacy of RNAi in amphibians and fish has always been controversial, great efforts have been invested towards establishing platforms that allow the acute and tissue-specific silencing of gene expression in mammals, whose potential for basic research and, in particular, therapy was immediately recognized. As one major problem for translating the use of RNAi from invertebrates to mammals, long dsRNAs in mammalian cells were found to trigger interferone response with overall unspecific shutdown of gene expression and cell death. Thus, an essential step was the realization that 20–22 nucleotides-long dsRNAs, rather than the much longer sequences previously used in invertebrates, do not trigger interferone response (Elbashir et al. 2001).

Several means to generate short dsRNA, referred to as short interfering RNA (siRNA), are commonly used, each with its specific advantages and limitations. For instance, siRNA can be chemically synthesized or prepared by (i) *in vitro* transcription from a DNA constructs followed by annealing of the resulting complementary RNA fragments (Yu et al. 2001), (ii) enzymatic digestion of a long dsRNA (esiRNA) (Yang et al. 2002, Myers et al. 2003, Kawasaki et al. 2003), or, as already mentioned, (iii) transcription from a DNA construct resulting in a shRNA (Brummelkamp et al. 2002, Paddison et al. 2002, Paul et al. 2002) (Figure 1). Much attention was also given to the criteria used for choosing a nucleotide sequence that should maximize specificity of RNAi while minimizing off target effects and many freeware algorithms are available to address this issue (Tilesi et al. 2009, Boutros et al. 2008). Finally, various authors have also suggested approaches that should be considered to validate the specificity of a RNAi-induced phenotype (Cullen 2006, Sarov et al. 2005).

Importantly, in parallel to the advent of RNAi, the scientific community has discovered a whole new mechanism through which cells physiologically regulate gene expression at the post-transcriptional level: the world of micro-RNAs. Discussing the role and physiology of micro-RNAs is beyond the purpose of this chapter, which exclusively focuses on techniques used by scientists to perform RNAi. Suffices to say that micro-RNAs are, in a snapshot, the physiologically-expressed equivalent of the artificially introduced siRNAs, which modulate translation and stability of a multitude of endogenous mRNAs (Inui et al. 2010, Chekulaeva et al. 2009, Winter et al. 2009).

However, despite remarkable progresses, the main limitation of RNAi technology remains tissue targeting i.e., the problem of delivering the relevant siRNAs to the relevant cell type. As we shall see, this problem has been addressed and partly solved, by many clever approaches.

EMBRYONIC DEVELOPMENT

Concomitant with the two first reports of RNAi in adult mice (discussed below) (McCaffrey et al. 2002, Lewis et al. 2002), a third study established a system to perform tissue-specific RNAi in post-implantation mouse embryos (Calegari et al. 2002), which could be achieved thank to the advent of a technology recently established by developmental biologists: the system of *in vivo* (or *in utero*) electroporation.

A few years before, chick embryologists have found that square-shaped electric pulses could be used to transfer DNA in animal cells with minimal tissue damage, a technique defined *in vivo* electroporation (Muramatsu et al.1997). In essence, the electric pulses create transitory pores in the plasma membrane of cells thereby allowing the migration of molecules into their cytoplasm. In particular for nucleic acids, which are negatively charged, electroporation provides not only the means to

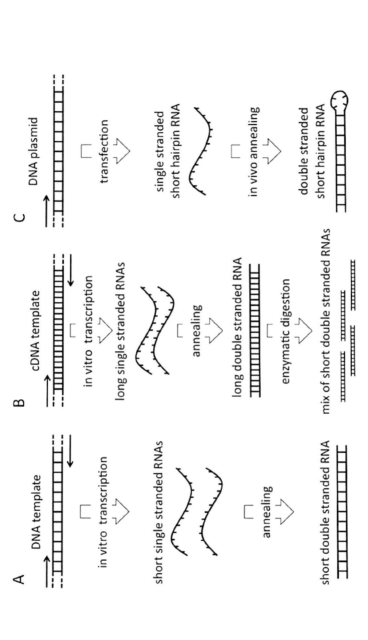

Figure 1. Methods for preparing siRNAs **(A)** Synthesis of a siRNA by *in vitro* transcription of a DNA fragment encoding for a 20–22 base-long sequence complementary to the target mRNA followed by annealing of the resulting short single-stranded RNAs. **(B)** Similar to A, *in vitro* transcription can be performed on a long cDNA encoding for the entire coding sequence of the target gene. Preparation of a mixture of siRNAs is then performed by enzymatic digestion of the long dsRNA. These siRNAs are also referred to as enzymatically-prepared siRNA, or esiRNAs. **(C)** Synthesis of a shRNA from a transfected DNA upon transcription of a RNA with a partial self-complementary sequence that triggers the formation of a loop.

"porate" cells, but also to trigger their migration by the directionality of the electric field, which is controlled positioning the electrodes in the appropriate anode-cathode orientation (Figure 2). This clever approach allows the tissue-specific manipulation of genes in living vertebrates with unprecedented speed and efficiency without the time-consuming generation of genetically manipulated organisms, which in many species has not yet been achieved.

Originally, this techniques has been established in the central nervous system of the chick embryo due to the ease of injecting solutions in the, relatively big, ventricular cavity and the possibility to access and manipulate avian embryos after opening a hole though the shell egg. Nevertheless, this technique was soon used to overexpress genes in the neural tube of mouse embryos kept, for one or two days, in a culture system reproducing *in utero* development (Calegari et al. 2002, Akamatsu et al. 1999, Osumi et al. 2001) (Figure 2). Even more remarkably, *in vivo* electroporation was also performed by injecting DNA through the uterine walls of anaesthetized pregnant mice into the lumen of the embryonic neural tube (Fukuchi-Shimogori et al. 2001, Saito et al. 2001, Tabata et al. 2001), thus, overcoming the two day limit of embryo culture systems. Moreover, even if the neural tube is the easier organ to target, *in vivo* electroporation was soon reported in many other tissues including skin, gut, muscle, heart, lung and retina of both embryos and adult organisms of several species, including humans (Calegari et al. 2004, Barnabe-Heider et al. 2008, LoTurco et al. 2009, Isaka et al. 2007).

As a fortunate coincidence, the finding that small siRNAs could be used in mammalian cells to trigger RNAi *in vitro* (Elbashir et al. 2001) temporally coincided with the development of *in vivo* electroporation as a means to deliver nucleic acids to developing mammals (Osumi et al. 2001, Fukuchi-Shimogori et al. 2001, Saito and Nakatsuji 2001, Tabata et al. 2001). This led to investigate the possibility to combine the two systems in order to achieve tissue-specific silencing of gene expression in developing mouse embryos.

In the first report (Calegari et al. 2002), a cocktail of RNaseIII-digested esiRNAs against GFP were injected in transgenic mouse embryos expressing GFP under the control of an endogenous promoter and electroporation performed to target neural stem cell of the lateral cortex, which triggered an almost complete reduction (ca. 90%) of endogenous gene expression as compared to the control controlateral cortex (Calegari et al. 2002). A similar approach was then used to knock-down various endogenous genes resulting in a number of publications elucidating important cell biological and genetic aspects of brain development. As one example, electroporation was the tool of choice to silence the expression

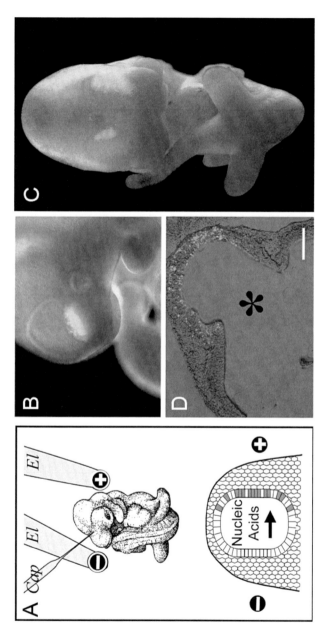

Figure 2. *In vivo* **electroporation of postimplantation mouse embryos (A)** (top) Drawing of a mouse embryo showing the injection of nucleic acids into the brain and positioning of the electrodes to perform electroporation, which triggers (bottom) the migration of nucleic acids into cells towards the anode (green). **(B** and **C)** Lateral (B) and frontal (C) views of a mouse embryo after electroporation with plasmids encoding for GFP and 24 hours of development in whole-embryo culture. **(D)** coronal section of a mouse embryo treated as in B–C showing individual neural stem cells of the midbrain expressing GFP (green).

Figure adapted with permission from Calegari et al., *Proc. Natl. Acad. Sci USA.* **99:**14236–39.

(Color image of this figure appears in the color plate section at the end of the book.)

of cell cycle regulators in neural stem cells revealing a novel mechanism of somatic stem cell differentiation by their cell cycle length (Lange et al. 2009, Salomoni et al. 2010, Lange et al. 2010). In addition, a number of studies have greatly contributed to revealing many aspects of mouse development by combining electroporation and RNAi, including the use of this tool for detecting micro-RNA activity (De Pietri Tonelli et al. 2006), a total of over 100 publications between its first use in 2002 (Calegari et al. 2002) and 2010 (Calegari et al. 2004, LoTurco et al. 2009, Matsuda et al. 2008).

It has to be mentioned that despite several advantages electroporation also has intrinsic limitations. In particular, transfection efficiency is rarely higher than 10–50% of cells in a given portion of tissue because to increase this efficiency higher voltages should be used, which induces cell death and tissue damage. In addition, electroporation is a transient expression system in which half-life of the ectopically introduced nucleic acids or their dilution due to cell division may limit the duration of the desired effect.

In part to overcome these limitations, a number of alternative were investigated to overexpress or knock-down genes in developing or postnatal mice after acute delivery of DNA or siRNA constructs. In particular, virus-mediated delivery of nucleic acids allows the integration of ectopic DNA in the host genome leading to constitutive expression of viral genes (Dunn et al. 2000), which can be used for the expression of shRNAs. This advantage is somehow compensated by a more laborious viral preparation, stricter limitations in DNA size due to encapsulation in the viral particle and a longer time required for the onset of transgene expression, which is in the order of days as compared to few hours needed after electroporation. While very effective and widely used in adult organisms, viral-mediated manipulation of gene expression in developing mouse embryos revealed less efficient and more troublesome than electroporation and is today only considered for very specific applications.

Another approach originally established in cell culture, lipofection, which relies on the fusion with the plasma membrane of liposomes containing nucleic acids in their lumen (or lipoplexes as heterogeneous association of lipids and nucleic acids), has also been applied to the mouse embryo (Hasssani et al. 2005) with, in particular for the vascular endothelium, remarkable success (Bollerot et al. 2006). Importantly, this approach is not dependent on the directionality of the electric field and, thus, can potentially target more cells than electroporation. Since efficacy of lipofection in the vascular endothelium appears to be superior to electroporation (Bollerot et al. 2006), it would be interesting to compare the performance of this approach in other tissues. Perhaps as one limitation of lipofection is the relatively high costs connected to the synthesis of the liposome compounds and their potential toxicity under certain conditions.

Although many other approaches revealed efficient in cell culture, such as conjugation to peptide fragments of ballistic gene delivery, no other system than the ones described above have, to my knowledge, been reported for triggering RNAi mouse embryos. In conclusion, electroporation is currently by far the most used tool for acute manipulation of gene expression in mouse embryos due to its versatility, simplicity and efficacy as demonstrated by the increase in the numbers of reports using this tool in the last decade.

ADULTHOOD

Due to the high potential of RNAi for therapy, greater efforts have been invested towards establishing systems of siRNA delivery in adult organisms.

As already mentioned, electroporation is widely used also for adult tissues (Barnabe-Heider et al. 2008, Isaka et al. 2007). However, when applied to adult organs, a limitation of electroporation arises that is the difficulty to diffuse sufficient amounts of nucleic acids to relatively big areas of tissue. For example, injection of DNA or siRNA is easy in organs containing cavities such as the neural tube, blood vessels, heart and lung, but efficiency of electroporation is dramatically reduced in tissues lacking cavities, such as the liver, skin and muscle, which is due to the fact that nucleic acids do not diffuse well in solid tissues. While this is not a major limitation for basic research, which can be performed on a limited number of cells, therapeutic applications using electroporation are unlikely to be effective in relatively big organs. Thus, the need to investigate alternative approaches.

The first use of RNAi in adult mouse was achieved independently by two laboratories by injecting in the blood stream a large volume of a physiological solution containing siRNAs, thereby systemically delivering the nucleic acids to various organs (McCaffrey et al. 2002, Lewis et al. 2002). In these studies, a plasmid for a reporter or viral gene was coinjected with siRNAs revealing specific RNAi against the target gene when a complementary, but not unrelated, siRNA was used (McCaffrey et al. 2002, Lewis et al 2002). These reports were important as a proof of principle for the use of acute RNAi in adult mammals and led to the first direct therapeutic application of RNAi in a mouse model of autoimmune hepatitis (Song et al. 2003).

However, many disadvantages are inherent in this approach, which limits its use in many other applications. In particular, in these studies the primarily targeted organ was the liver because of its elaborated vasculature, which allows efficient systemic delivery and accumulation of siRNAs. Since other organs are less vasculated and relevant cell types may be more distant form blood vessels, tissue specificity by systemic delivery of naked siRNA may be poor. Moreover, various nucleases are present in the blood stream that triggers degradation of nucleic acids and passive diffusion of siRNAs through the cell membrane is inefficient. In fact, in the studies described above very high amounts of siRNAs were used, which may lead to unspecific effects, including off-targeting, in other tissues.

Several variations to this common theme were introduced to achieve better tissue-specificity and reduce the amounts of siRNA used. The simplest was repetitively to administer loweramounts of siRNAs in the area of interest or implant polymers containing the naked siRNA, which is released constantly overtime (Gilmore et al. 2006, Novobransteva et al. 2008, Aigner 2006). A conceptually similar approach consisted in delivering naked siRNA to the respiratory tract and lungs by, simply, spray inhalation (Zhang et al. 2004, Massaro et al. 2004), which revealed to be effective for protecting the respiratory tract from viral infections (Bitko et al. 2005).

Besides the direct delivery of naked siRNAs, systems were developed that rely on their association with various molecules, including their direct chemical modification. As already described, lipofection is a method used to increase the efficiency of delivery of nucleic acids into cells. In two remarkable studies, intranasal or vaginal application of lipoplexes containing siRNAs against viruses was shown to protect animals from lethal viral infections (Bitko et al. 2005, Palliser et al. 2006). A conceptually similar approach was used in monkeys to target the liver by systemic injection of apolipoprotein-directed siRNA lipoplexes, which led to long-lasting reduction of apo-proteins and cholesterol secreted in the blood (Zimmermann et al. 2006).

A recent, potentially very powerful, chemical modification of siRNAs that improves their specific delivery is the use of RNA-aptamer/siRNA chimeric molecules. Briefly, an aptamer is a short oligonucleotide that, like an antibody, has the property to bind a specific target molecule. To find an aptamer for a given molecule, huge sequence variations of nucleic acid are generated (4^n different molecules for a sequence of n olygonucleotides). The target molecule is then fixed to a substrate and incubated with the oligonucleotide mixture. After washing, only olygonucleotides with a certain affinity to the target will remains and a polymerase chain reaction can be used to amplify them. The cycle is then repeated over and over in increasingly higher stringent conditions until a single aptamer, the best of all possible 4^n combinations, is selected, a process is referred to as systematic evolution of ligands by exponential enrichment, or SELEX (Tuerk et al. 1990, Ellington et al. 1990). To investigate the use of aptamers for RNAi, RNA-aptamers were selected for cell surface proteins of specific cell types and RNA-aptamer/siRNA chimeric molecules generated and delivered systemically asnaked or lipid-associated molecules, which revealed extremely effective in inhibiting tumor formation in animal models of prostate cancer (Chu et al. 2006, McNamara et al. 2006).

In addition to RNA-siRNA, also protein-siRNA linkage was investigated. In a recent study, specific targeting of neuronal cells in the brain has been achieved by associating siRNA to a 29 amino acids-long peptide from the rabies virus glycoprotein (Kumar et al. 2007). In short, the addition of this peptide allows crossing of the blood-brain barrier, a major obstacle for most drug-based therapies in the brain and internalization in neuronal cells. Using this method to deliver encephalitis specific antiviral siRNA sequences proved to be effective in protecting an astonishing 80% of mice from lethal encephalitis infection (Kumar et al. 2007). Moreover, antibodies-conjugated nanoparticles have also been used as carriers of siRNAs. In this case, the use of antibodies has the great potential of achieving the maximal tissue and cell specificity and, in fact, this technique has been used to target leukocytes and inhibit their proliferation by reducing the expression of the cell cycle regulator cyclinD1 (Peer et al. 2008). Specificity of antibodies versus aptamers may greatly vary depending on the specific molecule, cell and tissue to target. Permeability of the target tissues and internalization into a specific cell type should also be considered while choosing for the relatively big antibody-nanoparticle-siRNA as compared to the relatively small RNA-siRNA chimera. Clearly, these approaches are not mutually exclusive but represent a set of complementary strategies to extend the range of possible applications.

While the use of viruses for therapy is highly controversial due to concerns of genomic recombination the may lead to cancer, a number of chemical modifications of siRNAs, including binding to lipids, cholesterol (Castanotto et al. 2009), polyethylene-polymer chains (Zimmermann et al. 2006) and many other synthetic and natural substances are currently being investigated (Novobranstseva et al. 2008, de Fougerolles et al. 2007, Castanotto et al. 2009).

FIRST STEPS INTO THERAPY

There are few examples of basic research moving to the clinic in such a short time as for RNAi, barely five years from its first description (Fire et al. 1998) to its first use in a clinical trial in 2004. However, as a result of this short temporal gap, essentially all trials are still ongoing and, as of writing, only very few reports are available reporting on their efficacy (Novobranstseva et al. 2008, Tiemann et al. 2009, Whitehead et al. 2009).

In the first trial the aim was to reduce angiogenesis as a major cause of aged-related macular degeneration by using VEGF-targeted RNAi-aptamers. Clearly, the choice of the eye for focal administration made the trial more easily acceptable due to the ease of administration and the reduced risks of side effects to other organs. Following this first trial, many others started in a very rapid succession aiming at a multitude of diseases including kidney failure, various forms of cancer, various types of viral infection (HIV, adeno-associated, hepatitis and respiratory syncytial viruses) and others (Novobranstseva et al. 2008, Tiemann et al. 2009, Whitehead et al. 2009) (Table 1).

It is still very premature to draw any conclusion from these studies, which include very remarkable successes, such as a decreased viral infectivity upon RNAi treatment (DeVncenzo et al. 2010) and the first

Table 1. RNAi-based clinical trials.

Company	disease	target	status
Opko Health	amd	vegf	phase III
Sirna/Merck	amd	vegfr-1	phase II
Quark Pharma	amd	rtp801	phase II
Opko Health	dME	vegf	phase II
Anlylam Pharma	RSV	nucleocapsid	phase II
Benitec	HIV	Tat/rev	phase I
Nucleonics	HBV	four viral genes	phase I
Calando Pharma	solid tumors	rrm2	phase I
Quark Pharma	arf	p52	phase I
TransDerm	pachyonchina congenita	keratin 6a	phase I
Sanartis Pharma	hcv	mir-122	phase I
Duke university	metastatic melanima	mlp2/7	pilot study

AMD = age-related macular degeneration; DME = diabetic macular oedma; RSV = respiratory syncytial virus; HIV = human immunodeficiency virus; HVB = hepatitis-B virus; ARF = acute renal failure; HCV = hepatitis-C virus (Novobranstseva et al. 2008, Tiemann et al. 2009, Whitehead et al. 2009)

prove of systemic RNAi in the treatment of human cancer (Davis et al. 2010). However, also alarming findings paved the way, as in the case of a study on animal models of adeno-associated viral infection, which lead to lethality of a significant portion of the treated mice due to unpredicted off-target effects (Grimm et al. 2006). Moreover, with more basic research made in the field, it is becoming clear that exogenously delivered siRNAs may interfere with the action of endogenously expressed microRNAs as they both share common molecular machinery. Considering the many physiological roles of these small non coding RNAs, major concerns are raised about the specificity of siRNAs (Tiemann et al. 2009).

The use of RNAi for therapy has just started and uncertaininties and challenges ahead are many. Yet, the resources and initiatives of the scientists working in the field are at least as many.

CONCLUSION

In this chapter I have discussed the main strategies adopted to deliver siRNAs into living mammalian organisms by highlighting the great impact that these applications have in basic research and potential therapeutic applications. While the success of this technology is undoubted in the former area of investigation, more work is still necessary to assess the latter. Perhaps another message of this short report on RNAi, is that science has taught us to be ready for new, revolutionary findings, which, in a way or another, will have a great impact in our life.

ACKNOWLEDGEMENTS

FC is supported by the DFG-funded Center for Regenerative Therapies Dresden, the Medical Faculty of the Technical University Dresden and the Collaborative Research Center SFB655 of the DFG (subproject A20).

REFERENCES

Aigner, A. 2006. Gene silencing through RNA interference (RNAi) *in vivo*: strategies based on the direct application of siRNAs. *J. Biotechnol.* **124**:12–25.

Akamatsu, W., H.J. Okano, N. Osumi, T. Inoue and S. Nakamura. 1999. Mammalian ELAV-like neuronal RNA-binding proteins HuB and HuC promote neuronal development in both the central and the peripheral nervous systems. *Proc. Natl. Acad. Sci. USA.* **96**:9885–9890.

Barnabe-Heider, F., K. Meletis, M. Eriksson, O. Bergmann and H. Sabelstrom. 2008. Genetic manipulation of adult mouse neurogenic niches by *in vivo* electroporation. *Nat. Methods.* **5**:189–196.

Bitko, V., A. Musiyenko, O. Shulyayeva and S. Barik. 2005. Inhibition of respiratory viruses by nasally administered siRNA. *Nat. Med.* **11**:50–5.

Bollerot, K., D. Sugiyama, V. Escriou, R. Gautier and S. Tozer. 2006. Widespread lipoplex-mediated gene transfer to vascular endothelial cells and hemangioblasts in the vertebrate embryo. *Dev. Dyn.* **235**:105–114.

Boutros, M. and J. Ahringer. 2008. The art and design of genetic screens: RNA interference. *Nat. Rev. Genet.* **9**:554–566.

Brummelkamp, T.R., R. Bernards and R. Agami. 2002. A system for stable expression of short interfering RNAs in mammalian cells. *Science*. **296**:550–553.

Calegari, F, W. Haubensak, D. Yang, W.B. Huttner and F. Buchholz. 2002. Tissue-specific RNA interference in postimplantation mouse embryos with endoribonuclease-prepared short interfering RNA. *Proc. Natl. Acad. Sci. USA.* **99**:14236–14240.

Calegari, F, A.M. Marzesco, R. Kittler, F. Buchholz and W.B. Huttner. 2004. Tissue-specific RNA interference in post-implantation mouse embryos using directional electroporation and whole embryo culture. *Differentiation.* **72**:92–102.

Castanotto, D, J.J. Rossi. 2009. The promises and pitfalls of RNA-interference-based therapeutics. *Nature.* **457**:426-433.

Chekulaeva, M, W. Filipowicz. 2009 Mechanisms of miRNA-mediated post-transcriptional regulation in animal cells. *Curr. Opin. Cell. Biol.* **21**:452–460.

Chu, T.C, K.Y. Twu, A.D. Ellington and M. Levy. 2006. Aptamer mediated siRNA delivery. *Nucleic Acids Res.* **34**:73.

Collinet, C, M. Stoter, C.R. Bradshaw, N. Samusik and J.C. Rink. 2010. Systems survey of endocytosis by multiparametric image analysis. *Nature.* **464**:243–249.

Cullen, B.R. 2006. Enhancing and confirming the specificity of RNAi experiments. *Nat. Methods.* **3**:677–681.

Davis, M.E, J.E. Zuckerman, C.H. Choi, D. Seligson and A. Tolcher. 2010. Evidence of RNAi in humans from systemically administered siRNA via targeted nanoparticles. *Nature.* **464**:1067–1070.

de Fougerolles, A., H.P. Vornlocher, J. Maraganore and J. Lieberman. 2007. Interfering with disease:a progress report on siRNA-based therapeutics. *Nat. Rev. Drug Discov.* **6**:443–453.

De Pietri Tonelli, D., F. Calegari, J.F. Fei, T. Nomura and N. Osumi. 2006 Single-cell detection of microRNAs in developing vertebrate embryos after acute administration of a dual-fluorescence reporter/sensor plasmid. *Biotechniques.* **41**:727–732.

DeVincenzo, J., R. Lambkin-Williams, T. Wilkinson, J. Cehelsky and S. Nochur. 2010. A randomized, double-blind, placebo-controlled study of an RNAi-based therapy directed against respiratory syncytial virus. *Proc. Natl. Acad. Sci. USA.* **107**:8800–8805.

Dunn, K.J., B.O. Williams, Y. Li and W.J. Pavan. 2000 Neural crest-directed gene transfer demonstrates Wnt1 role in melanocyte expansion and differentiation during mouse development. *Proc. Natl. Acad. Sci. USA.* **97**:10050–10055.

Elbashir, S.M., J. Harborth, W. Lendeckel, A. Yalcin and K. Weber. 2001. Duplexes of 21-nucleotide RNAs mediate RNA interference in cultured mammalian cells. *Nature.* **411**:494–498.

Ellington, A.D., J.W. Szostak. 1990. *In vitro* selection of RNA molecules that bind specific ligands. *Nature.* **346**:818–22.

Fire, A., S. Xu, M.K. Montgomery, S.A. Kostas and S.E. Driver. 1998. Potent and specific genetic interference by double-stranded RNA in Caenorhabditis elegans. *Nature.* **391**:806–811.

Fraser, A.G., R.S. Kamath, P. Zipperlen, M. Martinez-Campos and M. Sohrmann. 2000. Functional genomic analysis of *C. elegans* chromosome I by systematic RNA interference. *Nature.* **408**:325–330.

Fukuchi-Shimogori, T., E.A. Grove. 2001. Neocortex patterning by the secreted signaling molecule FGF8. *Science.* **294**:1071–1074.

Gilmore, I.R., S.P. Fox, A.J. Hollins and S. Akhtar. 2006. Delivery strategies for siRNA-mediated gene silencing. *Curr. Drug Deliv.* **3**:147–145.

Gonczy, P., C. Echeverri, K. Oegema, A. Coulson and S.J. Jones. 2000. Functional genomic analysis of cell division in C. elegans using RNAi of genes on chromosome III. *Nature.* **408**:331–336.

Grimm, D., K.L. Streetz, C.L. Jopling, K. Storm and K. Pandey. 2006. Fatality in mice due to oversaturation of cellular microRNA/short hairpin RNA pathways. *Nature.* **441**:537–541.

Guo, S. and K.J. Kemphues. 1995. par-1, a gene required for establishing polarity in *C. elegans* embryos, encodes a putative Ser/Thr kinase that is asymmetrically distributed. *Cell.* **81**:611–620.

Hamilton, A.J. and D.C. Baulcombe. 1999. A species of small antisense RNA in posttranscriptional gene silencing in plants. *Science.* **286**:950–952.

Hassani, Z., G.F. Lemkine, P. Erbacher, K. Palmier and G. Alfama. 2005. Lipid-mediated siRNA delivery down-regulates exogenous gene expression in the mouse brain at picomolar levels. *J. Gene. Med.* **7**:198–207.

Inui, M., G. Martello and S. Piccolo. 2010. MicroRNA control of signal transduction. *Nat. Rev. Mol. Cell. Biol.* **11**:252–263.

Isaka, Y. and E. Imai. 2007. Electroporation-mediated gene therapy. *Expert Opin. Drug Deliv.* **4**:561–571.

Kawasaki, H., E. Suyama, M. Iyo and K. Taira. 2003. siRNAs generated by recombinant human Dicer induce specific and significant but target site-independent gene silencing in human cells. *Nucleic Acids Res.* **31**:981–987.

Kennerdell, J.R. and R.W. Carthew. 1998. Use of dsRNA-mediated genetic interference to demonstrate that frizzled and frizzled 2 act in the wingless pathway. *Cell.* **95**:1017–1026.

Kennerdell, J.R. and R.W. Carthew. 2000. Heritable gene silencing in Drosophila using double-stranded RNA. *Nat. Biotechnol.* **18**:896–898.

Kittler, R., G. Putz, L. Pelletier, I. Poser and A.K. Heninger. 2004. An endoribonuclease-prepared siRNA screen in human cells identifies genes essential for cell division. *Nature.* **432**:1036–1040.

Kumar, P., H. Wu, J.L. McBride, K.E. Jung and M.H. Kim. 2007. Transvascular delivery of small interfering RNA to the central nervous system. *Nature.* **448**:39–43.

Lange, C., F. Calegari. 2010. Cdks and cyclins link G1 length and differentiation of embryonic, neural and hematopoietic stem cells. *Cell. Cycle.* **9:**In press.

Lange, C., W.B. Huttner and F. Calegari. 2009. Cdk4/cyclinD1 overexpression in neural stem cells shortens G1, delays neurogenesis and promotes the generation and expansion of basal progenitors. *Cell. Stem Cell.* **5:**320–331.

Lewis, D.L., J.E. Hagstrom, A.G. Loomis, J.A. Wolff and H. Herweijer. 2002. Efficient delivery of siRNA for inhibition of gene expression in postnatal mice. *Nat. Genet.* **32:**107–108.

Lorenz, C., P. Hadwiger, M. John, H.P. Vornlocher and C. Unverzagt. 2004. Steroid and lipid conjugates of siRNAs to enhance cellular uptake and gene silencing in liver cells. *Bio. Org. Med. Chem. Lett.* **14:**4975–4977.

LoTurco, J., J.B. Manent and F. Sidiqi. 2009. New and improved tools for *in utero* electroporation studies of developing cerebral cortex. *Cereb Cortex.* 19 Suppl **1:**i120–125.

Massaro, D., G.D. Massaro and L.B. Clerch. 2004. Noninvasive delivery of small inhibitory RNA and other reagents to pulmonary alveoli in mice. *Am. J. Physiol. Lung Cell. Mol. Physiol.* **287:**L1066–1070.

Matsuda, T. and C.L. Cepko. 2008. Analysis of gene function in the retina. *Methods Mol. Biol.* **423:**259–278.

McCaffrey, A.P., L. Meuse, T.T. Pham, D.S. Conklin and G.J. Hannon. 2002. RNA interference in adult mice. *Nature.* **418:**38–39.

McNamara, J.O., E.R. Andrechek, Y. Wang, K.D. Viles and R.E. Rempel. 2006. Cell type-specific delivery of siRNAs with aptamer-siRNA chimeras. *Nat. Biotechnol.* **24:**1005–1015.

Muramatsu, T., Y. Mizutani, Y. Ohmori, J. Okumura. 1997. Comparison of three nonviral transfection methods for foreign gene expression in early chicken embryos *in vivo. Biochem. Biophys. Res. Commun.* **230:**376–380.

Myers, J.W., J.T. Jones, T. Meyer and J.E. Jr. Ferrell. 2003. Recombinant Dicer efficiently converts large dsRNAs into siRNAs suitable for gene silencing. *Nat. Biotechnol.* **21:**324–328.

Napoli, C., C. Lemieux and R. Jorgensen. 1990. Introduction of a Chimeric Chalcone Synthase Gene into Petunia Results in Reversible Co-Suppression of Homologous Genes in trans. *Plant Cell.* **2:**279–289.

Novobrantseva, T.I., A. Akinc, A. Borodovsky and A. de Fougerolles. 2008. Delivering silence:advancements in developing siRNA therapeutics. *Curr. Opin. Drug Discov. Devel.* **11:**217–224.

Osumi, N. and T. Inoue. 2001. Gene transfer into cultured mammalian embryos by electroporation. *Methods.* **24:**35–42.

Paddison, P.J., A.A. Caudy, E. Bernstein, G.J. Hannon and D.S. Conklin. 2002. Short hairpin RNAs (shRNAs) induce sequence-specific silencing in mammalian cells. *Genes. Dev.* **16:**948–958.

Palliser, D., D. Chowdhury, Q.Y. Wang, S.J. Lee and R.T. Bronson. 2006. An siRNA based microbicide protects mice from lethal herpes simplex virus 2 infection. *Nature.* **439:**89–94.

Paul, C.P., P.D. Good, I. Winer, D.R. Engelke. 2002. Effective expression of small interfering RNA in human cells. *Nat. Biotechnol.* **20:**505–508.

Peer, D., E.J. Park, Y. Morishita, C.V. Carman and M. Shimaoka. 2008. Systemic leukocyte-directed siRNA delivery revealing cyclin D1 as an anti-inflammatory target. *Science.* **319:**627–630.

Saito, T. and N. Nakatsuji. 2001. Efficient gene transfer into the embryonic mouse brain using *in vivo* electroporation. *Dev. Biol.* **240:**237–246.

Salomoni, P. and F. Calegari. 2010. *Cell.* cycle control of mammalian neural stem cells:putting a speed limit on G1. *Trends Cell. Biol.* 332–342.

Sarov, M. and A.F. Stewart. 2005. The best control for the specificity of RNAi. *Trends Biotechnol.* **23:**446–448.

Song, E., S.K. Lee, J. Wang, N. Ince and N. Ouyang. 2003. RNA interference targeting Fas protects mice from fulminant hepatitis. *Nat. Med.* **9:**347–351.

Tabata, H. and K. Nakajima. 2001. Efficient *in utero* gene transfer system to the developing mouse brain using electroporation: visualization of neuronal migration in the developing cortex. *Neuroscience.* **103:**865–872.

Tiemann, K. and J.J. Rossi. 2009. RNAi-based therapeutics-current status, challenges and prospects. *EMBO Mol. Med.* **1:**142–151.

Tilesi, F., P. Fradiani, V. Socci, D. Willems and F. Ascenzioni. 2009. Design and validation of siRNAs and shRNAs. *Curr. Opin. Mol. Ther.* **11:**156–164.

Timmons, L. and A. Fire. 1998. Specific interference by ingested dsRNA. *Nature.* 395:854. Tuerk, C. and L. Gold. 1990. Systematic evolution of ligands by exponential enrichment: RNA ligands to bacteriophage T4 DNA polymerase. *Science.* **249:**505–510.

Whitehead, K.A., R. Langer and D.G. Anderson. 2009. Knocking down barriers:advances in siRNA delivery. *Nat. Rev. Drug Discov.* **8:**129–138.

Wianny, F. and M. Zernicka-Goetz. 2000. Specific interference with gene function by double-stranded RNA in early mouse development. *Nat. Cell. Biol.* **2:**70–75.

Winter, J., S. Jung, S. Keller, R.I. Gregory, S. Diederichs. 2009. Many roads to maturity: microRNA biogenesis pathways and their regulation. *Nat. Cell. Biol.* **11:**228–234.

Yang, D., F. Buchholz, Z. Huang, A. Goga and C.Y. Chen. 2002. Short RNA duplexes produced by hydrolysis with Escherichia coli RNase III mediate effective RNA interference in mammalian cells. *Proc. Natl. Acad. Sci. USA.* **99:**9942–9947.

Yu, J.Y., S.L. DeRuiter and D.L. Turner. 2002. RNA interference by expression of short-interfering RNAs and hairpin RNAs in mammalian cells. *Proc. Natl. Acad. Sci. USA.* **99:**6047–6052.

Zhang, X., P. Shan, D. Jiang, P.W. Noble and N.G. Abraham. 2004. Small interfering RNA targeting heme oxygenase-1 enhances ischemia-reperfusion-induced lung apoptosis. *J. Biol. Chem.* **279:**10677–10684.

Zimmermann, T.S., A.C. Lee, A. Akinc, B. Bramlage and D. Bumcrot. 2006. RNAi-mediated gene silencing in non-human primates. *Nature.* **441:**111–114.

CHAPTER 2

Deciphering Post-transcriptional Gene Silencing Pathways Through Genetic Screens

Vincent Jauvion, Allison C. Mallory
and Hervé Vaucheret*

ABSTRACT

Small RNAs, 20- to 30-nucleotide in length, play important regulatory roles in many eukaryotic organisms. Except for ciliates, where they guide DNA elimination, small RNA regulate gene expression by either preventing or dampening transcription (Transcriptional Gene Silencing, TGS), or guiding RNA cleavage and translational repression (Post-Transcriptional Gene Silencing, PTGS), mechanisms collectively referred to as RNA silencing. In plants, the major role of TGS is to maintain transposable elements in a silent state, whereas PTGS primarily regulates genes involved in development and responses to biotic and abiotic stresses, such as viruses and salt stress. Consistent with it being an important viral defense mechanism, PTGS can spread cell-to-cell and systemically through a sequence-specific mobile signaling component, which sets up a defensive barrier at a distance from the site of initiation. Over the last 15 years, several forward and reverse genetic screens conducted in numerous laboratories have identified proteins necessary for the biogenesis and action of small RNA. Collectively, these screens revealed a diversity of small RNA pathways that individually require specific members of multigenic families. These screens also pointed to several instances of crosstalk among pathways, which, in certain cases, are important for establishing a homeostatic relationship among pathways.

Keywords: Arabidopsis, RNAi, PTGS, siRNA, miRNA.

---

* Corresponding author e-mail: herve.vaucheret@versailles.inra.fr

INTRODUCTION

PTGS was originally reported in transgenic plants as an undesired outcome occurring when researchers attempted to express to a high level the product of a sense transgene conferring a desirable trait. In several cases, plants expressing the introduced transgene could not be obtained, and, instead, only plants that lacked both the expression of the transgene and any gene sharing homology to the transgene were identified (Napoli et al. 1990, Smith et al. 1990, van der Krol et al. 1990). This observation puzzled researchers, but subsequent investigations led to the identification of double-stranded RNA (dsRNA) as the sequence-specific molecule inducing RNA silencing (Fire et al. 1998, Waterhouse et al. 1998). Following this discovery, several techniques leading to the direct and efficient production of dsRNA after transgene transcription, such as the expression of an inverted-repeat (IR) transgene, were developed to trigger silencing of a desired gene. In contrast to IR transgenes, the way by which sense transgenes produce dsRNA is not well understood, but unlike IR transgenes, sense transgenes require the activity of an RNA dependent RNA polymerase to induce silencing (see below). Based largely on this fundamental mechanistic difference, PTGS triggered by sense transgenes was renamed S-PTGS to distinguish it from IR-PTGS (Beclin et al. 2002). Both sense and inverted-repeat transgene-induced PTGS are powerful systems for identifying PTGS components through genetic screens. Moreover, the identification of small RNA as effectors of RNA silencing (Hamilton et al. 1999) and the discovery that several classes of endogenous small RNA exists have blurred the lines between the mechanistic differences among the various PTGS pathways and, thus, have necessitated precise methods to define the components of each small RNA pathway. Both forward and reverse genetic screens have had major roles in elucidating the protein requirements of these diverse small RNA pathways.

To accomplish genetic screens, various laboratory techniques are used to alter the genome sequence of an organism to determine how the loss of function or the change of function of a gene can impact a particular pathway, such as PTGS. In plants, the natural ability of the pathogen *Agrobacterium tumefaciens* is often exploited to transfer a DNA fragment of known sequence, usually containing a selectable marker, into the genome to disrupt gene function at random. In addition, chemical treatment with mutagenizing agents is routinely used to introduce point mutations, deletions or rearrangements in the DNA sequence. Once the genome disruptions are made, the positions of the mutations are ascertained through numerous techniques, including DNA sequencing and classical genome mapping, and, ultimately, the function of the interrupted gene determined. This suite of techniques has been used in both animals and plants by several laboratories to determine the components of the diverse RNA silencing systems. In the first section, we review the output of PTGS screens in plants, whereas the role and function of PTGS components will be presented in the subsequent section (current models of PTGS pathways).

OUTPUT OF PTGS SCREENS IN PLANTS

Identifying genes functioning in S-PTGS

The first reported S-PTGS forward genetic screen was based on the silencing of a transgene consisting of the strong viral 35S promoter fused to the bacterial *rolB* gene. Expression of the *rolB* gene at high levels causes developmental defects in plants, and, thus, after chemical mutagenesis of *p35s:rolB* Arabidopsis plants, the basis of this screen was to identify mutants exhibiting a wild-type phenotype due to silencing of the *p35S:rolB* transgene. The screen was a success and *enhancer of gene silencing (egs)* mutants that defined at least two loci were identified (Dehio et al. 1994), but unfortunately, these mutants were not further characterized and their identity remains unknown.

In contrast to the above forward genetic screen for enhancers of S-PTGS, a second S-PTGS screen was performed to identify mutants that suppressed PTGS of a transgene consisting of the viral 35S promoter fused to the bacterial *uidA* gene, which codes for β-glucuronidase (GUS), in an Arabidopsis line called *L1* (Elmayan et al. 1998). Immediately after germination, the *L1* line accumulates high levels of *GUS* mRNA and GUS protein, but, despite the maintenance of a very high rate of transcription, *GUS* mRNA and protein levels drop dramatically during the first week of growth due to PTGS of the *p35S:GUS* transgene, (Elmayan et al. 1998). After chemical mutagenesis of the *L1* line, *suppressor of gene silencing (sgs)* mutants that expressed GUS at a higher level than *L1* were identified. These mutants defined at least 15 *SGS* loci (Elmayan et al. 1998, Fagard et al. 2000, Mourrain et al. 2000, Morel et al. 2002, Boutet et al. 2003, Adenot et al. 2006, Elmayan et al. 2009, Jauvion et al. 2010), of which seven have been molecularly characterized and encode the RNA-dependent RNA polymerase SGS2/RDR6, the RNA-binding protein SGS3, the RNaseH enzyme SGS4/AGO1, the RNA methylase SGS5/HEN1, the RNA export protein SGS7/SDE5, the RNA trafficking protein SGS9/HPR1 and, unexpectedly, the DNA methyltransferase SGS6/MET1, which also has been implicated in TGS. Reverse genetics subsequently confirmed the importance of MET1 in S-PTGS and also implicated another TGS component, the chromatin-remodeling protein DDM1, in S-PTGS. These examples were the first indication that protein could be shared between TGS and S-PTGS (Morel et al. 2000).

To further investigate the requirements for S-PTGS, hypomorphic PTGS-deficient *ago1* mutants identified as suppressors of *L1* PTGS were subjected to a second round of mutagenesis to screen for mutants that restored *L1* PTGS. Seven mutants defining at least two loci were identified as enhancers of *L1* PTGS. In two mutants, the causal mutation was mapped to the nucleotidase/phoshatase FRY1, a protein that shares homology with a yeast and animal protein previously shown to regulate positively the activity of $5' \rightarrow 3'$ exoribonucleases (Gy et al. 2007). Consistent with FRY1 being a regulator of $5' \rightarrow 3'$ exoribonucleases in Arabidiopsis, reverse genetics implicated the $5' \rightarrow 3'$ exoribonucleases XRN2, XRN3 and XRN4 as suppressors of S-PTGS (Gy et al. 2007), extending previous findings that implicated XRN4 in S-PTGS (Gazzani et al. 2004). In addition to FRY1 and the XRNs, reverse genetics also revealed that *ago10* mutations restore *L1* S-PTGS in hypomorphic *ago1* mutants by increasing the amount of AGO1 protein (Mallory et al. 2009).

A second forward genetic approach screened for *silencing defective (sde)* mutants that reactivated the expression of a silenced GFP transgene in the *GxA* line, which carries both a *p35S:GFP* transgene (G) and a *p35S:PVX-GFP* viral amplicon transgene (A). In the absence of transgene A, GFP is not silenced and in the absence of transgene G, A is a weak inducer of GFP PTGS. By contrast, PTGS of GFP is efficiently triggered in the *GxA* line. After chemical mutagenesis of the *GxA* line, *sde* mutants that expressed GFP at a higher level than in the *GxA* line were identified. These mutants defined at least six loci and implicated the RNA-dependent RNA polymerase SDE1/SGS2/RDR6, the RNA-binding protein SDE2/SGS3, the RNA helicase SDE3, the largest subunit of PolIV SDE4/NRPD1, the RNA export protein SDE5 and the RNA-dependent RNA polymerase SDE6/RDR2 in PTGS (Dalmay et al. 2000, Dalmay et al. 2001, Herr et al. 2005, Hernandez-Pinzon et al. 2007). Extending these results, reverse genetics also implicated the second largest subunit of PolIV NRPD2 in *GxA* PTGS, confirming that both chromatin-related proteins and RNA-related proteins are important for S-PTGS (Herr et al. 2005).

Another amplicon system was used to identify mutants with an *enhanced silencing phenotype (esp)*. For this forward screen, two lines carrying a *p35S:PVX-PDS* amplicon and exhibiting weak PTGS of the endogenous *PHYTOENE DESATURASE (PDS)* gene were mutagenized and *esp* mutants exhibiting strong PDS PTGS were identified. These mutants defined at least five loci, four of which encode the RNA helicase splicing factor ESP3/PRP2 and the mRNA 3' formation proteins ESP1/CstF64-like, ESP4/Symplekin and ESP5/CPSF100, pointing to additional crosstalk between PTGS and RNA metabolism pathways (Herr et al. 2006).

Lastly, a third screen to identify mutants that suppressed PTGS was based on the Arabidopsis line *2a3*, which carries a *p35S:NIA2* transgene that triggers silencing of both the endogenous *NITRATE REDUCTASE1* (*NIA1*) and *NIA2* genes in addition to the *p35S:NIA2* transgene, due to sequence similarity. This transgene-induced homology based silencing is referred to as cosuppression. A reverse genetic screen using the *2a3* line revealed that the 15 *SGS* loci retrieved from the *L1* screen also controlled *2a3* cosuppression (Elmayan et al. 1998, Mourrain et al. 2000, Morel et al. 2002, Boutet et al. 2003, Jauvion et al. 2010), whereas a forward genetic screen using the *2a3* line identified additional *sgs* mutants that impacted *2a3* silencing but not *L1* silencing. These *2a3*-specific mutants defined at least three additional loci, including SDE3, which was previously identified in the *GxA* screen (Adenot et al. 2006, Jauvion et al. 2010).

To examine the similarities and differences between S-PTGS and IR-PTGS, the consequence of S-PTGS mutations on IR-PTGS was evaluated by introducing S-PTGS mutants into an Arabidopsis IR-PTGS line expressing a *GUS* dsRNA PTGS inducer and a *GUS* target mRNA under the control of the strong 35S promoter. This IR-PTGS system was shown to be insensitive to mutations affecting S-PTGS factors such as SGS2/RDR6 or SGS3 (Beclin et al. 2002), reinforcing the different genetic requirements of S-PTGS and IR-PTGS.

Identifying genes necessary for short-distance spread of silencing

Arabidopsis transgenic lines expressing and IR-transgene that generates dsRNA under the control of the *SUCROSE2* (*SUC2*) promoter, which drives expression specifically in the companion cells of the phloem, were generated to search specifically for mutants impaired in the spreading of PTGS. Tested dsRNA inducers included those that targeted *SULPHUR* (*SUL*) and *PHYTOENE DESATURASE* (*PDS*), which, when silenced, lead to bleaching of the leaf tissue. These IR-PTGS lines exhibited silencing of the *SUL* and *PDS* targets in a layer of 10–15 cells around the vasculature, due to the spreading of a mobile PTGS signal. Like the previously tested IR-GUS silenced line, silencing in the *pSUC:SUL* and *pSUC:PDS* lines was not impaired by mutations in SGS2/SDE1/RDR6 or SDE3, confirming the specific implication of SGS2/SDE1/RDR6, SGS3/SDE2 and SDE3 in S-PTGS (Himber et al. 2003).

Mutagenesis of the *pSUC2-SUL* line retrieved *silencing movement-deficient* (*smd*) mutants that defined at least four loci encoding the largest subunit of PolIV SMD1/NRPD1, the RNA-dependent RNA polymerase SMD2/RDR2, the chromatin-remodeling protein SMD3/CLSY1 and the RNaseIII enzyme DCL4, pointing to the implication of both chromatin-related proteins and RNA-related proteins in IR-PTGS, similar to S-PTGS.

Reverse genetics also implicated the RNaseH enzyme AGO1, the RNaseIII enzyme DCL1 and the RNA methylase HEN1 in IR-PTGS, indicating that at least some factors are required for both S-PTGS and IR-PTGS (Dunoyer et al. 2005, Dunoyer et al. 2007, Dunoyer et al. 2010b). Mutagenesis of the *SUC-SUL* line also identified one mutant exhibiting enhanced IR-PTGS called *enhanced silencing movement* (*esm*).

Like the mutagenesis of the *pSUC2-SUL* line, mutagenesis of two *pSUC2:PDS* lines (also referred to as JAP) retrieved mutants impaired in CLSY1, NRPD1 and RDR2 and implicated DCL4 and NRPD2 in systemic IR-PTGS. This screen also identified mutants impaired in the mRNA 3' end processing proteins FCA and FPA, the histone demethylase JMJ14 and the RNA trafficking protein TEX1. Moreover, reverse genetics revealed a role for the mRNA 3' end processing protein FY, the nucleosome/chromatin assembly factor FVE and histone demethylase FLD during IR-PTGS, confirming the implication of both chromatin-related proteins and RNA-related proteins in IR-PTGS and reinforcing the notion of a crosstalk between PTGS and general RNA metabolism (Baurle et al. 2007, Smith et al. 2007, Manzano et al. 2009, Searle et al. 2010, Yelina et al. 2010). In addition, reverse genetics revealed a role for the RNA trafficking protein HPR1 in systemic IR-PTGS, extending the list of proteins shared between S-PTGS and IR-PTGS (Jauvion et al. 2010).

Identifying genes involved in the endogenous miRNA pathway

Long before the discovery of microRNA (miRNA), small RNAs produced from endogenous *MIRNA* genes and the elucidation of their essential role in development, mutants exhibiting overlapping developmental defects were identified. These mutants, which turned out to essential players in the miRNA pathway, were impaired in a suite of proteins including the RNaseH enzyme AGO1, the closest paralog of AGO1 AGO10/ZLL/PNH, the Cap-binding proteins CBP20 and CBP80/ABH1, the RNaseIII enzyme DCL1/CAF/SIN/SUS, the RNA-binding protein DDL, the double-stranded RNA-binding protein DRB1/HYL1, the RNA methylase HEN1 the small RNA export protein HST or the zinc finger protein SE (Park et al. 2002, Han et al. 2004, Baumberger et al. 2005, Li et al. 2005, Park et al. 2005, Lobbes et al. 2006, Yang et al. 2006, Brodersen et al. 2008, Gregory et al. 2008, Kim et al. 2008, Laubinger et al. 2008, Yu et al. 2008).

A forward genetic screen searching for mutations that suppress the developmental defects of hypomorphic *hen1* mutants was carried out and mutants impaired in the TGS components NRPD1 and NRPD2 were identified (Yu et al. 2010). In addition, reverse genetics revealed that mutations in the TGS component RDR2 had a similar suppressive effect, suggesting that TGS and PTGS can compete for common components such as HEN1.

To further determine the proteins necessary for maintaining the integrity of the miRNA pathway, a genetic screen specifically dedicated to the identification of miRNA mutants was developed using a reporter line called *GFP171.1*. In this line, the 35S promoter drives the expression of a *GFP* mRNA that has a miR171 target site immediately downstream of the open reading frame (Parizotto et al. 2004, Brodersen et al. 2008). Thus, due to miR171-directed repression, GFP expression is suppressed. Mutagenesis of the *p35S:GFP171* line identified *microRNA biogenesis deficient* (*mbd*) mutants, which defined at least two loci, including DCL1 and HEN1 and *microRNA action deficient* (*mad*) mutants, which define at least six loci, including the microtubule severing protein MAD5/KTN1 that controls miRNA action at the level of translation (Brodersen et al. 2008). Reverse genetics also implicated the decapping complex factor VCS in miRNA-mediated control of translation (Brodersen et al. 2008).

Identifying genes involved in the endogenous tasiRNA pathway

Molecular and phenotypic characterization of S-PTGS deficient *sgs2/ rdr6* and *sgs3* mutants revealed the existence of a class of small RNA called trans-acting siRNA (tasiRNA), which derives from long non-coding RNA encoded by endogenous *TAS* genes (Peragine et al. 2004, Vazquez et al. 2004, Allen et al. 2005). These mutants exhibit an accelerated change from juvenile to adult vegetative phase, which is also observed in mutants impaired in the RNaseH enzyme AGO1 and AGO7/ZIP, the RNaseIII enzyme DCL1 and DCL4, the double-stranded RNA-binding proteins DRB1/HYL1 and DRB4, the RNA export protein SDE5, the Cyclophilin protein SQN and in *TAS3* tasiRNA precursor mutants (Hunter et al. 2003, Peragine et al. 2004, Yoshikawa et al. 2005, Adenot et al. 2006, Mallory et al. 2009, Smith et al. 2009, Jauvion et al. 2010). Based on the genetic requirements of the tasiRNA pathway and the presence of miRNA-binding sites in *TAS* precursors, it was deduced that the tasiRNA pathway requires the function of the miRNA and S-PTGS pathways, in addition to several *TAS*-specific components.

A forward genetic screen based on the suppression of the accelerated phase change defect of *ago7/zip* mutants was carried out and mutants impaired in the transcription factors ETT/ARF3 and ARF4, which are targeted by TAS3 tasiRNA (Hunter et al. 2006) were obtained. In addition, a forward genetic screen based on the suppression of *sqn* mutant developmental defects, which resemble hypomorphic *ago1* mutants, was carried out and mutants impaired in the F-box protein FBW2 (Earley et al. 2010) were retrieved. Reverse genetics also revealed that *fbw2* mutations suppress the developmental defects of hypomorphic *ago1* mutants, suggesting that SQN and FBW2 are positive and negative regulators of AGO1, respectively (Smith et al. 2009, Earley et al. 2010).

To further decipher the components of the tasiRNA pathway, a genetic screen dedicated to the identification of proteins specifically involved in the tasiRNA pathway was developed using the *TAS3-syn* reporter line carrying a *p35S:TAS3aPDS* transgene, which expresses a *TAS3a* mRNA in which one tasiRNA has been replaced by a sequence complementary to the endogenous *PDS* mRNA (Cuperus et al. 2009). Processing of this tasiRNA leads to repression of endogenous *PDS* mRNA, which can be monitored by the photobleaching phenotype typical of PDS silencing. Mutagenesis of the *p35S:TAS3aPDS* line identified mutants impaired in tasiRNA-directed PDS silencing and defined at least five loci, including *AGO7, DCL4, RDR6, SGS3* and *MIR390*, the miRNA required for *TAS3* tasiRNA production.

CURRENT MODELS OF PTGS PATHWAYS

Thus, forward and reverse genetic screens have been indispensable in the discovery of numerous genes controlling small RNA-directed regulation (Table 1). Based on the known function or proposed function of many of these genes, several models for small RNA-directed regulation have been proposed. Below, we place the identified genes in a functional context and develop our current understanding of how the multiple small RNA-directed pathways function separately or, sometimes, collectively in plants.

miRNA pathway

miRNAs are endogenous small RNAs processed from long primary transcripts (pri-miRNA) that are transcribed from non protein-coding *MIR* genes by PolII. Like classic protein-coding mRNA, pri-miRNA are capped at their 5' end, contain introns and are polyadenylated at their 3' end. One of the defining features of MIR transcripts is their sequence complementarity within a unique transcript, which permits the RNA to fold into an imperfect stem-loop structure that can be further processed into mature miRNA. The nuclear maturation and processing of pri-miRNA requires the activity of several proteins, including the Cap-binding proteins CBP20 and CBP80/ABH1 (Gregory et al. 2008, Kim et al. 2008, Laubinger et al. 2008), the zinc finger protein SE (Lobbes et al. 2006, Yang et al. 2006), the double-stranded RNA-binding protein DRB1/HYL1 (Han et al. 2004) and the RNaseIII DCL1 (Park et al. 2002), which

Table 1: Components of S-PTGS, IR-PTGS, miRNA and tasiRNA pathways.

Proteins are defined as actors or suppressors if they are required for or are antagonistic to the pathway, respectively. "–" indicates that impairment of these proteins does not affect the pathway. Only the effect on the PTGS part of the pathway is reported. If DNA methylation or 24-nt siRNA levels is affected without changing PTGS, a "–" is indicated".

	S-PTGS *L1*	S-PTGS *2a3*	S-PTGS *GxA*	IR-PTGS *IR71*	IR-PTGS *pSUC-PDS*	IR-PTGS *pSUC-SUL*	miRNA *MIR*	miRNA *GFP171.1*	tasiRNA *TAS*	tasiRNA *TAS3-syn*
AGO1	actor	actor		actor		actor	actor		actor	
AGO4				–	suppressor	–				
AGO7									actor	actor
AGO10	suppressor[1]						actor			
CBP20							actor			
CBP80							actor			
CLSY1				–	actor	actor	–			
DCL1				actor		actor	actor	actor	actor	
DCL2	actor			actor		actor	–			
DCL3				actor	suppressor	–	–			
DCL4				actor	actor	actor	actor[5]		actor	actor
DDL							actor			
DDM1	actor						–			
DRB1				–		–	actor		actor	
DRB4				suppressor		–	–		actor	
ESP1			suppressor[2]							
ESP3			suppressor[2]							
ESP4			suppressor[2]							

contd.... Table

Table...contd.

	S-PTGS	S-PTGS	S-PTGS	IR-PTGS	IR-PTGS	IR-PTGS	miRNA	miRNA	tasiRNA	tasiRNA
	L1	*2a3*	*GxA*	*IR71*	*pSUC-PDS*	*pSUC-SUL*	*MIR*	*GFP171.1*	*TAS*	*TAS3-syn*
ESP5			suppressor[2]							
FBW2							suppressor[3]	suppressor[3]	suppressor[3]	
FCA					actor					
FLD					actor					
FPA					actor					
FRY1	suppressor[1]						-			
FVE					actor					
FY					actor					
HEN1	actor	actor		actor		actor	actor	actor	actor	
HPR1	actor	actor			actor		-		actor	
HST				-		-	actor		-	
JMJ14					actor					
KTN1						actor	actor			
MET1	actor									
NRPD1			actor	-	actor	actor	suppressor[4]		-	
NRPD2			actor	-	actor		suppressor[4]		-	
RDR2			actor	-	actor	actor	suppressor[4]		-	
RDR6	actor	actor	actor		-	-	-		actor	actor

1: identified as a suppressor in a hypomorphic *ago1* background.
2: identified as a suppressor in a *PDS* amplicon background.
3: identified as a suppressor in a *sqn* background.
4: identified as a suppressor in a hypomorphic *hen1* background.
5: identified as an actor for young *MIR* genes only.

is the major RNaseIII responsible for cleaving the pri-miRNA into mature miRNAs (Figure 1). Mutations in these components result in decreased levels of mature miRNA accompanied by apparent stabilization of pri-miRNA. Accumulation of pri-miRNA also depends on the RNA-binding protein DDL, which interacts with DCL1 (Yu et al. 2008). DCL1 also forms a complex with HYL1 and SE (Fang et al. 2007, Fujioka et al. 2007), while SE likely associates with the CBP20-CBP80 complex (Laubinger et al. 2008). While dcL1 and *se* null alleles are embryolethal, *cbp20, cbp80, ddl* and HYL1 null alleles are viable, indicating that only some of these proteins play essential roles in miRNA maturation and suggesting that these protein complexes are dynamic structures.

Not all pri-miRNA are processed in the same manner. Indeed, whereas the first step of pri-miRNA processing usually involves DCL1-mediated cleavage at the base of the stem-loop structure of the fold-back precursor followed by elimination of the loop, the processing of some pri-miRNA, including *pri-MIR159* and *pri-MIR319*, begins with the cleavage of the loop and is followed by multiple DCL1-mediated cuts until the mature miRNA is released (Bologna et al. 2009). Moreover, whereas pri-miRNA processing generally liberates a single miRNA duplex, the processing of some pri-miRNA, including pri-MIR163, can liberate two adjacent miRNA, likely resulting from sequential DCL1 cuts (Kurihara et al. 2004). Several pri-miRNA also produce both DCL1-dependent canonical 21-nt species and DCL3-dependent 24-nt species, also called long miRNA (Dunoyer et al. 2004, Vazquez et al. 2008, Chellappan et al. 2010, Wu et al. 2010). Lastly, young *MIR* genes such as MIR822, MIR839 and MIR869 produce long and almost perfectly complementary stem-loops that are processed by DCL4 assisted by DRB4 instead of DCL1 and DRB1/HYL1 (Rajagopalan et al. 2006, Pouch-Pelissier et al. 2008, Ben Amor et al. 2009). Following excision from the pri-miRNA, the miRNA duplex is methylated at each 3' end by the methyltransferase HEN1 (Li et al. 2005). This methylation prevents 3' uridylation, which is a signal for degradation. Recently, a family of SMALL RNA DEGRADING NUCLEASE (SDN) enzymes implicated in small RNA degradation were identified by reverse genetics (Ramachandran et al. 2008), highlighting the importance of miRNA protection.

Following methylation by HEN1, miRNA remain in the nucleus or are exported to the cytoplasm by HST (Park et al. 2005). To execute silencing, one strand of the miRNA duplex associates with proteins of the AGO family (Vaucheret. 2008). AGO association depends, in part, on the nucleotide length of the miRNA and on the identity of the 5' nucleotide of the selected miRNA strand. The majority of miRNA are 21-nt long, have a U base at their 5' end and associate with AGO1, which exhibits RNaseH activity and executes silencing through

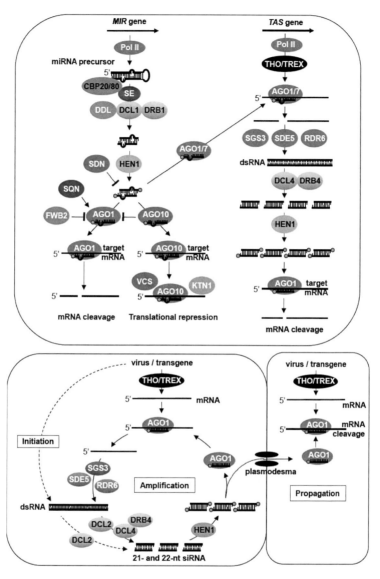

Figure 1: General components of miRNA, tasiRNA and S-PTGS pathways. Protein components represented by the same color participate in similar steps in the pathways. Additional steps that occur in only certain instances (nucleo-cytoplamic export of certains miRNA by HASTY, processing of young pri-miRNA by DCL4, etc.), or steps not directly related to PTGS (production of long miRNA by DCL3, DNA methylation guided by long miRNA loaded on AGO4, involvement of chromatin-related proteins in S-PTGS, etc.) are not represented on this figure but are detailed in the text.

(Color image of this figure appears in the color plate section at the end of the book.)

miRNA-guided target mRNA cleavage (Baumberger et al. 2005). Although its association with miRNA has not been directly demonstrated, AGO10/ZLL/PNH, the closest paralog of AGO1, is required for repressing target mRNA translation (Brodersen et al. 2008, Mallory et al. 2009). Whereas AGO1 alone appears sufficient to execute target mRNA cleavage (Baumberger et al. 2005), translational inhibition requires additional components such as the microtubule-severing enzyme KTN1 and the decapping complex member VCS (Figure 1) (Brodersen et al. 2008). In addition to, AGO1 and likely AGO10, other AGO proteins can associate with miRNA. Co-immunoprecipitation experiments revealed that 21-nt 5′ A and 21-nt 5′ C miRNA can associate with AGO2 and AGO5, respectively (Mi et al. 2008, Montgomery et al. 2008, Takeda et al. 2008). The functional relevance of these associations is not clear, as *ago2* and *ago5* mutants do not display strong developmental defects and silencing of miRNA targets by AGO2 and AGO5 has not been demonstrated. More recently, DCL3-dependent 24-nt miRNA species were shown to associate with AGO4 and direct DNA methylation of target genes (Chellappan et al. 2010, Wu et al. 2010). Whether AGO4/miRNA-mediated DNA methylation plays an important role in regulating target gene expression remains unknown.

tasiRNA pathway

Analysis of the transcriptome of miRNA mutants revealed that, whereas most miRNA targets encode proteins, a handful of non-coding RNA previously identified as tasiRNA precursors (Peragine et al. 2004, Vazquez et al. 2004) also are targeted by miRNA. This observation indicated that the tasiRNA pathway required the miRNA pathway for its initiation (Figure 1) (Allen et al. 2005), explaining why miRNA pathway mutants were also impaired in tasiRNA accumulation. TAS genes are transcribed by PolII and likely are exported/trafficked by the THO/TREX complex to AGO/miRNA catalytic centers (Jauvion et al. 2010, Yelina et al. 2010). Generation of tasiRNAs from the different *TAS* precursor RNA require different miRNA and different AGO protein. For example, a miR173/AGO1 complex cleaves *TAS1* and *TAS2* precursors RNAs, initiating tasiRNA production from the 3′ cleavage product, whereas a miR828/AGO1 complex initiates tasiRNA production by cleaving *TAS4* precursors. Both miR173 and miR828 are 22-nt long rather than 21-nt long, which appears to be necessary for the production of siRNA from AGO1-generated 3′ cleavage products (Chen et al. 2010, Cuperus et al. 2010). tasiRNA production from TAS3 requires dual targeting of the two miR390 complementary sites in the *TAS3* precursor by the miR390/AGO7 complex. The 3′ miR390 site is canonical and is cleaved by AGO7, whereas the 5′ miR390 site possesses several centrally located mismatches with miR390, which prevent the cleavage of the site but not miR390/AGO7 recognition (Axtell et al. 2006, Montgomery et al. 2008). The presence of this 5' non-cleavable AGO7-binding miR390 site is essential for the production of TAS3 tasiRNAs because this site cannot be replaced by a cleavable AGO7-binding miR390 site or a non-cleavable AGO1-binding miR171 site. The requirement for a noncleavable AGO7-binding miR390 site appears to be specific to the 5' site, because the 3' AGO7-binding miR390 site can be replaced by a cleavable AGO1-binding miR171 site without perturbing tasiRNA production (Montgomery et al. 2008).

After AGO-mediated slicing of TAS precursor RNAs, the cleavage fragments are protected against degradation by the RNA-binding SGS3 protein, which likely prevents their degradation by XRN4 and transformed into dsRNA by the RNA-dependent RNA polymerase RDR6 (Yoshikawa et al. 2005, Elmayan et al. 2009). The putative RNA export factor SDE5 is also essential for tasiRNA production, although its exact role remains uncertain (Hernandez-Pinzon et al. 2007, Jauvion et al. 2010). *TAS*-derived dsRNA are bound and sequentially processed, starting from the miRNA-guided cleavage site, by the RNase III DCL4/dsRNA binding protein DRB4 complex to produce a phased 21-nt tasiRNA population (Allen et al. 2005, Gasciolli et al. 2005, Xie et al. 2005, Yoshikawa et al. 2005, Adenot et al. 2006, Nakazawa et al. 2007). Analogous to miRNA, tasiRNA duplexes are methylated by HEN1 (Li et al. 2005) and one strand associates with AGO1 to guide cleavage of complementary tasiRNA targets (Figure 1) (Peragine et al. 2004, Vazquez et al. 2004).

Viral S-PTGS pathway

Whereas the miRNA and tasiRNA pathways rely on endogenous small RNA that act *in trans* to regulate partially complementary endogenous mRNA, S-PTGS pathways act *in cis* to destroy invading RNA from viruses, bacteria or transgenes (Ding et al 2007, Ruiz-Ferrer et al. 2009). The mechanistic basis of this plant immune system is the transformation of part of the invading RNA into dsRNA and its subsequent processing to siRNA, which, after amplification, associate with AGO proteins to direct the destruction of invading viral RNA. The S-PTGS pathway mostly relies on cellular components of the tasiRNA pathway, but additional proteins are sometimes required, depending on the nature of the invading pathogen or transgene.

Viruses produce dsRNA in a variety of ways. For example, dsRNA can be produced by viral RNA-dependent RNA polymerases during the replication of single-stranded RNA viruses. dsRNA can also result from convergent transcription of the two strands of DNA viruses. In addition, dsRNA can result from the folding of viral single-stranded

RNA, e.g., the leader region of the CaMV 35S RNA (Moissiard et al. 2006). Although all four Arabidopsis DCL proteins contribute to the production of viral siRNA, DCL4 appears to be the major actor (Blevins et al. 2006, Bouche et al. 2006, Deleris et al. 2006, Moissiard et al. 2006). DCL4 produces 21-nt viral siRNA that, when loaded onto AGO1, guide the cleavage of viral RNA. However, when the activity of DCL4 is compromised, for example by the TCV P38 protein, 22-nt and 24-nt viral siRNA are produced by DCL2 and DCL3, respectively (Figure 1) (Blevins et al. 2006, Bouche et al. 2006, Deleris et al. 2006). Whereas DCL2-dependent siRNA can efficiently guide the cleavage of viral RNA, DCL3-dependent siRNA do not appear to have this capacity. Due to the redundancy between DCL2 and DCL4, only *dcl2 dcl4* double mutants, but not the simple mutants, exhibit hyper-susceptibility to virus infection (Blevins et al. 2006, Bouche et al. 2006, Deleris et al. 2006, Wang et al. 2010). Despite the inability of DCL3-dependent siRNA to guide the cleavage of viral RNA, DCL3 seems to co-operate with DCL4 in the silencing of DNA viruses by guiding DNA methylation (Blevins et al. 2006). Although DCL1 does not appear to contribute directly to viral defense, it may have an indirect role by acting as a negative regulator of DCL4 (Qu et al. 2008) and in the case of geminiviruses and caulimoviruses, acting as a facilitator in the biogenesis of viral siRNAs (Blevins et al. 2006, Moissiard et al. 2006).

Infection of various PTGS mutants by CMV, a single-stranded RNA viruses, revealed that *ago1, hen1, rdr6, sgs3, sde3* and *sde5* mutants are hypersusceptible and, as such, accumulate higher level of CMV RNA than wildtype plants (Mourrain et al. 2000, Dalmay et al. 2001, Morel et al. 2002, Boutet et al. 2003, Hernandez-Pinzon et al. 2007). These results strongly suggest that, following viral RNA cleavage by AGO1, the RNA fragments are transformed into dsRNA by RDR6, similar to TAS cleavage products (Figure 1). Subsequent processing of viral dsRNA by DCL2 and DCL4 reinforces the production of viral siRNA and the destruction of viral RNA. Similar to DCL1, DCL2, DCL3 and DCL4, which exhibit partial redundancy, RDR1, RDR2 and RDR6 also have overlapping roles in viral siRNA biogenesis and virus silencing (Mourrain et al. 2000, Qu et al. 2005, Schwach et al. 2005, Donaire et al. 2008, Qi et al. 2009, Wang et al. 2010). Because PTGS primarily is an antiviral defense, most viruses have developed strategies to suppress PTGS at various levels and successfully infect plants (reviewed in (Ding 2010)).

Transgene S-PTGS pathway

How primary dsRNA is produced by sense transgenes during S-PTGS and why S-PTGS frequency varies from one transgenic line to another

(Palauqui et al. 1995, Vaucheret et al. 1995, Kunz et al. 1996) remains unclear. Several hypotheses have been evoked including locus-dependent unintended transcription of either antisense RNA or aberrant sense RNA that can serve as matrix for RDR6. These two hypotheses are consistent with the fact that mutations in the chromatin-remodeling protein DDM1 adversely affect S-PTGS frequency (Morel et al. 2000). However, once produced, dsRNA can be processed by one or several DCL, similar to viral dsRNA. In the case of the Arabidopsis *L1* locus, the availability of DCL2 seems to be the primary determinant initiating S-PTGS (Mlotshwa et al. 2008), probably because DCL2 produces 22-nt siRNA species that, after associating with AGO1, cleave target mRNA to generate cleavage products that can be protected against degradation by SGS3 and transformed into secondary dsRNA by RDR6 assisted by SDE5 (Mourrain et al. 2000, Morel et al. 2002, Boutet et al. 2003, Jauvion et al. 2010), similar to the role of 22-nt miRNAs in tasiRNA biogenesis. Because the *L1* locus produces an equal amount of 21-nt and 22-nt GUS siRNA (Mlotshwa et al. 2008), these secondary dsRNA likely are processed by DCL2 and DCL4 to produce secondary siRNA that, at least in the case of DCL2-dependent 22-nt siRNA, generate an amplification loop. This loop is essential for the spreading of S-PTGS. Indeed, S-PTGS is triggered locally and subsequently spreads from cell to cell through plasmodesmata or at long distance through the vasculature, indicating that a sequence-specific systemic silencing signal propagates S-PTGS throughout the plant (Palauqui et al. 1996, Palauqui et al. 1997, Voinnet et al. 1997, Voinnet et al. 1998). Because siRNA are mobile (Dunoyer et al. 2010a, Molnar et al. 2010), it is likely that the primary or the secondary siRNA produced in the cells where S-PTGS is initiated move to adjacent cells and, eventually, to the vasculature where they can guide S-PTGS (Figure 1).

The components identified in the *L1* screen likely are the core components of S-PTGS and represent the set of genes sufficient for strong silencers. In contrast, more weakly silenced lines, such as *2a3*, *GxA* and another 35S:GUS line called *Hc1* (Elmayan et al. 1998), require additional components such as the RNA helicase SDE3 that is dispensable for *L1* S-PTGS (Dalmay et al. 2000, Jauvion et al. 2010). Moreover, the efficiency of *GxA* S-PTGS is reduced in *nrpd1*, *nrpd2* and *rdr2* mutants (Herr et al., 2005), whereas *L1* S-PTGS is not affected by these mutations (M. Rivard, T. Elmayan, V.J. and H.V. unpublished) but is affected by *ddm1* and *met1* mutations (Morel et al. 2000). Because DDM1, MET1, NRPD1, NRPD2 and RDR2 affect the transcriptional state in a locus-specific manner, these different protein requirements likely reflect locus-specific rather than transgene-specific effects.

IR-PTGS pathway

Sense transgenes that produce a limited amount of initiating dsRNA require RDR6, SDE3, SDE5 and SGS3 to amplify the silencing signal. In contrast, IR transgenes that directly produce dsRNA do not require any of these components to produce siRNAs and silence their targets (Beclin et al. 2002, Himber et al. 2003, Dunoyer et al. 2005, Smith et al. 2007, Dunoyer et al. 2010b). Therefore, it is expected that IR transgenes have the same genetic requirement as endogenous IR loci, such as IR71 or IR2039, which produce 21-nt, 22-nt and 24-nt siRNA through DCL4, DCL2 and DCL3, respectively and silence endogenous targets through AGO1 (Henderson et al. 2006, Dunoyer et al. 2010b). However, IR-PTGS genetic screens based on the *pSUC2-SUL* and *pSUC2-PDS* lines not only revealed a requirement for AGO1 and DCL4 (Dunoyer et al. 2005, Smith et al. 2007, Dunoyer et al. 2010b), but also an unexpected requirement for the RNA-dependent RNA polymerase RDR2, the PolIV subunits NRPD1 and NRPD2 and a novel chromatin-remodeling protein CLSY1 for short-distance spreading of PTGS (Figure 2) (Dunoyer et al. 2005, Dunoyer et al. 2007, Smith et al. 2007, Dunoyer et al. 2010b). Because the amount of PDS siRNA in *pSUC2-PDS* lines is reduced in *clsy1*, *nrpd1* and *rdr2* mutants, it is reasonable to assume that, similar to the S-PTGS *GxA* line, the IR-PTGS *pSUC2-PDS* locus is under a transcriptional control that partly depends on 24-nt siRNA produced from this locus. Supporting this hypothesis, mutations in other chromatin-related proteins, including the nucleosome/chromatin assembly factor FVE and the histone demethylases FLD and JMJ14, affect IR-PTGS in the *pSUC2-PDS* line (Baurle et al. 2007, Searle et al. 2010). Alternatively, but not exclusively, an amplification step may also be required in IR-PTGS and CLSY1, PolIV and RDR2 may be required for movement or sensing of the primary signal in recipient cells (Dunoyer et al. 2010b). Supporting this hypothesis, the amount of SUL siRNA in the *pSUC2-SUL* line was unchanged in *clsy1, nrpd1* and *rdr2* mutants, despite a complete impairment of *pSUC2-SUL* IR-PTGS, suggesting that CLSY1, PolIV and RDR2 act downstream of the production of siRNA (Dunoyer et al. 2007).

HOMEOSTASIS OF SMALL RNA PATHWAYS

Most small RNA pathways involve specialized components. However, some components are shared among different pathways, opening the door to possible competition among pathways. For example, a suppressor screen recently shed light on a possible competition

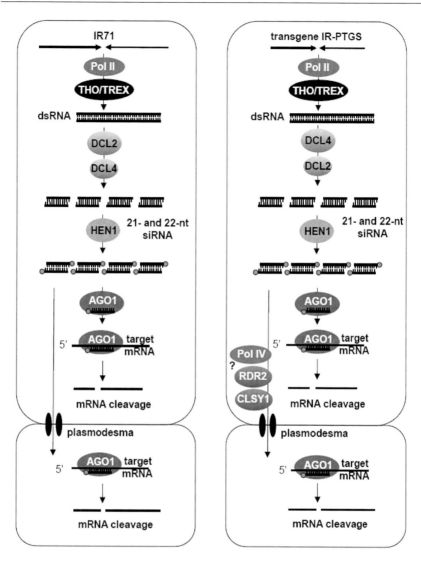

Figure 2: General components of IR-PTGS pathways. Protein components represented by the same color participate in similar steps in the pathways. The size of the ring around DCL2 and DCL4 in the endo IR and transgene IR pathways represents their hierarchical action. The question mark beside RDR2 and NRPD1 indicate that their exact step of action is not known. As in Figure 1, additional steps occurring in particular instances (involvement of 3′ RNA processing proteins in transgene IR-PTGS, involvement of chromatin-related proteins in IR-PTGS, etc), or steps not directly related to PTGS (production of 24-nt siRNA by DCL3, DNA methylation guided by 24-nt siRNA loaded on AGO4, etc.) are not shown on this figure but are detailed in the text.

(Color image of this figure appears in the color plate section at the end of the book.)

between siRNA and miRNA for methylation by HEN1. HEN1 methylates all type of known small RNA (Li et al. 2005) and this methylation is crucial to stabilize and protect small RNAs against uridylation and degradation by SDN (Ramachandran et al. 2008). *nrpd1, nrpd2* and *rdr2* mutations, which impair the production of endogenous 24-nt siRNAs, suppress the miRNA methylation defects of the hypomorphic *hen1-2* mutant (Yu et al. 2010). It is likely that the absence of these competing HEN1 substrates, which represent ca. 84% of the small RNA of a wildtype Arabidopsis cell, allows the partially active HEN1-2 protein to restore an almost wildtype level of miRNA methylation. Although HEN1 does not appear to be limiting in wildtype plants, the situation observed in the hypomorphic *hen1-2* mutant may mimic what happens during a burst of small RNA production, for example during virus infection. Indeed, virus-derived small RNA can represent up to 64% of the small RNA in an infected cell (Ding 2010). Because 24-nt siRNA are naturally more abundant than miRNA, virus-derived small RNA could compete with endogenous small RNA, including miRNA, so that virus infection could indirectly cause defects in miRNA methylation, which could be at least partly responsible for the developmental defects observed in virus-infected plants.

Like HEN1, AGO1 also is shared among different pathways. Compromising AGO1 by mutations (Bohmert et al. 1998) or by viral suppressors (Bortolamiol et al. 2007) leads to dramatic developmental consequences for the plant. In wildtype plants, AGO1 homeostasis is achieved by several regulatory loops, which allow miRNA and siRNA pathways to function correctly. AGO1 homeostasis is achieved through the miRNA pathway by miR168-guided cleavage of AGO1 mRNA (Vaucheret et al. 2004), miR168 preferential stabilization by AGO1 when AGO1 levels are high (Vaucheret et al. 2006) and translational repression of AGO1 by AGO10 (Mallory et al. 2009). Moreover, AGO1 homeostasis is achieved through the siRNA pathway by the production of a 22-nt miR168 species (Vaucheret 2009), which, after cleavage of AGO1 mRNA, allows the production of AGO1 siRNA that also contribute to the regulation of AGO1 mRNA level (Mallory et al. 2009). Supporting the importance of miR168 in the regulation of AGO1, expression of a miR168-resistant version of AGO1 causes developmental defects leading to the eventual death of the plant (Vaucheret et al. 2004). Recently, additional factors that regulate the level of AGO1 protein or activity were identified, including the cyclophilin protein SQN and F-box protein FBW2 (Smith et al. 2009, Earley et al. 2010). Another important protein of the miRNA pathway, DCL1, also is tightly regulated by regulatory loops involving miRNA. These regulatory loops include miR838-excision from DCL1 pre-mRNA (Rajagopalan et al. 2006) and miR162-guided cleavage of mature DCL1 mRNA (Xie et al.

2003). Because many miRNA are expressed in a tissue-specific manner or respond to endogenous or environmental stimuli, AGO1 and DCL1 regulatory loops likely allow cells to fine-tune the amount of AGO1 and DCL1 proteins so their levels are optimal for cellular function.

CROSSTALK BETWEEN PTGS AND RQC PATHWAYS

Many mRNA processing components and RNA quality control (RQC) proteins were found in PTGS genetic screens, acting either as effectors or suppressors of PTGS. For example, the Cap-binding proteins CBP20 and CBP80 and the splicing factor Zn-finger protein SE are involved in the miRNA pathway (Gregory et al. 2008, Laubinger et al. 2008), the 3′ RNA processing proteins FCA, FPA and FY are involved in IR-PTGS (Baurle et al. 2007, Manzano et al. 2009), while the members of the THO/TREX RNA trafficking complex HPR1, TEX1 and THO6 are involved in the tasiRNA, S-PTGS and IR-PTGS pathways (Jauvion et al. 2010, Yelina et al. 2010). Moreover, suppressor PTGS screens identified the putative splicing factor ESP3, the RNA 3′ end formation factors ESP1, ESP4 and ESP5 and the 5′→3′ exoribonucleases XRN2, XRN3 and XRN4 (Gazzani et al. 2004, Herr et al. 2006, Gy et al. 2007). These data are consistent with the impairment of IR-PTGS in plants lacking the non-sense mediated decay protein UPF1 (Arciga-Reyes et al. 2006) and the increase in S-PTGS efficiency when using transgenes lacking a polyadenylation signal (Luo et al. 2007). Altogether, these data indicate that RNA structure and localization are important determinants of PTGS efficiency.

These data also point to specific branchpoints between PTGS and RQC pathways. For example, *xrn2*, *xrn3* and *xrn4* mutations cause an increase in S-PTGS efficiency likely because siRNA-guided mRNA cleavage products can re-enter the S-PTGS pathway instead of being degraded. In contrast, *xrn2* and *xrn3* mutations cause an increase in the accumulation of miRNA processing products and *xrn4* mutations cause an increase in the accumulation of many miRNA-guided mRNA cleavage products without compromising the correct functioning of the miRNA pathway because these products cannot re-enter the miRNA or siRNA pathways (Souret et al. 2004, Gy et al. 2007). These results are consistent with the fact that 22-nt small RNA-derived mRNA cleavage products are protected from degradation by SGS3, allowing them to become substrates for RDR6 pathway and to escape degradation by XRN4 whereas 21-nt small RNA-derived mRNA cleavage products are degraded by XRN4 (Yoshikawa et al. 2005, Chen et al. 2010, Cuperus et al. 2010). How 22-nt small RNA-derived mRNA cleavage products selectively enter the PTGS pathway instead of the RQC pathway remains to be determined.

PTGS AS A TOOL

Knocking-out gene expression is an important goal in developmental biology, especially in higher plants where homologous recombination does not function in vegetative cells. In the late 80's, plant gene down-regulation was attempted using antisense transgenes, but this technique gave highly variable results. In the early 90's, the ability of sense transgenes to silence homologous endogenous genes by cosuppression (S-PTGS) offered a new tool for gene knock-out. However, the efficiency of this technique also appeared highly variable and unpredictable. After the discovery that dsRNA is the common inducer of PTGS, IR transgenes became popular and started to be used under the generic term "RNAi" to specifically knock-down gene expression. Despite a general success, the variability from gene to gene remained an unsolved problem (Kerschen et al. 2004). Moreover, it became obvious that IR transgenes lacked specificity in many instances, due to the production of hundreds of different siRNA, among which some could present sufficient homology with other endogenous genes to cause off-target silencing effects (Jackson et al. 2003). Indeed, up to five mismatches between a small RNA and its targets can be tolerated during AGO1-mediated cleavage in plants (Palatnik et al. 2003, Mallory et al. 2004). To circumvent this problem, artificial MIR genes were engineered by replacing a native miRNA sequence by a highly specific sequence (Parizotto et al. 2004, Vaucheret et al. 2004, Schwab et al. 2005). This method allowed the production of a single miRNA that, if correctly designed, could be a highly specific silencer. Moreover, artificial MIR genes can be expressed under tissue-specific or inducible promoters, allowing spatio-temporal down-regulation (Schwab et al. 2006). Artificial tasiRNA, replacement of one or multiple tasiRNA within a TAS gene, have also been used as highly specific silencers (de la Luz Gutierrez-Nava et al. 2008, Montgomery et al. 2008, Tretter et al. 2008, Felippes et al. 2009). However, the recent findings that siRNA can move from cell to cell and through the vasculature (Dunoyer et al. 2010a, Molnar et al. 2010), discourages to the use of artificial tasiRNA for tissue-specific down-regulation. Although artificial miRNA do not appear to be as mobile as artificial tasiRNA (Tretter et al. 2008, Schwab et al. 2009), examples of miRNA-mediated silencing in cells that do not express the corresponding miRNA (Chitwood et al. 2009, Nogueira et al. 2009, Marin et al. 2010) calls for caution when designing artificial miRNA for tissue-specific down-regulation. Moreover, the recent discovery that 22-nt small RNA engage cleavage products in the production of secondary siRNA (Cuperus et al. 2010) suggests that artificial miRNA could lead to the production of secondary siRNA, which could promote off-target effects. In these cases, experimental validation of the production of a single 21-nt miRNA and of the absence of production of secondary siRNA from cleavage products is highly recommended before using artificial miRNAs as silencers.

REFERENCES

Adenot, X., T. Elmayan, D. Lauressergues, S. Boutet, N. Bouche, V. Gasciolli and H. Vaucheret. 2006. DRB4-dependent TAS3 trans-acting siRNAs control leaf morphology through AGO7. *Curr. Biol.* **16**:927–932.

Allen, E., Z. Xie, A.M. Gustafson and J.C. Carrington. 2005. microRNA-directed phasing during trans-acting siRNA biogenesis in plants. *Cell.* **121**:207–221.

Arciga-Reyes, L., L. Wootton, M. Kieffer and B. Davies. 2006. UPF1 is required for nonsense-mediated mRNA decay (NMD) and RNAi in Arabidopsis. *Plant J.* **47**:480–489.

Axtell, M.J., C. Jan, R. Rajagopalan and D.P. Bartel. 2006. A two-hit trigger for siRNA biogenesis in plants. *Cell.* **127**:565–577.

Baumberger, N. and D.C. Baulcombe. 2005. Arabidopsis ARGONAUTE1 is an RNA Slicer that selectively recruits microRNAs and short interfering RNAs. *Proc. Natl. Acad. Sci. USA.* **102**:11928–11933.

Baurle, I., L. Smith, D.C. Baulcombe and C. Dean. 2007. Widespread role for the flowering-time regulators FCA and FPA in RNA-mediated chromatin silencing. *Science.* **318**:109–112.

Beclin, C., S. Boutet, P. Waterhouse and H. Vaucheret. 2002. A branched pathway for transgene-induced RNA silencing in plants. *Curr. Biol.* **12**:684–688.

Ben Amor, B., S. Wirth, F. Merchan, P. Laporte, Y. d'Aubenton-Carafa, J. Hirsch, A. Maizel, A. Mallory, A. Lucas, J.M. Deragon, H. Vaucheret, C. Thermes and Crespi. 2009. Novel long non-protein coding RNAs involved in Arabidopsis differentiation and stress responses. *Genome Res.* **19**:57–69.

Blevins, T., R. Rajeswaran, P.V. Shivaprasad, D. Beknazariants, A. Si-Ammour, H.S. Park, F. Vazquez, D. Robertson, F. Jr. Meins, T. Hohn and M.M. Pooggin. 2006. Four plant Dicers mediate viral small RNA biogenesis and DNA virus induced silencing. *Nucleic Acids Res.* **34**:6233–6246.

Bohmert, K., I. Camus, C. Bellini, D. Bouchez, M. Caboche and C. Benning. 1998. AGO1 defines a novel locus of Arabidopsis controlling leaf development. *EMBO J.* **17**:170–180.

Bologna, N.G., J.L. Mateos, E.G. Bresso and J.F. Palatnik. 2009. A loop-to-base processing mechanism underlies the biogenesis of plant microRNAs miR319 and miR159. *EMBO J.* **28**:3646–3656.

Bortolamiol, D., M. Pazhouhandeh, K. Marrocco, P. Genschik and V. Ziegler-Graff. 2007. The Polerovirus F box protein P0 targets ARGONAUTE1 to suppress RNA silencing. *Curr. Biol.* **17**:1615–1621.

Bouche, N., D. Lauressergues, V. Gasciolli and H. Vaucheret. 2006. An antagonistic function for Arabidopsis DCL2 in development and a new function for DCL4 in generating viral siRNAs. *EMBO J.* **25**:3347–3356.

Boutet, S., F. Vazquez, J. Liu, C. Beclin, M. Fagard, A. Gratias, J.B. Morel, P. Crete, X. Chen and H. Vaucheret. 2003. Arabidopsis HEN1: a genetic link between endogenous miRNA controlling development and siRNA controlling transgene silencing and virus resistance. *Curr. Biol.* **13**:843–848.

Brodersen, P., L. Sakvarelidze-Achard, M. Bruun-Rasmussen, P. Dunoyer, Y.Y. Yamamoto, L. Sieburth and O. Voinnet. 2008. Widespread translational inhibition by plant miRNAs and siRNAs. *Science.* **320**:1185–1190.

Chellappan, P., J. Xia, X. Zhou, S. Gao, X. Zhang, G. Coutino, F. Vazquez, W. Zhang and H. Jin. 2010. siRNAs from miRNA sites mediate DNA methylation of target genes. *Nucleic Acids Res.* **38**:6883–94.

Chen, H.M., L.T. Chen, K. Patel, Y.H. Li, D.C. Baulcombe and S.H. Wu. 2010. 22-nucleotide RNAs trigger secondary siRNA biogenesis in plants. *Proc. Natl. Acad. Sci. USA.* **107**:15269–74.

Chitwood, D.H., F.T. Nogueira, M.D. Howell, T.A. Montgomery, J.C. Carrington and M.C. Timmermans. 2009. Pattern formation via small RNA mobility. *Genes Dev.* **23**:549–554.

Cuperus, J.T., T.A. Montgomery, N. Fahlgren, R.T. Burke, T. Townsend, C.M. Sullivan and J.C. Carrington. 2009. Identification of MIR390a precursor processing-defective mutants in Arabidopsis by direct genome sequencing. *Proc. Natl. Acad. Sci. USA.* **107**:466–471.

Cuperus, J.T., A. Carbonell, N. Fahlgren, H. Garcia-Ruiz, R.T. Burke, A. Takeda, C.M. Sullivan, S.D. Gilbert, T.A. Montgomery and J.C. Carrington. 2010. Unique functionality of 22-nt miRNAs in triggering RDR6-dependent siRNA biogenesis from target transcripts in Arabidopsis. *Nat. Struct. Mol. Biol.* **17**:997–1003.

Dalmay, T., R. Horsefield, T.H. Braunstein and D.C. Baulcombe. 2001. SDE3 encodes an RNA helicase required for post-transcriptional gene silencing in Arabidopsis. *EMBO J.* **20**:2069–2078.

Dalmay, T., A. Hamilton, S. Rudd, S. Angell and D.C. Baulcombe. 2000. An RNA-dependent RNA polymerase gene in Arabidopsis is required for posttranscriptional gene silencing mediated by a transgene but not by a virus. *Cell.* **101**:543–553.

de la Luz Gutierrez-Nava, M., M.J. Aukerman, H. Sakai, S.V. Tingey and R.W. Williams. 2008. Artificial trans-acting siRNAs confer consistent and effective gene silencing. *Plant Physiol.* **147**:543–551.

Dehio, C. and J. Schell. 1994. Identification of plant genetic loci involved in a posttranscriptional mechanism for meiotically reversible transgene silencing. *Proc. Natl. Acad. Sci. USA.* **91**:5538–5542.

Deleris, A., J. Gallego-Bartolome, J. Bao, K.D. Kasschau, J.C. Carrington and O. Voinnet. 2006. Hierarchical action and inhibition of plant Dicer-like proteins in antiviral defense. *Science.* **313**:68–71.

Ding, S.W. 2010. RNA-based antiviral immunity. *Nat. Rev. Immunol.* **10**:632–44.

Ding, S.W. and O. Voinnet. 2007. Antiviral immunity directed by small RNAs. *Cell.* **130**:413–426.

Donaire, L., D. Barajas, B. Martinez-Garcia, L. Martinez-Priego, I. Pagan and C. Llave. 2008. Structural and genetic requirements for the biogenesis of tobacco rattle virus-derived small interfering RNAs. *J. Virol.* **82**:5167–5177.

Dunoyer, P., C. Himber and O. Voinnet. 2005. DICER-LIKE 4 is required for RNA interference and produces the 21-nucleotide small interfering RNA component of the plant cell-to-cell silencing signal. *Nat. Genet.* **37**:1356–1360.

Dunoyer, P., C.H. Lecellier, E.A. Parizotto, C. Himber and O. Voinnet. 2004. Probing the microRNA and small interfering RNA pathways with virus-encoded suppressors of RNA silencing. Plant *Cell.* **16**:1235–1250.

Dunoyer, P., C. Himber, V. Ruiz-Ferrer, A. Alioua, and O. Voinnet. 2007. Intra- and intercellular RNA interference in Arabidopsis thaliana requires components of the microRNA and heterochromatic silencing pathways. *Nat. Genet.* **39**:848–856.

Dunoyer, P., G. Schott, C. Himber, D. Meyer, A. Takeda, J.C. Carrington and O. Voinnet. 2010a. Small RNA duplexes function as mobile silencing signals between plant cells. *Science.* **328**:912–916.

Dunoyer, P., C.A. Brosnan, G. Schott, Y. Wang, F. Jay, A. Alioua, C. Himber and O. Voinnet. 2010b. An endogenous, systemic RNAi pathway in plants. *EMBO J.* **29**:1699–1712.

Earley, K., M. Smith, R. Weber, B. Gregory and R. Poethig. 2010. An endogenous F-box protein regulates ARGONAUTE1 in Arabidopsis thaliana. *Silence.* **1**:15.

Elmayan, T., X. Adenot, L. Gissot, D. Lauressergues, I. Gy and H. Vaucheret. 2009. A neomorphic sgs3 allele stabilizing miRNA cleavage products reveals that SGS3 acts as a homodimer. *FEBS J.* **276**:835–844.

Elmayan, T., S. Balzergue, F. Beon, V. Bourdon, J. Daubremet, Y. Guenet, P. Mourrain, J.C. Palauqui, S. Vernhettes, T. Vialle, K. Wostrikoff and H. Vaucheret. 1998. Arabidopsis mutants impaired in cosuppression. *Plant Cell.* **10**:1747–1758.

Fagard, M., S. Boutet, J.B. Morel, C. Bellini and H. Vaucheret. 2000. AGO1, QDE-2 and RDE-1 are related proteins required for post-transcriptional gene silencing in plants, quelling in fungi and RNA interference in animals. *Proc. Natl. Acad. Sci. USA.* **97**:11650–11654.

Fang, Y. and D.L. Spector. 2007. Identification of nuclear dicing bodies containing proteins for microRNA biogenesis in living Arabidopsis plants. *Curr. Biol.* **17**:818–823.

Felippes, F.F. and D. Weigel. 2009. Triggering the formation of tasiRNAs in Arabidopsis thaliana: the role of microRNA miR173. *EMBO Rep.* **10**:264–270.

Fire, A., S. Xu, M.K. Montgomery, S.A. Kostas, S.E. Driver and C.C. Mello. 1998. Potent and specific genetic interference by double-stranded RNA in *Caenorhabditis elegans. Nature.* **391**:806–811.

Fujioka, Y., M. Utsumi, Y. Ohba and Y. Watanabe. 2007. Location of a possible miRNA processing site in SmD3/SmB nuclear bodies in Arabidopsis. *Plant Cell. Physiol.* **48**:1243–1253.

Gasciolli, V., A.C. Mallory, D.P. Bartel and H. Vaucheret. 2005. Partially redundant functions of Arabidopsis DICER-like enzymes and a role for DCL4 in producing trans-acting siRNAs. *Curr. Biol.* **15**:1494–1500.

Gazzani, S., T. Lawrenson, C. Woodward, D. Headon and R. Sablowski. 2004. A link between mRNA turnover and RNA interference in Arabidopsis. *Science.* **306**:1046–1048.

Gregory, B.D., R.C. O'Malley, R. Lister, M.A. Urich, J. Tonti-Filippini, H. Chen, A.H. Millar and J.R. Ecker. 2008. A link between RNA metabolism and silencing affecting Arabidopsis development. *Dev. Cell.* **14**:854–866.

Gy, I., V. Gasciolli, D. Lauressergues, J.B. Morel, J. Gombert, F. Proux, C. Proux, H. Vaucheret and A.C. Mallory. 2007. Arabidopsis FIERY1, XRN2 and XRN3 are Endogenous RNA Silencing Suppressors. *Plant Cell.* **19**:3451–3461.

Hamilton, A.J. and D.C. Baulcombe. 1999. A species of small antisense RNA in posttranscriptional gene silencing in plants. *Science.* **286**:950–952.

Han, M.H., S. Goud, L. Song and N. Fedoroff. 2004. The Arabidopsis double-stranded RNA-binding protein HYL1 plays a role in microRNA-mediated gene regulation. *Proc. Natl. Acad. Sci. USA.* **101**:1093–1098.

Henderson, I.R., X. Zhang, C. Lu, L. Johnson, B.C. Meyers, P.J. Green and S.E. Jacobsen. 2006. Dissecting Arabidopsis thaliana DICER function in small RNA processing, gene silencing and DNA methylation patterning. *Nat. Genet.* **38**:721–725.

Hernandez-Pinzon, I., N.E. Yelina, F. Schwach, D.J. Studholme, D. Baulcombe and T. Dalmay. 2007. SDE5, the putative homologue of a human mRNA export factor, is required for transgene silencing and accumulation of trans-acting endogenous siRNA. *Plant J.* **50**:140–148.

Herr, A.J., M.B. Jensen, T. Dalmay and D.C. Baulcombe. 2005. RNA polymerase IV directs silencing of endogenous DNA. *Science.* **308**:118–120.

Herr, A.J., A. Molnar, A. Jones and D.C. Baulcombe. 2006. Defective RNA processing enhances RNA silencing and influences flowering of Arabidopsis. *Proc. Natl. Acad. Sci. USA.* **103**:14994–15001.

Himber, C., P. Dunoyer, G. Moissiard, C. Ritzenthaler and O. Voinnet. 2003. Transitivity-dependent and -independent cell-to-cell movement of RNA silencing. *EMBO J.* **22**:4523–4533.

Hunter, C., H. Sun and R.S. Poethig. 2003. The Arabidopsis heterochronic gene ZIPPY is an ARGONAUTE family member. *Curr. Biol.* **13**:1734–1739.

Hunter, C., M.R. Willmann, G. Wu, M. Yoshikawa, M. de la Luz Gutierrez-Nava and S.R. Poethig. 2006. Trans-acting siRNA-mediated repression of ETTIN and ARF4 regulates heteroblasty in Arabidopsis. Development **133**:2973–2981.

Jackson, A.L., S.R. Bartz, J. Schelter, S.V. Kobayashi, J. Burchard, M. Mao, B. Li, G. Cavet and P.S. Linsley. 2003. Expression profiling reveals off-target gene regulation by RNAi. *Nat. Biotechnol.* **21**:635–637.

Jauvion, V., T. Elmayan and H. Vaucheret. 2010. The Conserved RNA Trafficking Proteins HPR1 and TEX1 Are Involved in the Production of Endogenous and Exogenous Small Interfering RNA in Arabidopsis. *Plant Cell.* **22**:2697–2709.

Kerschen, A., and C.A. Napoli, R.A. Jorgensen and A.E. Muller. 2004. Effectiveness of RNA interference in transgenic plants. *FEBS Lett.* **566**:223–228.

Kim, S., J.Y. Yang, J. Xu, I.C. Jang, M.J. Prigge and N.H. Chua. 2008. Two cap-binding proteins CBP20 and CBP80 are involved in processing primary MicroRNAs. *Plant Cell. Physiol.* **49**:1634–1644.

Kunz, C., H. Schöb, M. Stam, J.M. Kooter and Jr.F. Meins. 1996. Developmentally regulated silencing and reactivation of tobacco chitinase transgene expression. *Plant J.* **10**:437–450.

Kurihara, Y. and Y. Watanabe. 2004. Arabidopsis micro-RNA biogenesis through Dicer-like 1 protein functions. *Proc. Natl. Acad. Sci. USA.* **101**:12753–12758.

Laubinger, S., T. Sachsenberg, G. Zeller, W. Busch, J.U. Lohmann, G. Ratsch and D. Weigel. 2008. Dual roles of the nuclear cap-binding complex and SERRATE in pre-mRNA splicing and microRNA processing in Arabidopsis thaliana. *Proc. Natl. Acad. Sci. USA.* **105**:8795–8800.

Li, J., Z. Yang, B. Yu, J. Liu and X. Chen. 2005. Methylation protects miRNAs and siRNAs from a 3'-end uridylation activity in Arabidopsis. *Curr. Biol.* **15**:1501–1507.

Lobbes, D., G. Rallapalli, D.D. Schmidt, C. Martin and J. Clarke. 2006. SERRATE: a new player on the plant microRNA scene. *EMBO Rep.* **7**:1052–1058.

Luo, Z. and Z. Chen 2007. Improperly terminated, unpolyadenylated mRNA of sense transgenes is targeted by RDR6-mediated RNA silencing in Arabidopsis. *Plant Cell.* **19**:943–958.

Mallory, A.C., H. Vaucheret. 2009. ARGONAUTE 1 homeostasis invokes the coordinate action of the microRNA and siRNA pathways. *EMBO Rep.* **10**:521–526.

Mallory, A.C., B.J. Reinhart, M.W. Jones-Rhoades, G. Tang, P.D. Zamore, M.K. Barton and D.P. Bartel. 2004. MicroRNA control of PHABULOSA in leaf development: importance of pairing to the microRNA 5' region. *EMBO J.* **23**:3356–3364.

Mallory, A.C., A. Hinze, M.R. Tucker, N. Bouche, V. Gasciolli, T. Elmayan, D. Lauressergues, V. Jauvion, H. Vaucheret and T. Laux. 2009. Redundant and specific roles of the ARGONAUTE proteins AGO1 and ZLL in development and small RNA-directed gene silencing. *PLoS Genet.* **5**:e1000646.

Manzano, D., S. Marquardt, A.M. Jones, I. Baurle, F. Liu and C. Dean. 2009. Altered interactions within FY/AtCPSF complexes required for Arabidopsis FCA-mediated chromatin silencing. *Proc. Natl. Acad. Sci. USA.* **106**:8772–8777.

Marin, E., V. Jouannet, A. Herz, A.S. Lokerse, D. Weijers, H. Vaucheret, L. Nussaume, M.D. Crespi and A. Maizel. 2010. miR390, Arabidopsis TAS3 tasiRNAs and their AUXIN RESPONSE FACTOR targets define an autoregulatory network quantitatively regulating lateral root growth. *Plant Cell.* **22**:1104–1117.

Mi, S., T. Cai, Y. Hu, Y. Chen, E. Hodges, F. Ni, L. Wu, S. Li, H. Zhou, C. Long, S. Chen, G.J. Hannon and Y. Qi. 2008. Sorting of small RNAs into Arabidopsis argonaute complexes is directed by the 5' terminal nucleotide. *Cell.* **133**:116–127.

Mlotshwa, S., G.J. Pruss, A. Peragine, M.W. Endres, J. Li, X. Chen, R.S. Poethig, L.H. Bowman and V. Vance. 2008. DICER-LIKE2 plays a primary role in transitive silencing of transgenes in Arabidopsis. *PLoS One.* **3**:e1755.

Moissiard, G. and O. Voinnet. 2006. RNA silencing of host transcripts by cauliflower mosaic virus requires coordinated action of the four Arabidopsis Dicer-like proteins. *Proc. Natl. Acad. Sci. USA.* **103**:19593–19598.

Molnar, A., C.W. Melnyk, A. Bassett, T.J. Hardcastle, R. Dunn and D.C. Baulcombe 2010. Small silencing RNAs in plants are mobile and direct epigenetic modification in recipient cells. *Science.* **328**:872–875.

Montgomery, T.A., M.D. Howell, Cuperus, D. Li, J.E. Hansen, A.L. Alexander, E.J. Chapman, N. Fahlgren, E. Allen and J.C. Carrington 2008. Specificity of ARGONAUTE7-miR390 interaction and dual functionality in TAS3 trans-acting siRNA formation. *Cell.* **133**:128–141.

Morel, J.B., C. Godon, P. Mourrain, C. Beclin, S. Boutet, F. Feuerbach, F. Proux and H. Vaucheret 2002. Fertile hypomorphic ARGONAUTE (*ago1*) mutants impaired in post-transcriptional gene silencing and virus resistance. *Plant Cell.* **14**:629–639.

Morel, J.B., P. Mourrain, C. Beclin and H. Vaucheret. 2000. DNA methylation and chromatin structure affect transcriptional and post-transcriptional transgene silencing in Arabidopsis. *Curr. Biol.* **10**:1591–1594.

Mourrain, P., C. Beclin, T. Elmayan, F. Feuerbach, C. Godon, J.B. Morel, D. Jouette, A.M. Lacombe, S. Nikic, N. Picault, K. Remoue, M. Sanial, T.A. Vo and H. Vaucheret. 2000. Arabidopsis SGS2 and SGS3 genes are required for posttranscriptional gene silencing and natural virus resistance. *Cell.* **101**:533–542.

Nakazawa, Y., A. Hiraguri, H. Moriyama and T. Fukuhara. 2007. The dsRNA-binding protein DRB4 interacts with the Dicer-like protein DCL4 *in vivo* and functions in the trans-acting siRNA pathway. *Plant Mol. Biol.* **63**:777–785.

Napoli, C., C. Lemieux and R. Jorgensen. 1990. Introduction of a Chimeric Chalcone Synthase Gene into Petunia Results in Reversible Co-Suppression of Homologous Genes in trans. *Plant Cell.* **2**:279–289.

Nogueira, F.T., D.H. Chitwood, S. Madi, K. Ohtsu, P.S. Schnable, M.J. Scanlon and M.C. Timmermans. 2009. Regulation of small RNA accumulation in the maize shoot apex. *PLoS Genet.* **5**:e1000320.

Palatnik, J.F., E. Allen, X. Wu, C. Schommer, R. Schwab, J.C. Carrington and D. Weigel. 2003. Control of leaf morphogenesis by microRNAs. *Nature.* **425**:257–263.

Palauqui, J.C. and H. Vaucheret. 1995. Field trial analysis of nitrate reductase co-suppression: a comparative study of 38 combinations of transgene loci. *Plant. Mol. Biol.* **29**:149–159.

Palauqui, J.C., H. Elmayan, J.M. Pollien and H. Vaucheret. 1997. Systemic acquired silencing: transgene-specific post-transcriptional silencing is transmitted by grafting from silenced stocks to non-silenced scions. *EMBO J.* **16**:4738–4745.

Palauqui, J.C., T. Elmayan, F.D. De Borne, P. Crete, C. Charles and H. Vaucheret. 1996. Frequencies, Timing and Spatial Patterns of Co-suppression of Nitrate Reductase and Nitrite Reductase in Transgenic Tobacco Plants. *Plant Physiol.* **112**:1447–1456.

Parizotto, E.A., P. Dunoyer, N. Rahm, C. Himber and O. Voinnet. 2004. *In vivo* investigation of the transcription, processing, endonucleolytic activity and functional relevance of the spatial distribution of a plant miRNA. *Genes Dev.* **18**:2237–2242.

Park, M.Y., G. Wu, A. Gonzalez-Sulser, H. Vaucheret and R.S. Poethig. 2005. Nuclear processing and export of microRNAs in Arabidopsis. *Proc. Natl. Acad. Sci. USA.* **102**:3691–3696.

Park, W., J. Li, R. Song, J. Messing and X. Chen. 2002. CARPEL FACTORY, a Dicer homolog and HEN1, a novel protein, act in microRNA metabolism in Arabidopsis thaliana. *Curr. Biol.* **12**:1484–1495.

Peragine, A., M. Yoshikawa, G. Wu, H.L. Albrecht and R.S. Poethig. 2004. SGS3 and SGS2/SDE1/RDR6 are required for juvenile development and the production of trans-acting siRNAs in Arabidopsis. *Genes Dev.* **18**:2368–2379.

Pouch-Pelissier, M.N., T. Pelissier, T. Elmayan, H. Vaucheret, D. Boko, M.F. Jantsch and J.M. Deragon. 2008. SINE RNA induces severe developmental defects in Arabidopsis thaliana and interacts with HYL1 (DRB1), a key member of the DCL1 complex. *PLoS Genet.* **4**:e1000096.

Qi, X., F.S. Bao and Z. Xie. 2009. Small RNA deep sequencing reveals role for Arabidopsis thaliana RNA-dependent RNA polymerases in viral siRNA biogenesis. *PLoS One.* **4**:e4971.

Qu, F., X. Ye and T.J. Morris. 2008. Arabidopsis DRB4, AGO1, AGO7 and RDR6 participate in a DCL4-initiated antiviral RNA silencing pathway negatively regulated by DCL1. *Proc. Natl. Acad. Sci. USA.* **105**:14732–14737.

Qu, F., X. Ye, G. Hou, S. Sato, T.E. Clemente and T.J. Morris. 2005. RDR6 has a broad-spectrum but temperature-dependent antiviral defense role in *Nicotiana benthamiana. J. Virol.* **79**:15209–15217.

R. Schwab, J.F. Palatnik, M. Riester, C. Schommer, M. Schmid and D. Weigel. 2005. Specific effects of microRNAs on the plant transcriptome. *Dev. Cell.* **8**:517–527.

Rajagopalan, R., H. Vaucheret, J. Trejo and D.P. Bartel. 2006. A diverse and evolutionarily fluid set of microRNAs in Arabidopsis thaliana. *Genes Dev.* **20**:3407–3425.

Ramachandran, V. and X. Chen. 2008. Degradation of microRNAs by a family of exoribonucleases in Arabidopsis. *Science.* **321**:1490–1492.

Ruiz-Ferrer, V. and O. Voinnet. 2009. Roles of plant small RNAs in biotic stress responses. *Annu. Rev. Plant Biol.* **60**:485–510.

Schwab, R., S. Ossowski, M. Riester, N. Warthmann and D. Weigel. 2006. Highly specific gene silencing by artificial microRNAs in Arabidopsis. *Plant Cell.* **18**:1121–1133.

Schwab, R., A. Maizel, V. Ruiz-Ferrer, D. Garcia, M. Bayer, M. Crespi, O. Voinnet and R.A. Martienssen. 2009. Endogenous TasiRNAs mediate non-cell autonomous effects on gene regulation in Arabidopsis thaliana. *PLoS One.* **4**:e5980.

Schwach, F., F.E. Vaistij, L. Jones and D.C. Baulcombe. 2005. An RNA-dependent RNA polymerase prevents meristem invasion by potato virus X and is required for the activity but not the production of a systemic silencing signal. *Plant Physiol.* **138**:1842–1852.

Searle, I.R., O. Pontes, C.W. Melnyk, L.M. Smith and D.C. Baulcombe. 2010. JMJ14, a JmjC domain protein, is required for RNA silencing and cell-to-cell movement of an RNA silencing signal in Arabidopsis. *Genes Dev.* **24**:986–991.

Smith, C.J., C.F. Watson, C.R. Bird, J. Ray, W. Schuch and D. Grierson. 1990. Expression of a truncated tomato polygalacturonase gene inhibits expression of the endogenous gene in transgenic plants. *Mol. Gen. Genet.* **224**:477–481.

Smith, L.M., O. Pontes, I. Searle, N. Yelina, F.K. Yousafzai, A.J. Herr, C.S. Pikaard and D.C. Baulcombe. 2007. An SNF2 protein associated with nuclear RNA silencing and the spread of a silencing signal between cells in Arabidopsis. *Plant Cell.* **19**:1507–1521.

Smith, M.R., M.R. Willmann, G. Wu, T.Z. Berardini, B. Moller, D. Weijers and R.S. Poethig. 2009. Cyclophilin 40 is required for microRNA activity in Arabidopsis. *Proc. Natl. Acad. Sci. USA.* **106**:5424–5429.

Souret, F.F., J.P. Kastenmayer and P.J. Green. 2004. AtXRN4 degrades mRNA in Arabidopsis and its substrates include selected miRNA targets. *Mol. Cell.* **15**:173–183.

Takeda, A., S. Iwasaki, T. Watanabe, M. Utsumi and Y. Watanabe. 2008. The mechanism selecting the guide strand from small RNA duplexes is different among argonaute proteins. *Plant Cell Physiol.* **49**:493–500.

Tretter, E.M., J.P. Alvarez, Y. Eshed and J.L. Bowman. 2008. Activity range of Arabidopsis small RNAs derived from different biogenesis pathways. *Plant Physiol.* **147**:58–62.

van der Krol, A.R., L.A. Mur, M. Beld, J.N. Mol and A.R. Stuitje. 1990. Flavonoid genes in petunia: addition of a limited number of gene copies may lead to a suppression of gene expression. *Plant Cell.* **2**:291–299.

Vaucheret, H. 2008. Plant ARGONAUTES. *Trends Plant Sci.* **13**:350–358. Vaucheret, H. 2009. AGO1 homeostasis involves differential production of 21-nt and 22-nt miR168 species by MIR168a and MIR168b. *PLoS One.* **4**:e6442.

Vaucheret, H., A.C. Mallory and D.P. Bartel. 2006. AGO1 homeostasis entails coexpression of MIR168 and AGO1 and preferential stabilization of miR168 by AGO1. *Mol. Cell.* **22**:129–136.

Vaucheret, H., J.C. Palauqui, T. Elmayan and B. Moffatt. 1995. Molecular and genetic analysis of nitrite reductase co-suppression in transgenic tobacco plants. *Mol. Gen. Genet.* **248**:311–317.

Vaucheret, H., F. Vazquez, P. Crete and D.P. Bartel. 2004. The action of ARGONAUTE1 in the miRNA pathway and its regulation by the miRNA pathway are crucial for plant development. *Genes Dev.* **18**:1187–1197.

Vazquez, F., T. Blevins, J. Ailhas, T. Boller and F.Jr. Meins 2008. Evolution of Arabidopsis MIR genes generates novel microRNA classes. *Nucleic Acids Res.* **36**:6429–6438.

Vazquez, F., H. Vaucheret, R. Rajagopalan, C. Lepers, V. Gasciolli, A.C. Mallory, J.L. Hilbert, D.P. Bartel and P. Crete. 2004. Endogenous trans-acting siRNAs regulate the accumulation of Arabidopsis mRNAs. *Mol. Cell.* **16**:69–79.

Voinnet, O. and D.C. Baulcombe. 1997. Systemic signalling in gene silencing. *Nature.* **389**:553.

Voinnet, O., P. Vain, S. Angell and D.C. Baulcombe. 1998. Systemic spread of sequence-specific transgene RNA degradation in plants is initiated by localized introduction of ectopic promoterless DNA. *Cell.* **95**:177–187.

Wang, X.B., Q. Wu, T. Ito, F. Cillo, W.X. Li, X. Chen, J.L. Yu and S.W. Ding. 2010. RNAi-mediated viral immunity requires amplification of virus-derived siRNAs in Arabidopsis thaliana. *Proc. Natl. Acad. Sci. USA.* **107**:484–489.

Waterhouse, P.M., M.W. Graham and M.B. Wang. 1998. Virus resistance and gene silencing in plants can be induced by simultaneous expression of sense and antisense RNA. *Proc. Natl. Acad. Sci. USA.* **95**:13959–13964.

Wu, L., H. Zhou, Q. Zhang, J. Zhang, F. Ni, C. Liu and Y. Qi. 2010. DNA methylation mediated by a microRNA pathway. *Mol. Cell.* **38**:465–475.

Xie, Z., K.D. Kasschau and J.C. Carrington. 2003. Negative feedback regulation of Dicer-Like1 in Arabidopsis by microRNA-guided mRNA degradation. *Curr. Biol.* **13**:784–789.

Xie, Z., E. Allen, A. Wilken and J.C. Carrington. 2005. DICER-LIKE 4 functions in trans-acting small interfering RNA biogenesis and vegetative phase change in Arabidopsis thaliana. *Proc. Natl. Acad. Sci. USA.* **102**:12984–12989.

Yang, L., Z. Liu, F. Lu, A. Dong and H. Huang. 2006. SERRATE is a novel nuclear regulator in primary microRNA processing in Arabidopsis. *Plant J.* **47**:841–850.

Yelina, N.E., L.M. Smith, A.M. Jones, K. Patel, K.A. Kelly and D.C. Baulcombe. 2010. Putative Arabidopsis THO/TREX mRNA export complex is involved in transgene and endogenous siRNA biosynthesis. *Proc. Natl. Acad. Sci. USA.* **107**:13948–13953.

Yoshikawa, M., A. Peragine, M.Y. Park and R.S. Poethig. 2005. A pathway for the biogenesis of trans-acting siRNAs in Arabidopsis. *Genes Dev.* **19**:2164–2175.

Yu, B., L. Bi, B. Zheng, L. Ji, D. Chevalier, M. Agarwal, V. Ramachandran, W. Li, T. Lagrange, J.C. Walker and X. Chen 2008. The FHA domain proteins DAWDLE in Arabidopsis and SNIP1 in humans act in small RNA biogenesis. *Proc. Natl. Acad. Sci. USA.* **105**:10073–10078.

Yu, B., L. Bi, J. Zhai, M. Agarwal, S. Li, Q. Wu, S.W. Ding, B.C. Meyers, H. Vaucheret and X. Chen. 2010. siRNAs compete with miRNAs for methylation by HEN1 in Arabidopsis. *Nucleic Acids Res.* **38**:5844–50.

CHAPTER 3

Implications of miRNA-directed Gene Silencing in Cancer

Duncan Ayers

ABSTRACT

The discovery of RNA interference (RNAi) pathways in the 1990s heralded the dawn to a new age of research, whereby minute strands of non-coding RNA were found to be master regulators of a myriad cellular processes in many organisms, including humans. Such non-coding RNA strands, now renowned as microRNAs (miRNAs), are the basis of a new plane of research—situated between the genomic and the transcriptomic levels. MiRNAs have been found to fine-tune the cell's arduous task of maintaining functional harmony within the ever changing micro-environment around it. However, if this miRNA orchestration is thwarted in any way, the delicate balance of inter-molecular interactions can be tipped to the point of development of a cacophony within the cell's processing ability, ultimately leading to the onset and development of disease conditions. Unsurprisingly, cancer biology is also very much dependent on the dysregulation of expression of individual and/or networks of miRNAs for tumour development, metastasis and acquisition of chemoresistance. Such miRNA dysregulations involve the increased expression of oncomirs and suppressed expression of tumour suppressor miRNAs. Since such miRNAs are confirmed to play crucial roles in tumour development and severity of condition, active research is highly focused on developing translational medicine therapeutics aimed at restoring such target miRNA expression levels to the norm. Additionally, the identification of unique miRNA expression signatures could be potentially exploited in the clinic for the development of novel, highly sensitive and vital diagnostic systems for the early detection and detailed characterisation of individual tumours within the cancer patient.

Keywords: microRNA, mRNA, cancer, oncomir, drug resistance, chemoresistance.

---

* *Corresponding author e-mail*: Duncan.Ayers@UGent.be

INTRODUCTION

The first ever documented discovery of the existence of miRNAs dates back to 1993, whereby researchers focusing on the development of the *C. elegans* nematode identified the gene *lin-4* which coded for the production of small non coding RNA products that affected growth at the larval stage (R.C. Lee et al. 1993). Since this ground-breaking revelation in the field of miRNA discovery, there are now over 1000 differing miRNAs identified in humans alone, all of which sequences are catalogued in a comprehensive, publically accessible database has known as miRBase (www. mirbase.org) (Griffiths-Jones et al. 2008). This chapter aims at giving a brief overview regarding the processing of the mature miRNA within the cell, the molecular functions which enable it to regulate finely the expression of its target mRNA/s, and will also delve into the roles that miRNAs play within major cancer biology processes such as oncogenesis, tumour suppression and resistance to chemotherapeutics.

miRNA Processing and Mode of Action

At the genomic levels, miRNAs are located in intronic areas of host genes or within intergenic regions (Rodriguez et al. 2004). Additionally, further research highlighted the fact that over 50% of all miRNAs expressed in humans fall within cancer-associated genomic regions or within fragile sites (Sevignani et al. 2006).

   The initial stages of miRNA processing begin with transcription of the miRNA-bearing host gene within the cell nucleus into a pri-miRNA by the enzyme RNA polymerase II (Pol II) (Davis and Hata 2010). The pri-miRNA is essentially a long primary transcript, which is consequently targeted for cleavage by the nuclear enzyme Drosha into a precursor RNA (pre-miRNA), an approximately 60–70 base long stem-loop structured RNA (Chhabra et al. 2010). The pre-miRNA is then transported out of the nucleus and into the cell cytoplasm by means of an Exportin-5/Ran-GTP enzyme complex (Chhabra et al. 2010). Once the pre-miRNA is released into the cytoplasm, the RNAse III enzyme Dicer binds onto its target pre-miRNA, resulting in the formation of a mature miRNA duplex (miRNA/miRNA*) of approximately 22 nucleotides in length but having two 3' overhangs at both ends (Kwak et al. 2010) (Merritt et al. 2010) (Michela Garofalo and Carlo M. Croce 2010). Finally, the mature miRNA duplex is then bound by the RNA-induced silencing complex (RISC), whereby the passenger mRNA* strand of the miRNA duplex is cleaved (Kwak et al. 2010). Consequently, the remaining miRNA strand of the mature miRNA acts as a guide for its bound RISC enzyme, with this strand being either totally or partially complementary to a unique sequence on the target mRNA (Kwak et al. 2010).

Two possible modes of action exist by which the miRNA/RISC complex manages to inhibit translation of the target mRNA. If there is perfect complementarity between the miRNA and its target mRNA sequence, the latter is cleaved by means of the endonucleolytic properties of the Argonaute 2 (AGO2) domain present in RISC (Wu and Belasco 2008). Alternatively, the miRNA binds to the 3' untranslated region (UTR) of the target mRNA with imperfect complementarity on the 3'UTR seed region (nucleotides #2–8), thus repressing translation of the target mRNA into its intended protein products (Wu and Belasco 2008).

miRNA Involvement in Cancer Biology

Throughout the last decade, ever more research has revealed the link between miRNA dysregulations and cancer conditions (Calin et al. 2002) (Metzler et al. 2004) (Foley et al. 2010) (Mestdagh et al. 2010) (Lowery et al. 2010) (Osada and Takahashi 2010). The influence of miRNAs in cancer can be two-fold; either act in a manner similar to oncogenes by enhancing the tumour's development and clinical aggressivity (oncomirs), or to regulate and diminish tumour development and progression (tumour suppressors). Additionally, specific miRNAs can be key players in rendering the tumours resistant to individual and/or multiple drugs commonly utilised in cancer chemotherapy regimes. The following sections will briefly highlight the importance of a selection of well characterised miRNAs which have been proven to influence cancer development.

Oncogenic miRNAs (oncomirs)

The research efforts leading to the identification of oncomirs have certainly been successful with the discovery of *miR–21* (Cai et al. 2004). This specific miRNA has been highlighted in various studies to be involved in multiple cancer conditions and may be deemed to be the first ever oncomir, or cancer-inducing miRNA, to be recognised from multiple independent cancer studies.

In breast cancer, *miR-21* was found to be over-expressed and linked to anti-apoptosis by modulating the expression of bcl-2 (Si et al. 2007). Evidence for its involvement in glioma tumourigenesis was also identified in varying studies, whereby *miR-21* expression was found to be highly upregulated (Moore and Zhang 2010) (Zhou et al. 2010). Moreover, further research revealed that one of the main modes of action of *miR-21* in glioma tumourigenesis was its anti-apoptotic effect by downstream regulation of caspase-3 and caspase-9 activities (Zhou et al. 2010). Additionally, another path by which over-expressed *miR-21* exerts its tumourigenic effects is by regulating tumour suppressor genes. Zhang and colleagues highlighted the role that *miR-21* plays in the negative modulation of the tumour suppressor phosphatase and tensin homolog (*PTEN*) gene in non-small cell lung cancer, with consequent exacerbation of tumour growth and invasion Zhang et al. 2010). Moreover, *miR-21* was also found to affect downstream expression of tissue inhibitor of metalloproteinase 3 (*TIMP3*) in breast cancer, with the resultant effect of enhanced tumour invasive properties through *TIMP3* modulation (Song et al. 2010). Similar studies have also implicated over-expression of *miR-21* to tumourigenesis and consequent poor patient prognosis in other cancer models such as prostate (Zhang et al. 2010), oral squamous cell carcinoma (Reis et al. 2010), colorectal cancer (Mudduluru et al. 2010), pre-B-cell lymphoma (Medina et al. 2010), osteosarcoma (Ziyan et al. 2010), pancreatic and gastric cancer respectively (Giovannetti et al. 2010)(Motoyama et al. 2010).

Other miRNAs of importance in the induction and exacerbation of human cancer biological pathways are *miR-221* and *miR-222*. These two miRNAs are both located and encoded within the X chromosome and their upregulated expression was found to be involved in prostate cancer proliferative properties, due to downstream modulation of the cell cycle inhibitor *p27(kip1)* (Galardi et al. 2007). This relationship between *p27* and *miR-221/222*was also observed in other cancer models such as hepatocellular carcinoma and chronic lymphocytic leukaemia (Fu et al. 2010) (Frenquelli et al. 2010). In a separate study, *miR-221* and *miR-222* were also implicated in glioblastoma anti-apoptotic survival through down regulation of *PUMA* (p53 upregulated modulator of apoptosis) expression (Chun-Zhi Zhang et al. 2010). The *PTEN* gene in gastric carcinoma is an additional downstream target following *miR-221/222* upregulation, leading to exacerbated cell proliferation and associated radio-resistance (Chun-Zhi et al. 2010). Similarly, in breast and ovarian cancers, *miR-221 / 222* were found to be involved in conferring exacerbation of cell proliferation and metastatic properties (Gianpiero Di Leva et al. 2010) (Wurz et al. 2010).

Apart from the influence of individual miRNAs in cancer biochemical pathways, entire networks of miRNAs can be simultaneously dysregulated at varying levels in order to modulate the expression of downstream targets. Amongst these networks of dysregulated miRNAs, the *miR-17-92* cluster is by far the most characterised for its involvement in tumourigenesis within multiple cancer conditions. This polycistronic cluster is composed of six individual miRNAs, namely *miR-17*, *miR-18a*, *miR-19a*, *miR-20a*, *miR-19b* and *miR-92a*, with the location of this cluster being the open reading frame 25 on chromosome 13 (C13orf25) (Xiang and Wu 2010). The amplification of the 13q31-q32 locus, where the *miR-17-92* cluster is located, was identified to be associated with a wide spectrum of solid tumours and lymphomas (Xiang and Wu 2010). Further research also highlighted that the *miR-17-92* cluster exerts its effects following its activation by C-MYC, which is a notorious oncogenic transcription factor that expresses an oncoprotein being highly dysregulated in a vast array of cancer conditions (Xiang and Wu 2010) (Olive et al. 2010). Interestingly, the recent study by Hong and colleagues revealed that the *miR-17-92* cluster also affects senescence pathways induced by the RAS oncogene, thus conferring tumourigenicity by targeting p21(WAF1), which is a key senescence effectors molecule (Hong et al. 2010). Other studies revealed that in specific cancer models, only individual components of the *miR-17-92* cluster were found to be dysregulated. In pancreatic cancer, for example, *miR-17-5p* alone was found to be linked to overall tumour aggressivity (Yu et al. 2010). The same miRNA was also associated with tumour cell migration properties in breast carcinoma cell lines (Hongling Li et al. 2010). Similarly, hepatocellular carcinoma was demonstrated to bear over-expression of *miR-92a* (Shigoka et al. 2010).

The *miR-106b~25* cluster also proves to be involved in multiple pathways that are relevant in tumour development, within a spectrum of human cancer conditions. This cluster was found to be over-expressed in prostate cancer, together with exacerbated expression of its host gene MCM7 (Poliseno et al. 2010). Additionally, the tandem over-expression of the cluster and its host gene were also found to initiate prostatic intraepithelial neoplasia in transgenic murine models, through PTEN targeting (Poliseno et al. 2010). A separate study focused on the identification of circulating miRNAs in plasma samples from 69 gastric cancer patients, prior to and following surgery for their condition (Tsujiura et al. 2010). The samples were found to contain elevated plasma levels of *miR-106a* and *miR-106b*, together with other miRNAs such as *miR-17-5p* and *miR-21* (Tsujiura et al. 2010). The *miR-106b~25* cluster was also found to be up-regulated by over two-fold expression in hepatocellular carcinoma tumour samples (Yang Li et al. 2009). Additional knock-down studies also confirmed the link between *miR-106b~25* over-expression and cell proliferation, together with anchorage—independent growth within this cancer cell line model (Yang Li et al. 2009). Interestingly, both the *miR-106b~25* and *miR-17-92* paralog clusters can concomitantly act upon the transforming growth factor β (TGF β) tumour suppressor signaling pathway for the ultimate goal of elevating tumorigenicity properties in a wide spectrum of cancers (Mestdagh et al., personal communication) (Petrocca et al. 2008).

Chemoresistance—inducing miRNAs

Another clinically relevant property manifested by most tumours is their development of resistance to a wide range of conventional and translational-based chemotherapeutic drugs (Ahmed et al. 2010). Since such drugs are the mainstay therapeutic tools, applied in chemotherapy cycles for cancer patient treatment across the globe, tumour resistance to these drugs must be rendered futile. Recent research has focused intensely on the biological and clinical implications of the roles which specific miRNAs could possibly play in inducing and regulating the drug resistance properties in a wide range of cancer models.

Individual miRNAs such as *miR-130a* and *miR-125b* were identified to be downregulated in ovarian cancer cells resistant to drugs such as paclitaxel and cisplatin (Sorrentino et al. 2008). Furthermore, *miR-130a* expression was found to be inversely associated with expression for the *M-CSF* ovarian cancer resistance gene in this study (Sorrentino et al. 2008). Interestingly, *miR-125b* is also implicated in other tumour models. This miRNA was recognized to downregulate the expression of Bcl-2 antagonist killer 1 (*BAK1*), which is a pro-apoptotic gene, and such a down regulation ultimately conferred paclitaxel chemoresistance properties to breast cancer cells (Zhou et al. 2010). In prostate cancer, *miR-125b* was also found to induce androgen withdrawal resistance in LNCaP cells through *BAK1* regulation (DeVere White et al. 2009).

In non-small cell lung cancer, cells which were resistant to cisplatin therapy were identified to have upregulation in expression of *miR-181a* and *miR-630*, with chemoresistance conferral occurring through multiple pathways such as decreased apoptosis and reduced DNA damage response mechanisms (Galluzzi et al. 2010).

Additionally, scientific literature has also reported the implication of well characterised oncogenic miRNAs which are also becoming recognized as chemoresistance miRNAs due to their occurrence (by dysregulated expression) in a wide spectrum of tumors. Examples of such miRNAs are *miR- 221* and *miR- 222*. These two miRNAs were identified to be up-regulated in breast cancer cells resistant to tamoxifen (Miller et al. 2008), non-small cell lung cancer cells resistant to tumor necrosis factor (TNF)-related apoptosis-inducing ligand (TRAIL) (Garofalo et al. 2008), and acute lymphocytic leukaemia cells resistant to fludarabine (*miR-221* only) (Moussay et al. 2010).

Tumour suppressor miRNAs

The influences of miRNAs are indeed multivariate and affect myriads of cellular processes and pathways within the cell. The above section briefly mentioned miRNAs involved in enhancing cancer development and consequent tumour aggressiveness in the clinical setting. However, other miRNAs have been identified and characterised to regulate or suppress such cancer-inducing pathways.

Of all presently characterised tumour suppressor miRNAs, the let-7 family (let-7-a1, a2, a3, b,c,d,e,f1,f2, g, i and miR-98) was the first human miRNA family to be identified, with its biological function of cellular differentiation being conserved across species (Roush and Slack 2008) (Peter 2009). Since its discovery, let-7 and its family members were found to be dysregulated, mainly through down regulation, in multiple cancer conditions including lung, glioblastoma, prostate, acute myeloid leukaemia and retinoblastoma (Osada and Takahashi 2010) (Soon-Tae Lee et al. 2010) (Dong et al. 2010) (Chandra et al. 2010) (Mu et al. 2010). The basis for the implication of let-7 in cancer as a tumour suppressor is that the RAS and C-MYC oncogenes, are both target genes of the former (Peter 2009)(Johnson et al. 2005). Additional target genes for let-7 include cell cycle regulators such as CDK6 and CDC25A, together with early embryonic genes such as HMGA2 and IMP-1 (Peter 2009). The let-7 family is also regulated through Lin-28 and Lin-28b, which are both RNA-binding proteins linked to regulation of mRNA translation (Newman and Hammond 2010).

Additional miRNAs playing crucial roles as key tumour suppressors is the miR-34 family, which is composed of three constituent miRNAs (miR-34a, b and c) (Christoffersen et al. 2010). This family of tumour suppressor miRNAs is regulated by p53 which is considered as an important signaling pathway to be affected in most cancer conditions (Christoffersen et al. 2010). Studies have demonstrated that miR-34a is downregulated in a myriad of cancer conditions including gliomas and neuroblastoma tumours with mutant-p53 status (Guessous et al. 2010) (Cole et al. 2008). Furthermore, miR-34a was proven to be a pro-apoptotic entity with essential oncogenes such as MYCN and E2F3 as its direct downstream expression regulatory targets in neuroblastoma (Welch et al. 2007) (Wei et al. 2008). Studies involving upregulation of miR-34a in nephrotoxicity, lung and other tumour models, have shown to lead to enhancements in cell cycle arrest and apoptosis (Wiggins et al. 2010) (Bhatt et al. 2010) (Hermeking 2010).

Other miRNAs which were recently found to be validated tumour suppressors are the miR-15a/16-1 cluster (Calin et al. 2008) (Gatt et al. 2010). This cluster was highlighted as a key regulator of tumour proliferative and invasive property in multiple myeloma and affecting expression of target genes including BCL2, MDM4, VEGFa and PI3KCa (Calin et al. 2008) (Gatt et al. 2010). MiR-205 was also found to be downregulated in prostate cancer (Majid et al. 2010). However, upon up-regulation, miR-205 was observed to induce transcriptomic expression of the interleukin (IL) tumour suppressor genes IL24 and IL32 through specific binding site activation within their promoter regions (Majid et al. 2010). Another study managed to shed further light on the regulatory pathways for C-MYC expression, which were found to involve miR-22 (Xiong et al. 2010). This study revealed that ectopic expression of miR-22 led to silencing of the c-myc binding protein (MYCBP), through direct binding of the miRNA to the 3' untranslated region of MYCBP (Xiong et al. 2010). Furthermore, MYCBP contains a spectrum of E-box-containing C-MYC target genes, whereby the latter were also found to be downregulated upon miR-22 over-expression (Xiong et al. 2010).

CONCLUSIONS

Clinical relevance of the identification of individual/networks of miRNAs directly linked with tumourigenesis (and related cancer pathways which render the tumour more clinically aggressive) are two-fold:

Firstly, the identification and proper validation of hallmark miRNAs which are proven to be key players in conferring individual and/or multi-drug chemoresistance properties, or oncogenic properties, to specific tumours may be utilised for the development and implementation of novel diagnostic miRNA expression profile screens. These screens may prove essential for early recognition of tumours suspected to possess oncogenic/chemoresistance properties, thus giving the clinician an accurate picture of the clinical manifestation in the individual cancer patient. Ultimately, the therapeutic avenues chosen by the clinician will be decided with the inclusion of pre-emptive strategies against such aggressive and/or chemoresistant tumours, based on the additional awareness of the miRNA profile of the individual tumour. The therapeutic options applied will thus maximise the chances of survival of the patient, with minimal suffering due to chemotherapy-induced adverse effects.

Secondly, such miRNAs are good candidates to be novel drug targets for future miRNA based therapies against aggressive tumours that are not responding to conventional chemotherapy. These miRNA—directed therapies are still in their infancy, though may in future prove to be a novel molecular therapeutics approach aimed at enhancing the efficacy of conventional chemotherapeutic treatments, with possible reduction in the dose and frequency of chemotherapy cycles through which the cancer patient has to undergo, ultimately reducing the level of distress and suffering from chemotherapy-induced adverse effects. Initial joint collaborations between researchers and the pharmaceutical industry have already led to promising results regarding miRNA therapies targeted against chronic Hepatitis C viral infection (Lanford et al. 2010).

It must be clearly stated that this chapter mentions but a small selection from the multitude of miRNAs which are referred to be implicated in cancer within scientific literature. This chapter serves to skim the surface on a novel and relevant research field which is attractive to researchers, clinicians and the pharma industries alike. This attraction lies in the proven premise that miRNA research has now opened new doors of perception; new perceptions into the origin of the cancer cell, and also provide us with novel tools with which to control its path. In the not too distant future, ever more miRNA based therapeutics and diagnostic tools will be at the disposal of the cancer clinician in order to eradicate, or at least attenuate drastically, the course of any individual tumour in the individual cancer patient afflicted with this devastating condition.

REFERENCES

Ahmed, N. et al. 2010. Epithelial mesenchymal transition and cancer stem cell-like phenotypes facilitate chemoresistance in recurrent ovarian cancer. *Current Cancer Drug Targets*. **10(3)**:268–278.

Bhatt, K. et al. 2010. MicroRNA-34a is induced via p53 during cisplatin nephrotoxicity and contributes to cell survival. *Molecular Medicine (Cambridge, Mass.)*. **16(9–10)**:409–416.

Cai, X., C.H. Hagedorn and B.R. Cullen. 2004. Human microRNAs are processed from capped, polyadenylated transcripts that can also function as mRNAs. *RNA (New York, N.Y.)*. **10(12)**:1957–1966.

Calin, G.A. et al. 2008. MiR-15a and miR-16-1 cluster functions in human leukemia. *Proceedings of the National Academy of Sciences of the United States of America*. **105(13)**:5166–5171.

Calin, G.A. et al. 2002. Frequent deletions and down-regulation of micro- RNA genes miR15 and miR16 at 13q14 in chronic lymphocytic leukemia. *Proceedings of the National Academy of Sciences of the United States of America*. **99(24)**:15524–15529.

Chandra, P. et al. 2010. Acute myeloid leukemia with t(9;11)(p21-22;q23): common properties of dysregulated ras pathway signaling and genomic progression characterize de novo and therapy-related cases. *American Journal of Clinical Pathology*. **133(5)**:686–693.

Chhabra, R., R. Dubey and N. Saini. 2010. Cooperative and individualistic functions of the microRNAs in the miR-23a~27a~24-2 cluster and its implication in human diseases. *Molecular Cancer*. **9**:232.

Christoffersen, N.R. et al. 2010. p53-independent upregulation of miR-34a during oncogene-induced senescence represses MYC. *Cell Death and Differentiation*. **17(2)**:236–245.

Chun-Zhi, Z. et al. 2010. MicroRNA-221 and microRNA-222 regulate gastric carcinoma cell proliferation and radioresistance by targeting PTEN. *BMC Cancer*. **10**:367.

Cole, K.A. et al. 2008. A functional screen identifies miR-34a as a candidate neuroblastoma tumor suppressor gene. *Molecular Cancer Research: MCR*. **6(5)**:735–742.

Davis, B.N. and A. Hata. 2010. Mechanisms of control of microRNA biogenesis. *Journal of Biochemistry.* Available at: http://www.ncbi.nlm.nih.gov/pubmed/20833630 [Accessed September 26, 2010].

DeVere White, R.W. et al. 2009. MicroRNAs and their potential for translation in prostate cancer. *Urologic Oncology.* **27(3)**:307–311.

Di Leva, G. et al. 2010. MicroRNA cluster 221–222 and estrogen receptor alpha interactions in breast cancer. *Journal of the National Cancer Institute.* **102(10)**:706–721.

Dong, Q. et al. 2010. MicroRNA let-7a inhibits proliferation of human prostate cancer cells *in vitro* and *in vivo* by targeting E2F2 and CCND2. *PloS One.* **5(4)**:e10147.

Foley, N.H. et al. 2010. MicroRNA-184 inhibits neuroblastoma cell survival through targeting the serine/threonine kinase AKT2. *Molecular Cancer.* **9**:83.

Frenquelli, M. et al. 2010. MicroRNA and proliferation control in chronic lymphocytic leukemia: functional relationship between miR-221/222 cluster and p27. *Blood.* **115(19)**:3949–3959.

Fu, X. et al., 2010. Clinical significance of miR-221 and its inverse correlation with p27(Kip1) in hepatocellular carcinoma. *Molecular Biology Reports.* Available at: http://www.ncbi.nlm.nih.gov/pubmed/20146005 [Accessed October 19, 2010].

Galardi, S. et al. 2007. miR-221 and miR-222 expression affects the proliferation potential of human prostate carcinoma cell lines by targeting p27Kip1. *The Journal of Biological Chemistry.* **282(32)**:23716–23724.

Galluzzi, L. et al. 2010. miR-181a and miR-630 regulate cisplatin-induced cancer cell death. *Cancer Research.* **70(5)**:1793–1803.

Garofalo, M. et al. 2008. MicroRNA signatures of TRAIL resistance in human non-small cell lung cancer. *Oncogene.* **27(27)**:3845–3855.

Garofalo, M. and C.M. Croce. 2010. microRNAs: Master Regulators as Potential Therapeutics in Cancer. *Annual Review of Pharmacology and Toxicology.* Available at: http://www.ncbi.nlm.nih.gov/pubmed/20809797 [Accessed September 26, 2010].

Gatt, M.E. et al. 2010. MicroRNAs 15a/16-1 function as tumor suppressor genes in multiple myeloma. *Blood.* Available at: http://www.ncbi.nlm.nih.gov/pubmed/20962322 [Accessed November 3, 2010].

Giovannetti, E. et al. 2010. MicroRNA-21 in pancreatic cancer: correlation with clinical outcome and pharmacologic aspects underlying its role in the modulation of gemcitabine activity. *Cancer Research.* **70(11)**:4528–4538.

Griffiths-Jones, S. et al. 2008. miRBase: tools for microRNA genomics. *Nucleic Acids Research,* 36(Database issue), D154–158.

Guessous, F. et al. 2010. microRNA-34a is tumor suppressive in brain tumors and glioma stem cells. *Cell Cycle (Georgetown, Tex.),* 9(6). Available at: http://www.ncbi.nlm.nih.gov/pubmed/20190569 [Accessed November 2, 2010].

Hermeking, H. 2010. The miR-34 family in cancer and apoptosis. *Cell Death and Differentiation,* **17(2)**:193–199.

Hong, L. et al. 2010. The miR-17-92 Cluster of MicroRNAs Confers Tumorigenicity by Inhibiting Oncogene-Induced Senescence. *Cancer Research.* Available at: http://www.ncbi.nlm.nih.gov/pubmed/20851997 [Accessed October 21, 2010].

Johnson, S.M. et al., 2005. RAS is regulated by the let-7 microRNA family. *Cell.* **120(5):**635–647.

Kwak, P.B., S. Iwasaki and Y. Tomari. 2010. Review Article: The microRNA pathway and cancer. *Cancer Science.* Available at: http://www.ncbi.nlm.nih.gov/pubmed/20726859 [Accessed September 26, 2010].

Lanford, R.E. et al. 2010. Therapeutic silencing of microRNA-122 in primates with chronic hepatitis C virus infection. *Science (New York, N.Y.).* **327(5962):**198–201.

Lee, S. et al. 2010. Let-7 microRNA inhibits the proliferation of human glioblastoma cells. *Journal of Neuro-Oncology.* Available at: http://www.ncbi.nlm.nih.gov/pubmed/20607356 [Accessed November 1, 2010].

Li, H. et al. 2010. miR-17-5p promotes human breast cancer cell migration and invasion through suppression of HBP1. *Breast Cancer Research and Treatment.* Available at: http://www.ncbi.nlm.nih.gov/pubmed/20505989 [Accessed October 21, 2010].

Li, Y. et al. 2009. Role of the miR-106b-25 microRNA cluster in hepatocellular carcinoma. *Cancer Science.* **100(7):**1234–1242.

Lowery, A.J. et al. 2010. Dysregulated miR-183 inhibits migration in breast cancer cells. *BMC Cancer.* **10(1):**502.

Majid, S. et al. 2010. MicroRNA-205-directed transcriptional activation of tumor suppressor genes in prostate cancer. *Cancer.* Available at: http://www.ncbi.nlm.nih.gov/pubmed/20737563 [Accessed November 3, 2010].

Medina, P.P., M. Nolde and F.J. Slack. 2010. OncomiR addiction in an *in vivo* model of microRNA-21-induced pre-B-cell lymphoma. *Nature.* **467(7311):**86–90.

Merritt, W.M., M. Bar-Eli and A.K. Sood, 2010. The dicey role of Dicer: implications for RNAi therapy. *Cancer Research.* **70(7):**2571–2574.

Mestdagh, P. et al. 2010. MYCN/c-MYC-induced microRNAs repress coding gene networks associated with poor outcome in MYCN/c-MYC-activated tumors. *Oncogene.* **29(9):**1394–1404.

Metzler, M. et al. 2004. High expression of precursor microRNA-155/BIC RNA in children with Burkitt lymphoma. *Genes, Chromosomes and Cancer.* **39(2):**167–169.

Miller, T.E. et al. 2008. MicroRNA-221/222 confers tamoxifen resistance in breast cancer by targeting p27Kip1. *The Journal of Biological Chemistry.* **283(44):**29897–29903.

Moore, L.M. and W. Zhang. 2010. Targeting miR-21 in glioma: a small RNA with big potential. *Expert Opinion on Therapeutic Targets.* **14(11):**1247–1257.

Motoyama, K. et al., 2010. Clinicopathological and prognostic significance of PDCD4 and microRNA-21 in human gastric cancer. *International Journal of Oncology.* **36(5):**1089–1095.

Moussay, E. et al. 2010. Determination of genes and microRNAs involved in the resistance to fludarabine *in vivo* in chronic lymphocytic leukemia. *Molecular Cancer.* **9:**115.

Mu, G. et al. 2010. Correlation of overexpression of HMGA1 and HMGA2 with poor tumor differentiation, invasion, and proliferation associated with let-7 down-regulation in retinoblastomas. *Human Pathology*. **41(4)**:493–502.

Mudduluru, G. et al. 2010. Curcumin regulates miR-21 expression and inhibits invasion and metastasis in colorectal cancer. *Bioscience Reports*. Available at: http://www.ncbi.nlm.nih.gov/pubmed/20815812 [Accessed October 19, 2010].

Newman, M.A. and S.M. Hammond. 2010. Lin-28: an early embryonic sentinel that blocks Let-7 biogenesis. *The International Journal of Biochemistry and Cell Biology*. **42(8)**:1330–1333.

Olive, V., I. Jiang and L. He. 2010. mir-17-92, a cluster of miRNAs in the midst of the cancer network. *The International Journal of Biochemistry and Cell Biology*. 42(8):1348–1354.

Osada, H. and T. Takahashi. 2010. Review Article: let-7 and miR-17-92: Small-sized major players in lung cancer development. *Cancer Science*. Available at: http://www.ncbi.nlm.nih.gov/pubmed/20735434 [Accessed September 26, 2010].

Peter, M.E. 2009. Let-7 and miR-200 microRNAs: guardians against pluripotency and cancer progression. *Cell Cycle (Georgetown, Tex.)*. **8(6)**:843–852.

Petrocca, F., A. Vecchione and C.M. Croce. 2008. Emerging Role of miR-106b-25/miR-17-92 Clusters in the Control of Transforming Growth Factor â Signaling. *Cancer Research*. **68(20)**:8191–8194.

Poliseno, L. et al. 2010. Identification of the miR-106b~25 microRNA cluster as a proto-oncogenic PTEN-targeting intron that cooperates with its host gene MCM7 in transformation. *Science Signaling*. **3(117)**:ra29.

Reis, P.P. et al. 2010. Programmed cell death 4 loss increases tumor cell invasion and is regulated by miR-21 in oral squamous cell carcinoma. *Molecular Cancer*. **9**:238.

Rodriguez, A. et al. 2004. Identification of mammalian microRNA host genes and transcription units. *Genome Research*. **14(10A)**:1902–1910.

Roush, S. and F.J. Slack. 2008. The let-7 family of microRNAs. *Trends in Cell Biology*. 18(10):505–516.

Sevignani, C. et al. 2006. Mammalian microRNAs: a small world for fine-tuning gene expression. *Mammalian Genome: Official Journal of the International Mammalian Genome Society*. **17(3)**:189–202.

Shigoka, M. et al. 2010. Deregulation of miR-92a expression is implicated in hepatocellular carcinoma development. *Pathology International*. **60(5)**:351–357.

Si, M. et al. 2007. miR-21-mediated tumor growth. *Oncogene*. **26(19)**:2799–2803.

Song, B. et al. 2010. MicroRNA-21 regulates breast cancer invasion partly by targeting tissue inhibitor of metalloproteinase 3 expression. *Journal of Experimental and Clinical Cancer Research: CR*. **29**:29.

Sorrentino, A. et al. 2008. Role of microRNAs in drug-resistant ovarian cancer cells. *Gynecologic Oncology*. **111(3)**:478–486.

Tsujiura, M. et al. 2010. Circulating microRNAs in plasma of patients with gastric cancers. *British Journal of Cancer*. **102(7)**:1174–1179.

Wei, J.S. et al. 2008. The MYCN oncogene is a direct target of miR-34a. *Oncogene.* **27(39)**:5204–5213.

Welch, C., Y. Chen and R.L. Stallings. 2007. MicroRNA-34a functions as a potential tumor suppressor by inducing apoptosis in neuroblastoma cells. *Oncogene.* **26(34)**:5017–5022.

Wiggins, J.F. et al. 2010. Development of a lung cancer therapeutic based on the tumor suppressor microRNA-34. *Cancer Research.* **70(14)**:5923–5930.

Wu, L. and J.G. Belasco. 2008. Let me count the ways: mechanisms of gene regulation by miRNAs and siRNAs. *Molecular Cell.* **29(1)**:1–7.

Wurz, K. et al. 2010. MiR-221 and MiR-222 alterations in sporadic ovarian carcinoma: Relationship to CDKN1B, CDKNIC and overall survival. *Genes, Chromosomes and Cancer.* **49(7)**:577–584.

Xiang, J. and J. Wu. 2010. Feud or Friend? The Role of the miR-17-92 Cluster in Tumorigenesis. *Current Genomics.* **11(2)**:129–135.

Xiong, J., Q. Du and Z. Liang, 2010. Tumor-suppressive microRNA-22 inhibits the transcription of E-box-containing c-Myc target genes by silencing c-Myc binding protein. *Oncogene.* **29(35)**:4980–4988.

Yu, J. et al. 2010. MicroRNA miR-17-5p is overexpressed in pancreatic cancer, associated with a poor prognosis and involved in cancer cell proliferation and invasion. *Cancer Biology and Therapy*, 10(8). Available at: http://www.ncbi.nlm. nih.gov/pubmed/20703102 [Accessed October 21, 2010].

Zhang, C. et al. 2010. MiR-221 and miR-222 target PUMA to induce cell survival in glioblastoma. *Molecular Cancer.* **9**:229.

Zhang, H. et al. 2010. Serum miRNA-21: Elevated levels in patients with metastatic hormone-refractory prostate cancer and potential predictive factor for the efficacy of docetaxel-based chemotherapy. *The Prostate.* Available at: http:// www.ncbi.nlm.nih.gov/pubmed/20842666 [Accessed October 19, 2010].

Zhang, J. et al. 2010. MicroRNA-21 (miR-21) represses tumor suppressor PTEN and promotes growth and invasion in non-small cell lung cancer (NSCLC). *Clinica Chimica Acta; International Journal of Clinical Chemistry.* **411(11–12)**:846–852.

Zhou, M. et al. 2010. MicroRNA-125b confers the resistance of breast cancer cells to paclitaxel through suppression of pro-apoptotic Bcl-2 antagonist killer 1 (Bak1) expression. *The Journal of Biological Chemistry.* **285(28)**:21496–21507.

Zhou, X. et al. 2010. Reduction of miR-21 induces glioma cell apoptosis via activating caspase 9 and 3. *Oncology Reports.* **24(1)**:195–201.

Ziyan, W. et al. 2010. MicroRNA-21 is involved in osteosarcoma cell invasion and migration. *Medical Oncology (Northwood, London, England).* Available at: http:// www.ncbi.nlm.nih.gov/pubmed/20480266 [Accessed October 19, 2010].

CHAPTER 4

MicroRNA Interference: Concept and Technologies

Zhiguo Wang* and Xiaobin Luo

ABSTRACT

Discovery of microRNAs (miRNAs) has revolutionized our understanding of the mechanisms of gene expression regulation. miRNAs are universally expressed in mammalian cells, involving in nearly every aspects of life of organisms; to date, ~850 miRNAs have been identified in humans and >5000 miRNAs have been found in vertebrates. miRNAs play important roles in the fundamental biological processes such as cell growth, differentiation and death and are implicated in a wide spectrum of disease including developmental malformations, cancer and cardiovascular disease, neuronal disorders such as Alzheimer's disease, metabolic disturbance such as diabetes mellitus and viral disease. Aberrant miRNA expression has been documented in human disease and in animal models as well, with evidence for causative roles in tumorigenesis and other pathological processes. The expression profiles of miRNAs demonstrate unique disease-associated signatures. These properties and functions of miRNAs make them attractive viable therapeutic targets through interfering with their expression. Both gain-of-function and loss-of-function technologies are necessary tools for understanding miRNAs. The miRNA-Targeting strategy resulting in gain-of-function of miRNAs is an essential way for gain-of-knowledge about miRNA targets and functions. Ever since the elucidation of functions and implications of miRNA, many creative and innovative strategies and methodologies have been invented and widely and successfully applied to miRNA research as gain-of-function or loss-of-function approaches and even to new drug development. Stimulated by these promising advances, we proposed the concept of microRNA interference (miRNAi): manipulating the function, stability, biogenesis of expression of miRNAs to interfers with the expression of their target protein-coding mRNAs to alter the cellular functions. The aim of this book is to provide brief descriptions of the concept of miRNAi and the related miRNAi technologies. It is our expectation

* *Corresponding author e-mail*: wangz.email@gmail.com

that from this chapter, readers would be able to acquire the basic knowledge of miRNAs, the new concepts pertinent of miRNAi, and insight into the principles of various miRNAi technologies.

Keywords: miRNAi, gene expression, miRNA interference, post-transcriptional regulation.

INTRODUCTION

miRNAs, endogenous noncoding regulatory mRNAs of around 22-nucleotides long, have rapidly emerged as one of the key governors of the gene expression regulatory program in cells of varying species (Latronico et al. 2007, Latronico and Condorelli 2009, Wang et al. 2008, Wang 2010, Yang et al. 2010, Brennecke et al. 2005, Jackson and Standart 2007, Pillai et al. 2007). Accumulating evidence suggests that miRNAs constitutes a novel, universal mechanism for fine regulation of gene expression in all organisms, "tuning" the cellular phenotype during delicate processes. Owing to their ever-increasing number in the mammalian cells and their ever-increasing implication in the control of the fundamental biological processes (such as development, cell growth and differentiation, cell death, etc.), miRNAs have now become a research subject capturing major interest of scientists worldwide. Moreover, with recent studies revealing the macro roles of miRNAs in the pathogenesis of adult humans, we have now entered a new era of miRNA research. The exciting findings in this field have inspired us with a premise and a promise that miRNAs will ultimately be taken to the heart for therapy of human disease.

In the past years, we have witnessed a rapid evolving of many creative, innovative, inventive techniques and methodologies that are of great promising in miRNA research and applications (Wang et al. 2008, Wang 2010). These technologies have demonstrated their efficacy and reliability in producing gain-of-function or loss-of-function of miRNAs, providing new tools for elucidating miRNA functions and opening up a new avenue for the development of new agents targeting miRNAs for therapeutic aims. These stimulating advances prompted us to propose the concept of microRNA interference (miRNAi) (Wang et al. 2008, Wang 2010): manipulating the function, stability, biogenesis or expression of miRNAs to interferes with the expression of their target protein-coding mRNAs to alter the cellular functions. The aim of this review article is to provide introductive descriptions of the strategies and methodologies for interfering miRNA expression, biogenesis and function and their applications in miRNA research and new drug design using miRNAs as therapeutic targets.

BIOGENESIS AND ACTION OF miRNA

The genes for miRNAs, like the protein-coding genes, are located in the chromosomes as an integral part of the complex genome. Some of miRNAs are encoded by unique genes and others are embedded into the intronic regions of protein-coding genes. They are initially transcribed by RNA polymerase as long primary miRNAs (pri-miRNAs) with typical stem-loop structures (Figure 1). Pri-miRNAs are quickly processed by the nuclear RNase endonuclease-III Drosha to remove the branches to become precursor miRNAs (pre-miRNAs) of around 60–100 nts in length. Pre-miRNAs are then transported from the nucleus to the cytosol by exportin-5, where they are further processed by another RNase endonuclease-III Dicer to generate mature miRNAs of around 22-nt long (Kim 2005, Lee et al. 2002, Hutvagner and Simard 2008).

The first step for mature miRNAs to take their effects in gene regulation is to incorporate into the protein complex, followed by binding to the 3' untranslated region (3'UTR) of its target mRNA through a Watson-Crick basepairing mechanism with its 5'–end 2–8 nts exactly complementary to recognition motif within the target. Hence, the actual function of a miRNA is to serve as the target-recognition component of the complex. This 5'–end 2–8 nt region is termed "seed sequence" or "seed site" for it is critical and sufficient for miRNA actions (Lewis et al. 2003, Lewis et al. 2005). Partial complementarity with the rest of the sequence of a miRNA creates bulges and mismatches in the miRNA:mRNA heteroduplex and plays a role in producing post-transcriptional regulation of gene expression, presumably by stabilizing the miRNA:mRNA interaction. Because of the lax require-ment for miRNA:mRNA interaction, a majority, if not all, of protein-coding genes contain the target motifs for miRNA actions.

It is important to note that on one hand each miRNA has the potential to target multiple genes and on the other hand each gene may be a target for multiple miRNAs. The target genes of a given miRNA may encode proteins that have different or even opposing functions (e.g., cell growth vs. cell death) and the net outcome of integral actions of a miRNA on its multiple target genes defines the function of that miRNA. Clearly, the action of miRNAs is non-gene-specific.

DEREGULATION OF miRNA EXPRESSION IN HUMAN DISEASE

To date, >800 different miRNAs have been identified in humans and >5000 in animal species. Expression of miRNAs in mammalian species under normal condition is genetically programmed with certain spatial (depending on cell-, tissue-, or organ-type) and temporal (depending on developmental stage) patterns (Wang and Yang 2010). On one hand, some miRNAs are expressed in cell/tissue/organ-restricted manners and others may be ubiquitously expressed. On the other hand, expression of miRNAs is dynamic, changing along with the lifespan of an organism from fetal development to aging process. These spatial heterogeneities and temporal differences are critical for the involvement of miRNAs in the fine regulation of versatile cellular functions and cell lineage decisions with right timings in right places. Perhaps, the most important feature relevant to miRNA interference is the fact that expression of miRNAs is deregulated, abnormally upregulated or downregulated, in diseased states and different diseases and different stages of a disease is associated with different miRNA expression profiles: the miRNA signature. For example, available studies indicate that all cancers are connected together by miRNA profiles, suggesting that various cancers may share similar associations at the miRNA level, in which some strong onco-miRNAs or miRNA tumor suppressors may play key roles (Lu et al. 2005, Xi et al. 2006). A potential correlation exists between miRNA tissue specificity and disease, which may be of value in predicting specific disease-related miRNAs by combining the miRNA tissue specificity values. Thus, if a disease occurs specifically in a given tissue, the miRNAs specifically expressed in that tissue will have a great potential to be related to that disease. Disease-associated miRNAs show various dysfunctions, such as mutation, up-regulation, deleted and down-regulation. The elucidation of miRNA transcriptomes between diseased and normal tissues or between different disease types, stages and grades, gives the chance to identify the miRNAs most probably involved in human disease and to establish new diagnostic and prognostic markers. Importantly, it also provides the rationale for miRNA interference.

THE CONCEPT OF "miRNA INTERFERENCE (miRNAi)"

RNA interference (RNAi) is a well-known strategy for gene silencing; this strategy takes the advantages of the capability of small double-stranded RNA molecules (siRNAs) to bind RNA-induced silencing complex (RISC) on one hand and to bind target genes (mRNAs) on the other hand (Xia et al. 2002, Pusharaj et al.2008, Golden et al. 2008). Through such dual interactions, siRNAs elicit powerful knockdown of gene expression by degrading their target mRNAs. Two key characteristics of the RNAi strategy are that the only target of RNAi is mRNAs and that the only outcome of RNAi is silencing of mRNAs. In other words, the RNAi strategy uses siRNAs to interfere directly with mRNAs (mostly protein-coding genes) to silence gene expression.

In light of this concept, we proposed a new concept: microRNA interference (miRNAi) (Wang et al. 2008, Wang 2010). We define miRNAi as the strategies of manipulating the function, stability, biogenesis, or expression of miRNAs so as to interfere with expression of protein-coding mRNAs (Table 1). This concept is based on the thoughts outline below. (1) The fundamental mechanism of miRNA actions is miRNA:mRNA interaction or binding. A key to interfere with miRNA actions is to disrupt the miRNA:mRNA interaction. In order to achieve this aim, we can either manipulate miRNAs or mRNAs to alter the miRNA:mRNA interaction. For miRNAs, we can either mimic miRNA actions to enhance the miRNA:mRNA interaction or to inhibit miRNAs to break the miRNA:mRNA interaction. Additionally, we can also manipulate mRNA to interrupt the miRNA:mRNA interaction. (2) miRNA biogenesis involves many steps amenable to artificial interventions that can either enhance or inhibit the process to alter miRNA levels. (3) miRNAs, like protein-coding genes, are tightly regulated in their expression by transcriptional factors. Altering expression of transcription factors through miRNAi can alter expression of miRNAs. In theory, any of the components along the miRNA expression and biogenesis pathways can be interfered to alter the level and function of miRNAs thereby expression of protein-coding genes (Figure 1).

Table 1. Comparison between RNAi and miRNAi.

	RNAi	*miRNAi*
Associated technologies	siRNA only	many (see Table 2)
Nature of interfering molecules	artificial only	artificial and endogenous
Direct target	mRNAs	miRNAs and mRNAs
Site of action	coding region of mRNAs	all sites along miRNA biogenesis and 3'UTR of mRNAs
Outcome of gene expression	silencing only	silencing or enhancing

THE CONCEPT OF miRNAi TECHNOLOGIES (miRNAi-Tech)

We further proposed the strategies for interfering with miRNAs be called miRNAi technologies (miRNAi-Tech) (Wang et al. 2008, Wang 2010). miRNAi-Tech can be categorized based on the following six different perspectives (Table 2).

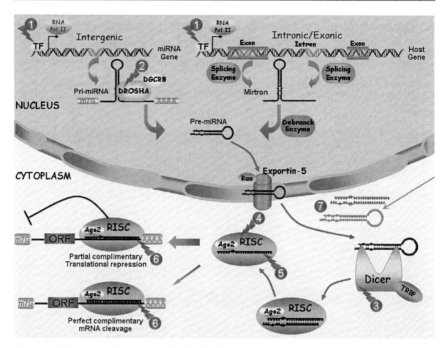

Figure 1. Schematic illustration of miRNA biogenesis pathway (see detailed description in text) and the sites for miRNA interference along the pathway. The arrows and the numbers indicate the sites of miRNAi action. RNA Pol II: RNA polymerase II; TF: transcription factor; intergenic miRNAs: miRNA-coding genes located in between protein-coding genes; intronic miRNAs (miRNA-coding genes located within introns of their host protein-coding genes; mitron: the intron within a protein-coding gene (the host gene) which contains miRNA sequences; Drosha: the nuclear RNase III ribonuclease; Dicer: cytoplasmic RNase III ribonuclease; pri-miRNA: primary miRNA; pre-miRNA: precursor miRNA; RISC: RNA-induced silencing complex. u: the site for interfering transcription of endogenous miRNAs by altering TFs; v the site for interfering miRNA biogenesis by silencing Drosha; w the site for interfering miRNA biogenesis by knocking down or knocking out Dicer; x the site for interfering miRNA action by inhibiting RISC; y the site for interfering miRNA action and stability by targeting mature miRNA using antisense (such as AMO, MT-AMO, miRNA sponge, antagomiR, and LNA-antimiR); z the site for interfering miRNA action in a gene-specific manner by using miR-Mimic and miR-Mask technologies; and exogenous canonical miRNAs or pre-miRNAs delivered into cells to achieve miRNA-gain-of-function.

(Color image of this figure appears in the color plate section at the end of the book.)

(1) According to the mechanisms of actions, miRNAi-Tech can be divided into two major strategies: miRNA-targeting and targeting-miRNAs strategies.

The "miRNA-targeting strategy of miRNAi" refers to the approaches producing "gain-of-function" of miRNAs to enhance gene targeting so as to alter the gene expression and cellular function.

Table 2. General Introduction to miRNAi Technologies.

Mechanism of action	miRNA-targeting	producing "miRNA-gain-of-function" to enhance "mRNA-loss-of-function"
	Targeting-miRNAs	producing "miRNA-gain-of-function" to enhance "mRNA-loss-of-function"
Mode of action	AMO:miRNA interaction	anti-miRNA oligomer inhibiting miRNA level or activity
	ASO:mRNA interaction	anti-mRNA oligomer inhibiting miRNA function
Outcome of miRNA	miRNA-gain-of-function	enhancement of miRNA function
	miRNA-loss-of-function	inhibiting miRNA expression, biogenesis, or function
Outcome of target gene	mRNA-gain-of-function	causing derepression of protein-coding genes
	mRNA-loss-of-function	enhancing repression of protein-coding genes
miRNA specificity	miRNA-specific	targeting a specific miRNA or a specific group of miRNAs
	Non-miRNA-specific	targeting the whole population of miRNAs
Gene specificity	Non-gene-specific	a miRNA targeting multiple genes with its binding site
	Gene-specific	a miRNA targeting a particular gene

The "targeting-miRNA strategy of miRNAi" refers to the approaches leading to "loss-of-function" of miRNAs via inhibiting miRNA expression and/or action to alter the gene expression and cellular function.

(2) In terms of the outcome of actions, miRNAi-Tech can be grouped into "miRNA-gain-of-function" (enhancement of miRNA function) and "miRNA-loss-of-function" (inhibition of miRNA expression, biogenesis, or function).

The "miRNA-gain-of-function" approach is in general achieved by overexpression of the endogenous miRNAs and forced expression of exogenous miRNAs, which results in enhanced miRNA targeting.

The "miRNA-loss-of-function" approach can be conferred through targeting-miRNAs by knockdown or knockout of miRNA expression and inhibition of miRNA action.

(3) With respect to the specificity of target miRNAs, miRNAi-Tech can be miRNA-specific or non-miRNA-specific.

The miRNA-specific miRNAi-Tech interferes only with miRNA function to produce derepression of target protein-coding genes. They belong to the "targeting-miRNA" strategy and are miRNA-specific but non-gene-specific.

The non-miRNA-specific miRNAi-Tech interferes with the biogenesis of miRNAs such as inhibition of Dicer or Drosha, affecting the levels of the whole population of miRNAs, but not a particular miRNA. These approaches are therefore neither miRNA-specific nor gene-specific.

(4) As to the target genes of miRNAs, miRNAi-Tech comprise gene-specific and non-gene-specific approaches, which can be either "gain-of-function" of miRNAs or "loss-of-function" of miRNAs.

The gene-specific miRNAi-Tech act on the target gene of a given miRNA but not on the miRNA per se. This is achieved by enhancing or removing the actions of miRNAs.

The non-gene-specific miRNAi-Tech directs their actions to miRNAs without interactions with the target genes of the miRNAs. Such actions are miRNA-specific but not gene-specific given that a miRNA can target multiple protein-coding genes. Most of the currently available miRNAi-Tech belongs to non-gene-specific approaches.

(5) From the perspective of target protein-coding genes (mRNAs), miRNAi-Tech may result in either "mRNA-gain-of-function" (derepression of genes) or "mRNA-loss-of-function" (enhancement or establishment of repression).

The "targeting-miRNA" strategy causes "mRNA-gain-of-function" by relieving the repressive actions induced by the targeted miRNAs. The "mRNA-gain-of-function" miRNAi-Tech could be gene-specific or non-gene-specific, but they must be miRNA-specific.

The "miRNA-targeting" strategy produces "mRNA-loss-of-function" effects by enhancing the repressive actions of the miRNAs. The "mRNA-loss-of-function" miRNAi-Tech could be gene-specific or non-gene-specific, but they must be miRNA-specific.

(6) Taking the mode of actions into consideration, miRNAi-Tech can act by creating AMO:miRNA interaction (AMO: anti-miRNA antisense oligomer) or by creating ASO:mRNA interaction (ASO: anti-mRNA antisense oligomer) to disrupt normal miRNA:mRNA interaction.

The AMO:miRNA-interacting miRNAi-Tech suppress miRNA function using AMO approaches. They are all miRNA-specific and non-gene-specific, belonging to the targeting-miRNAs strategies and the miRNA-loss-of-function and mRNA-gain-of-function class.

The ASO:mRNA-interacting miRNAi-Tech interrupt miRNA function using ASO approaches. They belong to the targetingmiRNAs strategies and the miRNA-loss-of-function and mRNA-gain-of-function class.

Based on above classifications, some miRNAs are both miRNA- and gene-specific, some are miRNA-specific but non-gene-specific and others are neither miRNA-specific nor gene-specific (Table 2).

GENERAL INTRODUCTION TO miRNAi TECHNOLOGIES

miRNA-Targeting miRNAi-Technologies

An array of technologies has been developed for achieving gain-of-function of miRNAs by forced expression or overexpression. These include Synthetic Canonical miRNA technology, miRNA Mimic technology, Multi-miRNA Hairpins technology and miRNA Transgene technology (Wang 2010).

The Synthetic Canonical miRNA technology is an indispensable and most essential approach for miRNA gain-of-function in the fundamental research of miRNAs and of miRNA-related biological processes. The SC-miRNA technology involves an application of synthetic miRNAs that are identical in sequence to their counterpart endogenous miRNAs (Yang et al. 2007, Xiao et al. 2007a, b, Luo et al. 2007, Luo et al. 2008, Lu et al. 2009). It has been used to force expression of miRNAs of interest to investigate the pathophysio-logical outcome of upregulation of the miRNAs. The approach can also be used to normalize the level of down regulated miRNAs to alleviate the pathological conditions as a result of downregulation of those miRNAs. Synthetic Canonical miRNAs (SC-miRNAs) are exogenously applied miRNAs in the form of mature double-stranded constructs or of precursor hairpin constructs that are identical in sequence to their counterpart endogenous miRNAs. Once introduced into cells, SC-miRNAs produce seemingly identical gene regulation and cellular functions as their endogenous miRNA counterparts do. The SC-miRNA approach belongs to the "miRNA-targeting" and "miRNA-gain-of-function" strategy.

The miRNA Mimics technology (miR-Mimic) (Xiao et al. 2007b) takes the advantage of non-natural double-stranded miRNA-like RNA fragments that can mimic the action of canonical miRNAs. It can be used when a particular protein-coding gene needs to be knocked down. miR-Mimics and siRMAs are quite similar in terms of their applications to basic research and the potential to disease treatment.

It is an innovative approach for gene silencing. This approach is to generate non-natural double-stranded miRNA-like RNA fragments. A miR-Mimic is a double-stranded RNA fragment designed to have its 5'-end 2–8 nts bearing an exact complementary motif (equivalent to seed site) to the selected sequence in the 3'UTR of an mRNA unique to that target gene. Once introduced into cells, this RNA fragment, mimicking an endogenous miRNA, can bind specifically to its target gene and produce post-transcriptional repression of the gene. Unlike endogenous miRNAs, miR-Mimics act in a gene-specific fashion. The miR-Mimic approach belongs to the "miRNA-targeting" and "miRNA-gain-of-function" strategy and is primarily used as an exogenous tool to study gene function by targeting mRNA through miRNA-like actions in mammalian cells. The technology was developed by our research group (Department of Medicine, Montreal Heart Institute, University of Montreal) in 2007 (Xiao et al. 2007b).

The Multi-miRNA Hairpins technology uses a single artificial construct in the form of double-stranded RNA to produce multiple mature miRNAs once introduced into cells (Sun et al. 2006, Xia et al. 2006). Multi-miRNA Hairpins technology refers to a single artificial construct that can produce multiple mature miRNAs to improve gene repression over single miRNAs offering expression silencing of multiple genes. It takes the advantage of expression vehicles that are able to give high efficiency and high fidelity of producing multiple mature miRNAs providing both an enhanced single-gene knockdown and linked multi-gene knockdowns. This technology was concurrently developed in 2006 by Zhu's laboratory (Department of Developmental and Molecular Biology, Albert Einstein College of Medicine) (Sun et al. 2006) and by Xu's laboratory (Department of Biochemistry and Molecular Pharmacology, University of Massachusetts Medical School) (Xia et al. 2006). The Multi-miRNA Hairpins technology belongs to the "miRNA-targeting" and "miRNA-gain-of-function" strategy.

The miRNA Transgene technology is an efficient approach to create stable overexpression of miRNAs of interest to study the role of miRNAs under *in vivo* conditions. Other miRNA-targeting or miRNA-gain-of-function technologies can only create transient expression of miRNAs, which may be inadequate when the research or therapy requires long-lasting. There are in general two different approaches for establishing miRNA transgene animals. One is the conventional transgene approach that has been widely used for protein-coding genes. It requires production of mice by incorporating a non-native segment of DNA containing a pre-miRNA-coding sequence of interest into mice's germ line, which retains the ability to overexpress that miRNA in the transgenic mice. This approach was applied to miRNA research first in 2006 by Costinean et al. (2006) and Peng et al. (2006) in 2006, then in 2007 by Lu et al. (2007) and recently by us as well (Lu et al. 2010). The second is an innovative approach utilizing artificial intronic miRNAs generated by inserting a transposon into the intron of a protein-coding gene, which was originally developed by Ying's laboratory in 2003 (Lin et al. 2003, Lin and Ying 2006). The miRNA-transgene technology belongs to the "miRNA-targeting" and "miRNA-gain-of-function" strategy and is primarily used for studying miRNA target genes, cellular function and pathological role in animal models.

TARGETING-miRNA miRNAi-TECHNOLOGIES

Targeting-miRNA technologies include Anti-miRNA Antisense Inhibitor Oligoribonucleotides (AMO) technology, Multiple-Target AMO technology (MT-AMO technology), miRNA Sponge technology, miRNA-Masking Antisense Oligonucleotides technology and MiRNA Knockout technology (Wang 2010) (Table 3).

The AMO technology has been the most commonly and frequently used targeting-miRNA/loss-of-function approach to knock down a miRNA of interest. This method offers a miRNA-specific way to wipe out the function of the targeted miRNAs and can be used to normalize levels of miRNAs that are abnormally upregulated in the diseases state. One of the indispensable approaches in miRNA research as well as in miRNA therapy is inhibition or loss-of-function of miRNAs. Multiple steps in pathway for miRNA biogenesis could be targeted for inhibition of miRNA production and maturation. Thus far, nonetheless, the anti-miRNA antisense inhibitor oligoribonucleotides (AMO) technology used to target mature miRNAs has found most value for these applications. Standard AMO is a single-stranded 2'-O-methyl (2'-OMe)-modified oligoribonucleotide fragment exactly antisense to its target miRNA. The AMO technology was initially established in 2004 by Tuschl's laboratory (Laboratory of RNA Molecular Biology, The Rockefeller University) (Meister et al. 2004) and by Zamore's laboratory (Department of Biochemistry and Molecular Pharmacology, University of Massachusetts Medical School) (Hutvágner et al. 2004). The idea was however originated from Boutla et al. (2003) who used antisense 2'-deoxyoligonucleotides to sequence-specifically inactivate miRNAs in microinjected *D. melanogaster* embryos. Since then, AMO technology has undergone many important modifications to enhance the efficiency and specificity of miRNA interference. These include a cholesterol moiety-conjugated 2'-OMe modified AMOs called antagomiR (Krutxfeldt et al. 2005, 2007), locked nucleic acid (LNA)-modified AMOs (Ørom

Table 3. Summary of miRNAi Technologies and Their Applications.

Name of miRNAi Technology	Characteristics	Applications
miRNA-targeting miRNA-gain-of-function mRNA-loss-of-function	Enhancing miRNA action to enhance gene silencing	
Synthetic Canonical miRNA(SC-miRNA)	miRNA-specific non-gene-specific	(1) transient miRNA-gain-of-function in miRNA research; (2) miRNA replacement therapy
miRNA Mimic (miR-Mimic)	non-miRNA-specific gene-specific	(1) transient miRNA-gain-of-function in a gene-specific manner for miRNA research; (2) miRNA replacement therapy
Multi-miRNA Hairpins	miRNA-specific non-gene-specific	(1) transient miRNA-gain-of-function in miRNA research; (2) miRNA replacement therapy
Viral Vectors (carrying SC-miRNAs)	miRNA-specific non-gene-specific	(1) long-lasting miRNA-gain-of-function in miRNA research; (2) for *in vitro* and *in vivo* applications; (3) miRNA replacement therapy
miRNA Transgene	miRNA-specific non-gene-specific	(1) whole-life miRNA-gain-of-function in miRNA research; (2) controllable or conditional overexpression of miRNA; (3) *in vivo* animal model; (4) miRNA replacement therapy
Targeting-miRNA miRNA-loss-of-function mRNA-gain-of-function	Knockdown or knockout of miRNA to relieve gene silencing	
Anti-miRNA Antisense Oligonucleotides (AMO)	miRNA-specific non-gene-specific	(1) to knockdown target miRNA for validating the miRNA targets and function under *in vitro* conditions; (2) to achieve upregulation of the cognate target protein; (3) to reverse the pathological process associated with miRNA upregulation
AntagomiR (AMO conjugated with cholesterol moiety to gain the capability of penetrating cells)	miRNA-specific non-gene-specific	(1) to knockdown target miRNA for validating the miRNA targets and function under *in vivo* conditions; (2) to achieve upregulation of the cognate target protein; (3) to reverse the pathological process associated with miRNA upregulation
LNA-antimiR (AMO modified with locked nucleic acids to improve thermal	miRNA-specific non-gene-specific	(1) to knockdown target miRNA for validating the miRNA targets and function under *in vitro* and *in vivo*

Name of miRNAi Technology	Characteristics	Applications
stability, miRNA-binding affinity and specificity)		conditions; **(2)** to achieve upregulation of the cognate target protein; **(3)** to reverse the pathological process associated with miRNA upregulation
Multiple-Target Anti-miRNA Antisense Oligonucleotides (MT-AMO)	miRNA-specific non-gene-specific	**(1)** to knockdown multiple miRNAs from same or different seed families; **(2)** to achieve upregulation of multiple target proteins; **(3)** to reverse the pathological process associated with miRNA upregulation
miRNA Sponge	miRNA-specific non-gene-specific	**(1)** to knockdown multiple miRNAs from a same seed family; **(2)** to achieve upregulation of the cognate target proteins; **(3)** to reverse the pathological process associated with miRNA upregulation
miRNA-Masking Antisense Oligonucleotides (miR-Mask)	miRNA-specific gene-specific	**(1)** to block action of a miRNA without knocking down the miRNA; **(2)** to achieve upregulation of the cognate target protein; **(3)** to reverse the patho logical process associated with miRNA upregulation
miRNA Knockout (miR-KO)	miRNA-specific non-gene-specific	**(1)** to acquire permanent removal of a target miRNA; **(2)** to allow for controllable or conditional miRNA silencing; **(3)** to allow for studying miRNA-loss-of-function *in vivo* whole-animal context
Dicer Inactivation	non-miRNA-specific non-gene-specific	**(1)** to achieve a global loss-of-function of literally all cellular miRNAs; **(2)** to allow for controllable or conditional miRNA silencing; **(3)** to allow for studying miRNA-loss-of-function *in vivo* whole-animal context

et al. 2006, Davis et al. 2006), 2'-O-methoxyethyl (2'-MOE) (Davis et al. 2006, Esau et al. 2006), 2'-flouro (2'-F) (Davis et al. 2006), phosphorothioate backbone modification and a peptide nucleic acid (PNA)-modified AMOs (Fabani and Gait 2008). The AMO technology belongs to the "targeting-miRNA" and "miRNA-loss-of-function" strategy.

The AntagomiR technology is a method initial developed by Krutzfeldt et al. (2005) to confer AMOs the capability of penetrating cells by conjugating AMOs with cholesterol moieties. Conjugation of a cholesterol to the 3'end of a 2'-OMe modified AMO with two or three phosphorothioate modifications on each end has been reported to facilitate *in vivo* delivery of the AMO targeting miR-122 into the liver (Krutzfeldt et al. 2005). After three days of daily intravenous administration of 80 mg/kg AMO to normal mice, increased levels of miR-122 target gene mRNAs were observed in the liver, as well as reduced plasma cholesterol and an apparent degradation of the miRNA. A more recent study claimed that 3'-cholesterol-modified AMOs have enhanced potency, allowing miRNA inhibition for at least 7 d from a single transfection (Horwich and Zamore 2008). Since the development of the antagomiR technology, it has been documented by numerous studies to be highly effective in knocking down target miRNAs with long-lasting efficacies under *in vivo* conditions (Krutzfeldt et al. 2007, 2005 Esau et al. 2006, Pineau et al. 2010, van Solingen et al. 2009).

The Locked Nucleic Acid (LNA)-antimiR technology. LNA modified AMOs, which are commercially available, also showed significantly better activity than the 2'-OMe (Elmén et al. 2005). LNA is defined as oligonucleotides containing one or more LNA monomers, the 2'-O, 4'-C-methylene-β-D-ribofuranosyl nucleosides. A major structural characteristic of LNA is the close resemblance to the natural nucleic acids because methylene group-linked O2' and the C4' atoms introduce a conformational lock of the molecule into a near perfect N-type conformation adopted by RNA (Saenger 1984, Jepsen and Wengel 2004). This modification confers a higher thermal stability, a greater discriminative power and a longer half-life with proven efficacy for *in vivo* application to knockdown miRNAs (Ørom et al. 2006, Davis et al. 2006, Elmen et al. 2005, Grünweller and Hartmann 2007, Lanford et al. 2010, Mook et al. 2007, Veedu and Wengel 2010, Stenvang et al. 2008, Naguibmeva et al. 2006, Wang et al. 2009, Kocerha et al. 2009). A fully modified LNA sequence has been reported to be fully resistant to the 3'-exonuclease (Frieden et al. 2003). End-blocked sequence, i.e. LNA-DNA-LNA gapmers display a high stability in human serum compared to similar 2'-OMe modified sequences (Kurreck et al. 2002). Recently, Elmén et al. (2008) reported a systemically administered 16-nt, unconjugated LNA-antimiR towards miR-122 leads to specific, dose-dependent silencing of miR-122 in the liver and shows no hepatotoxicity in mice. Moreover, the efficacy of LNA/OMe mxiMer in anti-miRNA activity has also been tested (Fabani and Gait 2008).

The multiple-target AMO (MT-AMO) technology is an innovative strategy which confers a single AMO fragment the capability of targeting multiple miRNAs. This modified AMO is single-stranded 2'-*O*-methyl-modified oligoribonucleotides carrying multiple AMO units which are engineered into a single unit and are able to simultaneously silence multiple target miRNAs or multiple miRNA seed families. Studies suggest the MT-AMO is an improved approach for miRNA target finding and miRNA function validation; it not only enhances the effectiveness of targeting miRNAs but also confers diversity of actions. It has been successfully used to identified target genes and cellular function of several oncogenic miRNAs and of the muscle-specific miRNAs (Lu et al. 2009). This novel strategy may find its broad application as a useful tool in miRNA research for exploring biological processes involving multiple miRNAs and multiple genes and the potential as a miRNA therapy for human disease such as cancer and cardiac disorders. This technology was developed by our research group (Lu et al. 2009). The MT-AMO technology belongs to the "targeting-miRNA" and "miRNA-loss-of-function" strategy. The Multiple-Target AMO technology is designed as a simple and straightforward approach for studying the biological processes involving multiple miRNAs.

The miRNA Sponge technology is an innovative approach used to generate RNAs containing multiple, tandem binding sites for a miRNA seed family of interest being able to target all members of that miRNA seed family. When vectors encoding the miRNA sponges are transiently transfected into cultured cells, they depress miRNA targets as strongly as the conventional AMOs. The major advancement of this technique over the AMO technique is that it can better inhibit functional classes of miRNAs than do AMOs that are designed to block single miRNA sequences. The main principle of the miRNA Sponge technology is identical to that the MT-AMO technology described above: targeting multiple miRNAs. The miRNA Sponge technology was established by Sharp's laboratory in 2007 (Ebert et al. 2007, Hammond 2007). Similar to the AMO approach, miRNA Sponge technology belongs to the "targeting-miRNA" and "miRNA-loss-of-function" strategy.

The miRNA-Masking Antisense Oligonucleotides (miR-Mask) technology is a supplement to the AMO technique; while AMO is indispensable for studying the overall function of a miRNA, the miR-Mask might be more appropriate for studying the specific outcome of regulation of the target gene by the miRNA (Xiao et al. 2009, Choi et al. 2007). A standard miR-Mask is single-stranded 2'-*O*-methyl- or LNA-modified oligoribonucleotide (or other chemically modified) that is a 22-nt antisense to a protein-coding mRNA as a target for an endogenous miRNA of interest. Instead of binding to the target miRNA like an AMO, a miR-Mask does not directly interact with its target miRNA but binds to the binding site of that miRNA in the 3'UTR of the target mRNA by fully complementary mechanism. In this way, the miR-Mask covers up the access of its target miRNA to the binding site so as to derepress its target gene (mRNA) via blocking the action of its target miRNA. The anti-miRNA action of a miR-Mask is gene-specific because it is designed to be fully complementary to the target mRNA sequence of a miRNA. The anti-miRNA action of a miR-Mask is also miRNA-specific as well because it is designed to target the binding site of that particular miRNA. The miR-Mask approach is a valuable supplement to the AMO technique; while AMO is indispensable for studying the overall function of a miRNA, the miR-Mask might be more appropriate for studying the specific outcome of regulation of the target gene by the miRNA. This technology was first established by my research group in 2007 (Xiao et al. 2009) and a similar approach with the same concept was subsequently reported by Schier's laboratory (Choi et al. 2007). Similar to the AMO approach, miR-Mask technology belongs to the "targeting-miRNA" and "miRNA-loss-of-function" strategy.

The miRNA Knockout (miR-KO) technology aims to generate mouse lines with genetic ablation of specific miRNAs or targeted disruption of miRNA genes. This approach allows for investigations of miRNA function related to the development of particular biological processes and/or pathological conditions in an *in vivo* context and in a permanent manner. miR-KO technology aims to generate mouse lines with genetic ablation of specific miRNAs or targeted disruption of miRNA genes. For some applications, this knockout strategy has demonstrated its superiority over the knockdown strategies such as antisense to miRNA, which are primarily *in vitro*, transient, local targeting-miRNA or miRNA-loss-of-function approaches. The first applications of knockout models to study miRNA function in development were performed in flies in 2005 Kwon et al (2005) and by Sokol and Ambros (2005). Later, the miR-KO techniques were introduced to mouse models by Zhao et al. (2007) and Thai et al. (2007). It seems that this strategy is increasingly appreciated and favored by recent studies (Kuhnert et al. 2008, Wang et al.2008), though limitations exist.

Dicer Inactivation technology aims to disrupt the maturation of miRNAs by inhibiting Dicer functions through knocking down or knocking out the Dicer gene. This approach has been widely used to study the global requirement of miRNAs for certain fundamental biological and pathological processes (Yang et al. 2005, da Costa Martins et al. 2008, Andl et al. 2006, Bernstein et al. 2003, Chen et al. 2008, Harris et al. 2006). Targeting miRNAs via interrupting the miRNA biogenesis pathway has become an important alternative to the direct targeting of mature miRNAs. In theory, every single one of the components along the miRNA biogenesis could be manipulated to disrupt the pathway. Dicer, which is critical for processing of pre-miRNAs into their mature form, has been a popular target helping to assess the global requirement for miRNAs in mammalian biology. Numerous studies using the Dicer inactivation approach have been documented. While this approach provides an invaluable means of studying the global requirement of miRNA for certain biological and pathophysiological processes, it has its inherent limitations. The Dicer inactivation technology is in general a miRNA-loss-of-function strategy; nonetheless, as already noticed by some studies, it can give rise to paradoxical miRNA gain-of-function outcomes.

CONCLUSION

miRNAs are universally expressed in mammalian cells, involving in nearly every aspects of life of organisms. They play important roles in the fundamental biological processes such as cell growth, differentiation and death and are implicated in a wide spectrum of disease including developmental malformations, cancer and cardiovascular disease, neuronal disorders such as Alzheimer's disease, metabolic disturbance such as diabetes mellitus and viral disease. Many pathological conditions are caused by deregulated expression of miRNAs. miRNAs have been considered a part of the epigenetic program in organisms. These properties and functions of miRNAs make them attractive viable therapeutic targets through interfering with their expression. Both gain-of-function and loss-of-function technologies are necessary tools for understanding miRNAs. The miRNA-Targeting strategy resulting in gain-of-function of miRNAs is an essential way for gain-of-knowledge about miRNA targets and functions. On the other hand, the Targeting-miRNA technologies leading to loss-of-function of miRNAs is also an indispensable strategy for miRNA research. These two approaches are mutually complementary for elucidating miRNA biology and pathophysiology. The miRNAi-Tech has opened new opportunities for creative and rational designs of a variety of combinations integrating varying nucleotide fragments for various purposes and provided exquisite tools for functional analysis related to identification and characterization of targets of miRNAs and their functions in gene controlling program.

In addition, the miRNAi-Tech also offers the strategies and tools for designing new agents for gene therapy of human diseases associated with miRNA deregulation (Table 3). These agents will possess a backbone structure as of oligoribonucleotides or oligodeoxyribonucleotides. Like other types of nucleic acids for gene therapy, such as antisense oligodeoxynucleotides, decoy oligodeoxynucleotides, siRNA, triplex-forming oligodeoxynucleotides, aptamer and DNAzyme, the oligomer fragments generated using miRNAi-tech can be chemically modified to improve stability and constructed into plasmids for easier delivery into organisms to treat diseases. Particularly noteworthy is the development of antagomiR and LNA-antimiR which can be delivered into organisms to knockdown target miRNAs for *in vivo* whole animal applications providing new opportunities for drug development.

The rationale for using miRNAs as potential therapeutic targets is based on the following facts. (1) Aberrant miRNA expression is a common feature of human multigenic diseases and could be used as a prognostic biomarker for disease diagnosis; (2) Particular miRNAs are involved in, as causal factors or key contributors, particular types of human diseases and could serve as novel therapeutic targets for disease treatment; (3) The level of miRNA expression responds to physiological stimuli and is susceptible to drug intervention; (4) Small size (18–26 nucleotides in length) of miRNAs makes them relatively easy to manipulate; and (5) A miRNA can be viewed as a regulator of a cellular function or a cellular program, a concept about miRNA actions to be discussed in the next section (Wang et al. 2008, Wang 2010).

ACKNOWLEDGMENTS

This work was supported by the Canadian Institute of Health Research and Heart and Stroke Foundation of Quebec to Z Wang.

REFERENCES

Andl, T., E.P. Murchison, F. Liu, Y. Zhang, M. Yunta-Gonzalez, J.W. Tobias, C.D. Andl, J.T. Seykora, G.J. Hannon and S.E. Millar. 2006. The miRNA-processing enzyme dicer is essential for the morphogenesis and maintenance of hair follicles. *Curr. Biol.* **16**:1041–1049.

Bernstein, E., S.Y. Kim, M.A. Carmell, E.P. Murchison, H. Alcorn, M.Z. Li, A.A. Mills, S.J. Elledge, K.V. Anderson and G.J. Hannon. 2003. Dicer is essential for mouse development. *Nat. Genet.* **35**:215–217.

Boutla, A., C. Delidakis and M. Tabler. 2003. Developmental defects by antisense mediated inactivation of micro-RNAs 2 and 13 in Drosophila and the identification of putative target genes. *Nucleic Acids Res.* **31**:4973–4980.

Brennecke, J., A. Stark, R.B. Russell and S.M.Cohen. 2005. Principles of microRNA-target recognition. *PLoS. Biol.* **3**:404–418.

Chen, J.F., E.P. Murchison, R. Tang, T.E. Callis, M. Tatsuguchi, Z. Deng, M. Rojas, S.M. Hammond, M.D. Schneider, C.H. Selzman, G. Meissner, C. Patterson, G.J. Hannon and D.Z. Wang. 2008. Targeted deletion of Dicer in the heart leads to dilated cardiomyopathy and heart failure. *Proc. Natl. Acad. Sci. USA.* **105**:2111–2116.

Choi, W.Y., A.J. Giraldez and A.F. Schier. 2007. Target protectors reveal dampening and balancing of Nodal agonist and antagonist by miR-430. *Science.* **318**:271–274.

Costinean, S., N. Zanesi, Y. Pekarsky, E. Tili, S. Volinia, N. Heerema and C.M. Croce. 2006. Pre-B cell proliferation and lymphoblastic leukemia/high-grade lymphoma in E(mu)-miR155 transgenic mice. *Proc. Natl. Acad. Sci. USA.* **103**:7024–7029.

da Costa Martins, P.A., M. Bourajjaj, M. Gladka, M. Kortland, R.J. van Oort, Y.M. Pinto, J.D. Molkentin and L.J. De Windt. 2008. Conditional dicer gene deletion in the postnatal myocardium provokes spontaneous cardiac remodeling. *Circulation.* **118**:567–1576.

Davis, S., B. Lollo, S. Freier and C. Esau. 2006. Improved targeting of miRNA with antisense oligonucleotides. *Nucleic Acids Res.* **34**:2294–2304.

Ebert, M.S., J.R. Neilson and P.A. Sharp. 2007. MicroRNA sponges: competitive inhibitors of small RNAs in mammalian cells. *Nat. Methods,* **4**:721–726.

Elmén, J., M. Lindow, S. Schütz, M. Lawrence, A. Petri, S. Obad, M. Lindholm, M. Hedtjärn, H.F. Hansen, U. Berger, S. Gullans, P. Kearney, P. Sarnow, E.M. Straarup and S. Kauppinen. 2008. LNA-mediated microRNA silencing in non-human primates. *Nature.* **452**:896–899.

Elmén, J., H. Thonberg, K. Ljungberg, M. Frieden, M. Westergaard, Y. Xu, B. Wahren, Z. Liang, H. Ørum, T. Koch and C. Wahlestedt. 2005. Locked nucleic acid (LNA) mediated improvements in siRNA stability and functionality. *Nucleic Acids Res.* **33**:439–447.

Esau, C., S. Davis, S.F. Murray, X.X. Yu, S.K. Pandey, M. Pear, L. Watts, S.L. Booten, M. Graham, R. McKay, A. Subramaniam, S. Propp, B.A. Lollo, S. Freier, C.F. Bennett, S. Bhanot and B.P. Monia. 2006. miR-122 regulation of lipid metabolism revealed by *in vivo* antisense targeting. *Cell. Metab.* **3**:87–98.

Fabani, M.M. and M.J. Gait. 2008. miR-122 targeting with LNA/2′-O methyl oligonucleotide mixmers, peptide nucleic acids (PNA) and PNA-peptide conjugates. *RNA.* **14**:336–346.

Frieden, M., S.M. Christensen, N.D. Mikkelsen, C. Rosenbohm, C.A. Thrue, M. Westergaard, H.F. Hansen, H. Ørum and T. Koch. 2003. Expanding the design horizon of antisense oligonucleotides with alpha-L-LNA. *Nucleic Acids Res.* **31**:6365–6372.

Golden, D.E., V.R. Gerbasi and E.J. Sontheimer. 2008. An inside job for siRNAs. *Mol. Cell.* **31**:309–312.

Grünweller, A. and R.K. Hartmann. 2007. Locked nucleic acid oligonucleotides: the next generation of antisense agents? *BioDrugs.* **21**:235–243.

Hammond, S.M. 2007. Soaking up small RNAs. *Nat. Methods.* **4**:694–695.

Harris, K.S., Z. Zhang, M.T. McManus, B.D. Harfe and X. Sun. 2006. Dicer function is essential for lung epithelium morphogenesis. *Proc. Natl. Acad. Sci. USA.* **103**:2208–2213.

Horwich, M.D. and P.D. Zamore. 2008. Design and delivery of antisense oligonucleotides to block microRNA function in cultured Drosophila and human cells. *Nature. Protocols.* **3**:1537–1549.

Hutvagner, G. and M.J. Simard. 2008. Argonaute proteins: key players in RNA silencing. *Nat. Rev. Mol. Cell. Biol.* **9**:22–32.

Hutvágner, G., M.J. Simard, C.C. Mello and P.D. Zam. 2004. Sequence-Specific Inhibition of Small RNA Function. *PLoS Biol.* **2**:465–475.

Jackson, R.J. and N. Standart. 2007. How do microRNAs regulate gene expression? *Sci. STKE.* **23**:243–249.

Jepsen, J.S. and L. Wengel. 2004. LNA-antisense rivals siRNA for gene silencing. *Curr. Opin. Drug Discovery Dev.* **7**:188–194.

Kim, V.N. 2005. MicroRNA biogenesis: co-ordinated cropping and dicing. *Nat. Rev. Mol. Cell. Biol.* **6**:376–385.

Kocerha, J., S. Kauppinen and C. Wahlestedt. 2009. microRNAs in CNS disorders. *Neuromolecular Med.* **11**:162–172.

Krutzfeldt, J., S. Kuwajima, R. Braich, K.G. Rajeev, J. Pena, T. Tuschl, M. Manoharan and M. Stoffel. 2007. Specificity, duplex degradation and subcellular localization of antagomirs. *Nucleic Acids Res.* **35**:2885–2892.

Krutzfeldt, J., N. Rajewsky, R. Braich, K.G. Rajeev, T. Tuschl, M. Manoharan and M. Stoffel. 2005. Silencing of microRNAs *in vivo* with 'antagomirs'. *Nature.* **438**:685–689.

Kuhnert, F., M.R. Mancuso, J. Hampton, K. Stankunas, T. Asano, C.Z. Chen and C.J. Kuo. 2008. Attribution of vascular phenotypes of the murine Egfl7 locus to the microRNA miR-126. *Development.* **135**:3989–3993.

Kurreck, J., E. Wyszko, C. Gillen and V.A. Erdmann. 2002. Design of antisense oligonucleotides stabilized by locked nucleic acids. *Nucleic Acids Res.* **30**:1911–1198.

Kwon, C., Z. Han, E.N. Olson and D. Srivastava. 2005. MicroRNA1 influences cardiac differentiation in Drosophila and regulates Notch signaling. *Proc. Natl. Acad. Sci. USA.* **102**:18986–18991.

Lanford, R.E., E.S. Hildebrandt-Eriksen, A. Petri, R. Persson, M. Lindow, M.E. Munk, S. Kauppinen and H. Ørum. 2010. Therapeutic silencing of microRNA-122 in primates with chronic hepatitis C virus infection. *Science.* **327**:198–201.

Latronico, M.V. and G. Condorelli. 2009. MicroRNAs and cardiac pathology. *Nat. Rev. Cardiol.* **6**:419–429.

Latronico, M.V.G., D. Catalucci and G. Condorelli. 2007. Emerging Role of MicroRNAs in Cardiovascular Biology. *Circ. Res.* **101**:1225–1236.

Lee, Y., K. Jeon and J.T. Lee, et al. 2002. MicroRNA maturation: stepwise processing and subcellular localization. *EMBO J.* **21**:4663–4670.

Lewis, B.P., C.B. Burge and D.P. Bartel. 2005. Conserved seed pairing, often flanked by adenosines, indicates that thousands of human genes are microRNA targets. *Cell.* **120**:15–20.

Lewis, B.P., I.H. Shih, M.W. Jones-Rhoades, D.P. Bartel and C.B. Burge. 2003. Prediction of mammalian microRNA targets. *Cell*, **115**:787–798.

Lin, S.L. and S.Y. Ying. 2006. Gene silencing *in vitro* and *in vivo* using intronic microRNAs. *Methods Mol. Biol.* **342**:295–312.

Lin, S.L., D. Chang, D.Y. Wu and S.Y. Ying. 2003. A novel RNA splicing-mediated gene silencing mechanism potential for genome evolution. *Biochem. Biophys. Res. Commun.* **310**:754–760.

Lu, J., G. Getz, E.A. Miska, E. Alvarez-Saavedra, J. Lamb, D. Peck, A. Sweet-Cordero, B.L. Ebert, R.H. Mak, A.A. Ferrando, J.R. Downing, T. Jacks, H.R. Horvitz and T.R. Golub. 2005. MicroRNA expression profiles classify human cancers. *Nature.* **435**:834–838.

Lu, Y., J. Xiao, H. Lin, Y. Bai, X. Luo, Z. Wang and B. Yang. 2009. A single anti-microRNA antisense oligodeoxyribonucleotide (AMO) targeting multiple microRNAs offers an improved approach for microRNA interference. *Nucleic Acids Res.* **37**:e24.

Lu, Y., J.M. Thomson, H.Y. Wong, S.M. Hammond and B.L. Hogan. 2007. Transgenic over-expression of the microRNA miR-17-92 cluster promotes proliferation and inhibits differentiation of lung epithelial progenitor cells. *Dev. Biol.* **310**:442–453.

Lu, Y., Y. Zhang, N. Wang, Z. Pan, X. Gao, F. Zhang, Y. Zhang, H. Shan, X. Luo, J. Xiao, G. Qiao, Y. Li, Y. Bai, L. Sun, N. Ma, Y. Hui, C. Xu, Z. Wang and B. Yang. 2010. Control of experimental atrial fibrillation by microRNA-328. **122**:2378–2387.

Luo,X., H. Lin, J. Xiao, Y. Zhang, Y. Lu, B. Yang and Z. Wang. 2008. Down regulation of miRNA-1/miRNA-133 contributes to re-expression of pacemaker channel genes HCN2 and HCN4 in hypertrophic heart. *J. Biol. Chem.* **283**:20045–20052.

Luo, X., H. Lin, Y. Lu, B. Li, J. Xiao, B. Yang and Z. Wang. 2007. Transcriptional activation by stimulating protein 1 and post-transcriptional repression by muscle-specific microRNAs of I_{Ks}-encoding genes and potential implications in regional heterogeneity of their expressions. *J. Cell. Physiol.* **212**:358–367.

Meister, G., M. Landthaler, Y. Dorsett and T. Tuschl. 2004. Sequence-specific inhibition of microRNA- and siRNA-induced RNA silencing. *RNA.* **10**:544–550.

Mook, O.R., F. Baas, M.B. de Wissel and K. Fluiter. 2007. Evaluation of locked nucleic acid-modified small interfering RNA *in vitro* and *in vivo. Mol. Cancer Ther.* **6**:833–843.

Naguibneva, I., M. Ameyar-Zazoua, N. Nonne, A. Polesskaya, S. Ait-Si-Ali, R. Groisman, M. Souidi, L.L. Pritchard and A. Harel-Bellan. 2006. An LNA-based loss-of-function assay for micro-RNAs. *Biomed. Pharmacother.* **60**:633–638.

Ørom, U.A., S. Kauppinen and A.H. Lund. 2006. LNA-modified oligonucleotides mediate specific inhibition of microRNA function. *Gene.* **372**:137–141.

Peng, S., J.P. York and P. Zhang. 2006. A transgenic approach for RNA interference-based genetic screening in mice. *Proc. Natl. Acad. Sci. USA.* **103**:2252–2256.

Pillai, R.S., S.N. Bhattacharyya and W. Filipowicz. 2007. Repression of protein synthesis by miRNAs: how many mechanisms? *Trends Cell. Biol.* **17**:118–126.

Pineau, P., S. Volinia, K. McJunkin, A. Marchio, C. Battiston, B. Terris, V. Mazzaferro, S.W. Lowe, C.M. Croce and A. Dejean. 2010. miR-221 overexpression contributes to liver tumorigenesis. *Proc. Natl. Acad. Sci. USA.* **107**:26426–26429.

Pushparaj, P.N., J.J. Aarthi, J. Manikandan and S.D. Kumar. 2008. siRNA, miRNA and shRNA: *in vivo* applications. *J. Dent. Res.* **87**:992–1003.

Saenger, W. 1984. Principles of nucleic acid structure. New York, USA: Springer.

Sokol, N.S. and V. Ambros. 2005. Mesodermally expressed Drosophila microRNA-1 is regulated by Twist and is required in muscles during larval growth. *Genes Dev.* **19**:2343–2354.

Stenvang, J., A.N. Silahtaroglu, M. Lindow, J. Elmen and S. Kauppinen. 2008. The utility of LNA in microRNA-based cancer diagnostics and therapeutics. *Semin Cancer Biol.* **18**:89–102.

Sun, D., M. Melegari, S. Sridhar, C.E. Rogler and L. Zhu. 2006. Multi-miRNA hairpin method that improves gene knockdown efficiency and provides linked multi-gene knockdown. *Biotechniques.* **41**:59–63.

Thai, T.H., D.P. Calado, S. Casola, K.M. Ansel, C. Xiao, Y. Xue, A. Murphy, D. Frendewey, D. Valenzuela, J.L. Kutok, M. Schmidt-Supprian, N. Rajewsky, G. Yancopoulos, A. Rao and K. Rajewsky. 2007. Regulation of the germinal center response by microRNA-155. *Science.* **316**:604–608.

van Solingen, C., L. Seghers, R. Bijkerk, J.M. Duijs, M.K. Roeten, A.M. van Oeveren-Rietdijk, H.J. Baelde, M. Monge, J.B. Vos, H.C. de Boer, P.H. Quax, T.J. Rabelink and A.J. van Zonneveld. 2009. Antagomir-mediated silencing of endothelial cell specific microRNA-126 impairs ischemia-induced angiogenesis. *J. Cell. Mol. Med.* **13**:1577–1585.

Veedu, R.N. and J. Wengel. 2010. Locked nucleic acids: promising nucleic acid analogs for therapeutic applications. *Chem. Biodivers.* **7**:536–542.

Wang, S., A.B. Aurora, B.A. Johnson, X. Qi, J. McAnally, J.A. Hill, J.A. Richardson, R. Bassel-Duby and E.N. Olson. 2008. The endothelial-specific microRNA miR-126 governs vascular integrity and angiogenesis. *Dev. Cell.* **15**:261–271.

Wang, X., P. Liu, H. Zhu, Y. Xu, C. Ma, X. Dai, L. Huang, Y. Liu, L. Zhang and C. Qin. 2009. miR-34a, a microRNA up-regulated in a double transgenic mouse model of Alzheimer's disease, inhibits bcl2 translation. *Brain Res. Bull.* **80**:268–273.

Wang, Z. 2009. MicroRNA-Interference Technologies. New York, USA: Springer-Verlag.

Wang, Z. 2010. The role of microRNA in cardiac excitability. *J. Cardiovasc. Pharmacol.*, 2010;doi: 10.1097/FJC.0b013e3181edb22c.

Wang, Z. and B. Yang. 2010. 1st Edn. MicroRNA Expression Detection Methods. New York, USA: Springer-Verlag.

Wang, Z., X. Luo, Y. Lu and B. Yang. 2008. miRNAs at the heart of the matter. *J. Mol. Med.* **86**:772–783.

Xi, Y., A. Formentini, M. Chien, D.B. Weir, J.J. Russo, J. Ju, M. Kornmann and J. Ju. 2006. Prognostic values of microRNAs in colorectal cancer. *Biomark Insights.* **2:**113–121.

Xia, H., Q. Mao, H.L. Paulson and B.L. Davidson. 2002. siRNA-mediated gene silencing *in vitro* and *in vivo. Nat. Biotechnol.* **20:**1006–1010.

Xia, X.G., H. Zhou and Z. Xu. 2006. Multiple shRNAs expressed by an inducible pol II promoter can knock down the expression of multiple target genes. *Biotechniques.* **41:**64–68.

Xiao, J., X. Luo, H. Lin, Y. Zhang, Y. Lu, N. Wang, Y.Q. Zhang, B. Yang and Z. Wang. 2007a. MicroRNA miR-133 represses HERG K⁺ channel expression contributing to QT prolongation in diabetic hearts. *J. Biol. Chem.* **282:**12363–12367.

Xiao, J., B. Yang, H. Lin, Y. Lu, X. Luo and Z. Wang. 2007b. Novel approaches for gene-specific interference via manipulating actions of microRNAs: examination on the pacemaker channel genes HCN2 and HCN4. *J. Cell. Physiol.* **212:**285–292.

Yang, B., H. Lin, J. Xiao, Y. Lu, X. Luo, B. Li, Y. Zhang, C. Xu, Y. Bai, H. Wang, G. Chen and Z. Wang. 2007. The muscle-specific microRNA miR-1 causes cardiac arrhythmias by targeting GJA1 and KCNJ2 genes. *Nat. Med.* **13:**486–491.

Yang, W.J., D.D. Yang, S. Na, G.E. Sandusky, Q. Zhang and G. Zhao. 2005. Dicer is required for embryonic angiogenesis during mouse development. *J. Biol. Chem.* **280:**9330–9335.

Zhao, Y., J.F. Ransom, A. Li, V. Vedantham, M. von Drehle, A.N. Muth, T. Tsuchihashi, M.T. McManus, R.J. Schwartz and D. Srivastava. 2007. Dysregulation of cardiogenesis, cardiac conduction and cell cycle in mice lacking miRNA-1-2. *Cell.* **129:**303–311.

CHAPTER 5

Silencing the Disease Messengers: Progress and Prospects in Developing Nucleic Acid based Therapeutics

Rakha H. Das,* and Nandini Verma

ABSTRACT

The conception that all enzymes are proteins was challenged few decades ago by the surprising discovery of cleavage and ligation activity of the group I intron. Since then numerous categories of catalytic nucleic acids have emerged, principally exploiting an anti-sense oligonucleotide that can hybridize with a specific mRNA to block translation and its subsequent degradation. However, the phenomenon of ribonucleic acid (RNA) interference (RNAi) was rather a recent breakthrough which brought a new expansion of research interest in catalytic nucleic acids. Besides a promising non-protein catalytic gene manipulating tool whose potential can be exploited in understanding physiological and pathological processes, RNAi accounts for a versatile candidature for the development of nucleic acid based therapeutic agents. Owing to the specificity of post-transcriptional gene silencing (PTGS) in cells, the RNAi technique is now established in *in vitro* systems and presently more exploration focusing its derivation into the *in vivo* application are being carried out. An altogether novel class of catalytic nucleic acid made entirely of DNA has evolved through *in vitro* selection with its enormous potential for mediating gene inactivation. These RNA-cleaving DNA enzymes (deoxyribozyme, DNAzyme) rapidly cleaved their way to successful PTGS applications because of their terrific RNA cleavage activity, target flexibility and stability demonstrated by their ability of gene suppression both *in vitro* and *in vivo*. So far, several genes have been targeted by nucleic acid-based PTGS tools in animal models of various diseases to scrutinize and determine the intricate

* *Corresponding author e-mail*: rakhahdas@yahoo.com

processes at the level of molecular signaling events. By and large, these PTGS studies generated candidate molecules which have been tested in the clinical trials and few have successfully proceeded for conversion into drugs, instigating a new epoch of gene-specific medication. These developments can prospectively overcome the restraints in current therapeutic regimes. This chapter delineates current advancements in PTGS technology, primarily on short interfering RNA and DNA enzymes, as these are primed to hold central stage in transferring the anti-gene therapies to the clinic.

Keywords: Post-transcriptional gene silencing, RNAi, DNAzyme, gene therapy.

INTRODUCTION

Regulation of gene expression is an intricate process involving acetylation and deacetylation of histones, methylation and demethylation of cytosine and *trans-* and *cis-* acting elements in addition to several transcription factors. Microarray analysis of cellular mRNAs indicate several genes both up- and down-regulated in various pathological conditions of diseases. Over and under expressions of one or one set of genes often effects the expression of other set of genes. Molecules that can specifically silence expression of a gene are powerful research tools and perhaps, prospective therapeutics [Anderson 1998]. Much effort has been put into the development of such molecules and has resulted in the creation of different classes of potential therapeutic agents. Single-stranded nucleic acids have the ability to alter gene-expression by binding to and consequently interfering with the stability or function of a target mRNA transcript. The major classes of antisense agents currently used by investigators for sequence-specific mRNA knockdown are RNA interference (RNAi) associated with small interference RNA (siRNA) and microRNA (miRNA), ribozymes, DNAzymes, oligonucleotide decoys and antisense oligonucleotides (ODNs), (Figure 1). Selective gene silencing by nucleic acid enzymes has provided researchers with a new strategy to block gene expression to either understand the function of target genes or drug target validation as potential therapeutics.

RNAi

RNA interference is an intracellular mechanism which regulates genes. This class includes two types of small RNA molecules, miRNA and siRNA [Lavery and King 2003]. This mechanism operates against the long, double-stranded RNA is introduced into the cell that is processed by the cytoplasmic RNase III enzyme Dicer into 21-23nt siRNAs. These siRNAs are then become associated to form the multi-component RNA-induced silencing complex (RISC) that mediates unwinding of the duplex and uses the antisense strand to recognize and degrade the complementary mRNA (Figure 1A). The specific down-regulation of target mRNAs is achievable by the use of short synthetic siRNAs and short hairpin RNAs (shRNAs) [Dykxhoorn et al. 2003]. However, some non-specific effects of siRNA have also been observed together with its immunogenicity which leads to the activation of complement pathway.

MicroRNAs are short RNA (about 22 nucleotides long) post-transcriptional regulatory molecules which recognize and bind to complementary sequences in the three prime untranslated regions (3' UTRs) of target messenger RNA transcripts (mRNAs) that causes the gene silencing effect [Trang et al. 2008]. The human genome encodes over 1000 miRNAs, targeting around 60% of mammalian genes. In eukaryotic organisms miRNAs are well conserved and are vital component of genetic regulation.

Ribozymes and DNAzymes: The Non-identical Catalytic Twins

Ribozymes and DNAzymes, are single-stranded nucleic acids having a conserved sequence that bring about the cleavage of the target RNA upon binding (Figure 1B) [Bramlage et al. 1998, Breaker 1997]. Both of these nucleic acid enzymes have three essential components: (i) a highly conserved nucleotide catalytic domain; (ii) base-pairing arms flanking the catalytic domain for binding to the target RNA; and (iii) a recognition cleavage site of the target RNA sequence. However, as their names indicate, DNAzymes are DNA oligonucleotide while ribozymes are composed of RNA. There are five naturally occurring classes of ribozymes that include hairpin, hammerhead, group I intron, ribonuclease P and hepatitis delta virus ribozyme [Bramlage et al. 1998]. The hammerhead and hairpin ribozymes are most extensively studied due to their small size and versatility. The complementary sequence directed base-pairing of ribozyme and substrate RNA to form a complex secondary structure having catalytic centre required for the cleavage. The substrate is cleaved 3' to the recognition sequence in the presence of divalent metal ions such as Mg^{2+} or Mn^{2+} generating two products with 2'-3' cyclic phosphate and a 5'-OH termini. After cleavage of the target message, the ribozyme may be released and can participate in further reactions, hence acting catalytically.

Unlike ribozymes, the DNAzymes have not been observed in nature, instead have been isolated in the 1990s following many rounds of *in vitro* selection. The RNA-cleaving 10–23 DNAzyme possess the ability to cleave target RNA molecules under physiological conditions in a manner similar to ribozymes provided there is a purine-pyrimidine dinucleotide in the sequence of the target RNA [Breaker 1997]. This can be accomplished at very high kinetic efficiency

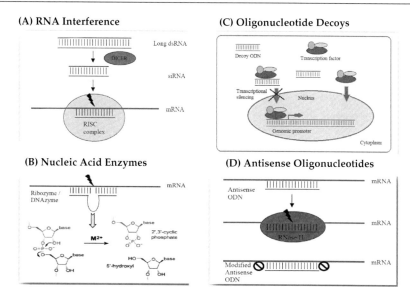

Figure 1. Mechanisms of post-transcriptional gene silencing by nucleic acid agents. (A) In RNA interference a long, double-stranded RNA is processed to siRNA by DICER. The siRNA uses the RISC complex to degrade the target mRNA. **(B)** Nucleic acid enzymes like ribozyme or DNAzyme cleave the target mRNA directly owing to their catalytic activity. **(C)** Decoy ODN silences the expression of specific genes by binding to the transcription factors responsible for their transcription. **(D)** An antisense oligonucleotide hybridizes to the target mRNA which is then either degraded by RNase H or blocks translation by steric hindrance.

(Color image of this figure appears in the color plate section at the end of the book.)

with rates equal to or greater than that of ribozymes and endo-ribonucleases. The activity of the DNAzyme is characterized by progression through multiple rounds of mRNA substrate binding, cleavage and release of products. The mode of action of the 10–23 DNAzyme combines the advantageous features of both ribozymes and antisense DNA. DNAzymes are significantly more stable than ribozymes due to the deoxyribose backbone and are relatively inexpensive and easy to synthesize. The DNAzyme also has the added advantage of functioning against a wider selection of cleavage sites in a chosen target.

Decoys

In studies involving transcription factors to investigate the molecular mechanisms, knockout of transcription factor is rather less preferred as it can result into undesired secondary effects and perplexed outcomes. Suppressing the activity of the transcription factor using decoy oligonucleotides can decrease these secondary effects. ODN decoys are synthetic double-stranded DNA containing *cis*-acting elements that bind to the transcription factor(s), therefore, averting the interaction of the nuclear protein with consensus sequence in the genomic promoter (Figure 1C) [Mann 2005].

ODN decoys, however, are not strictly 'gene-specific' agents; they have been shown to be extremely specific for their target transcription factor. A number of transcription factors including E2F and NF-κB have been successfully targeted using ODN decoys in animal models of various diseases [Nogoy 2007]. An essential concern in the use of decoys is their possible interference in the normal physiological responses [Mann 2005] due to pleiotropy of transcription factors.

Antisense

Antisense oligonucleotides are short sequences of single-stranded DNA that are typically between 18–25nt in length. The inhibition associated with this type of antisense occurs in the cytoplasm and affect translation of the target mRNA by either steric hindrance or RNase H mediated cleavage (Figure 1D) [Lavery and King 2003, Crook and Bennett 1996]. The major route of antisense action and the most widely understood mechanism involves RNase H mediated cleavage of the target mRNA recognized as RNA-DNA heteroduplex. Similarly, antisense RNA is a single-stranded RNA that is complementary to a messenger RNA (mRNA) strand transcribed within cells that can physically obstruct the translation machinery [Lavery and King 2003, Crook and Bennett 1996].

Gene Silencing: A Class of Emerging Therapeutics

As our understanding of the pathogenesis of human diseases is persistently progressing, a parallel evolution in the therapeutic developments is also accompanied for a better and specific combat. More precisely, the latest advancements in genomics have generated a substantial transformation in the conventional therapeutic opinion to concentrate on the intricate molecular and cellular processes which in turn are under regulation of a cluster of key genes. It is at this intersection where the nucleic acid molecules are emerging as prevailing tools in facilitating the characterization of key biochemical pathways by dissecting the molecular functions. With the ability of nucleic acids to obstruct specifically the expression of a targeted gene they have also obtained considerable attention as a prospective class of therapeutic agents.

The conception of a gene-specific therapy emerged nearly three decades ago when the efficacy of nucleic acids in altering the gene expression was demonstrated for the first time [Paterson et al. 1977]. In 1978 the earliest application of antisense molecules was directed to inhibit the Rous sarcoma viral replication [Stephenson and Zamecnik 1978]. Of late there has been a substantial advancement in nucleic acid therapeutics having several gene targeted and which are under metamorphosis to develop into a clinical reality. This therapeutics encompassed natural, modified, as well as synthetic molecules both in *in vitro* and *in vivo* interventions in diverse patho-physiological contexts with different degrees of success. The first gene therapy study was commenced in 1990 for curing adenosine deaminase deficiency [Blaese et al. 1995], subsequently a number of clinical gene therapy trials for diseases caused by genetic disorders, viral infections and malignancies were performed [Anderson 1995, Miller 1992]. The present gene therapy approaches are chiefly focused on either the substitution of faulty genes or repression of a pathological gene target.

RNA interference: Current Status as a Contemporary Therapeutics

The efficiency and specificity of siRNA-mediated target degradation reported in most studies opens the door to its possible use against human diseases. As compared to other nucleic acid based strategies, RNAi has evolved much more rapidly as a candidate therapeutic agent. A principal advantage of RNAi against other nucleic acid-based methodology with respect to their medicinal relevance is that it employs cellular machinery which efficiently assists targeting of complementary transcripts, often leading in extremely effective down-regulation of gene expression. Despite a high expectancy regarding this significant biological process for sequence specific gene regulation, there are several hurdles and apprehensions which need to be addressed including its off-target effects, immunogenicity and delivery prior to flourishing RNAi as a therapeutic reality.

Apart from initial *in vitro* investigations to judge the functional proficiency of RNAi against several target genes, in the past decade a number of *in vivo* studies have been carried out in various animal models of human diseases including genetic, neurological, cardiovascular, viral and cancer in order to elucidate its therapeutic extrapolation [Aigner 2006, Vaishnav et al. 2010]. Numerous RNAi based molecules eventually also entered into various phases of clinical trials in assorted diseases together with age-related macular degeneration (AMD), diabetes, hypercholesterolaemia, respiratory disease, hepatitis, human immunodeficiency virus infection and cancer in the past decade and a few RNAi therapeutic programs are under clinical practice (Table 1).

Table 1. Current Clinical Trials of RNAi-Based Therapeutics*.

Drug	Molecule Type	Program (clinical stage)	Target	Sponsor	Indication	Status
ALN-RSV	siRNA	phase IIb	RSV nucleocapsid	Alnylam/Cubist/ Kyowa Kirin	Adult RSV infection	Ongoing
PF-04523655	siRNA	phase II	RTP801	Pfizer/Quark	AMD, diabetic macular edema	Ongoing
QPI 1002	siRNA	phase II	p53	Quark	Acute kidney injury, delayed graft function	recruiting participants
Excellair	siRNA	phase II	Syk kinase	Zabecor	Asthma	Ongoing
ALN-VSP	siRNA	phase I	VEGF, KSP	Alnylam	Primary and secondary liver cancer	Ongoing
CALAA-01	siRNA	phase I	RRM2	Calando	Cancer	recruiting participants
Atu-027	siRNA	phase I	PKN3	Silence	Cancer (GI, lung other)	recruiting participants
SYL040012	siRNA	phase I	β2 adrenergic receptor	Sylentis	Glaucoma	Ongoing
ALN-TTR	siRNA	phase I	TTR	Alnylam	TTR amyloidosis	Ongoing
Bevasiranib	siRNA	phase III	VEGF-A	Opko	AMD	Terminated
AGN211745	siRNA	phase II	VEGFRI	Allergan/SIRNA	AMD	Terminated
ApoB SNALP	siRNA	phase I	ApoB	Tekmira	Hypercholesterolemia	Completed
TD101	siRNA	phase I	Mutant K6a	Transderm	Pachyonychia congenita	Completed
Bcr-abl	siRNA	phase I	Bcr-abl oncogene	Univ. Duisberg-Essen	CML	Completed
RTP801i-14	siRNA	Phase I	RTP801 (DDIT4)	Quark Pharmaceuticals	AMD	Ongoing

contd. ... Table

Table ... contd.

Drug	Molecule Type	Program(clinical stage)	Target	Sponsor	Indication	Status
SPC3649	LNA-anti miRNA	Phase I	miR-122	Santaris Pharma	Hepatitis C	Ongoing
Proteasome siRNA	siRNA	Phase I	Immuno-proteasome beta subunits (LMP2, LMP7, MECL1)	Duke University	Metastatic melanoma	recruiting participants
Anti-tat/rev	shRNA	Pilot feasible study	Tat/rev	City of Hope National medical center	AIDS lymphoma	Unknown

* All data from corporate websites, press releases and http://www.clinicaltrials.gov

AMD age-related macular degeneration, RSV respiratory syncytial virus; AMD = age-related macular degeneration; CML = chronic myeloid leukemia; GI = gastrointestinal; KSP = kinesin spindle protein; PKN = protein kinase N3; RRM2 = ribonucleotide reductase M2; TTR = transthyretin; VEGF = vascular endothelial growth factor; VEGFRI = vascular endothelial growth factor receptor I.

Nearly half of them involve local/topical delivery to the eye, respiratory tract or skin. The others are systemic programs targeting kidney, liver, hepatic and extrahepatic cancer or leukocytes. After the evaluation of previous oligonucleotide trials which were terminated in failures in early clinical development owing to unclear evidences of perception in humans, majorly due to poor target validation and lack of clear post-delivery readouts [Vaishnav et al. 2010], the current programs are being developed to ascertain the unmet medical need. A large number of clinical trials are based on the intravitreal treatment of AMD. The initial and most advanced clinical investigation was the phase III trial of bevasiranib (siRNA to VEGF) which was introduced as continuance therapy subsequent to anti-VEGF treatment with Lucentis (antibody to VEGF). Unfortunately, this trial was terminated prematurely since the preliminary data fall short to achieve the expected primary endpoint. The other two ongoing trials involve VEGFA165b siRNA targeting VEGF A165 isoform and a chemically modified siRNA, AGN211745 (Sirna-027), against VEGF receptor-1. Surprisingly, some uncertainties were observed on the mechanism of anti-VEGF siRNA therapeutics for treating AMD according to a recent study which indicated that intravitreal injections of all siRNA of 21-nucleotides or longer, irrespective of the nucleotide sequences or their deliberated targets (i.e., whether or not the siRNA targets VEGF or VEGFR), were able to suppress neovascularization in mice without causing off-target RNAi or interferon-alpha/beta activation. Moreover, this study demonstrated that this in general siRNA-mediated suppression of neovascularization was due to non-specific stimulation of the cell surface Toll-like receptor-TLR3 pathway, its adaptor toll-receptor-associated activator of interferon (TRIF) and induction of interferon-gamma and interleukin- 12 [Kleinman et al. 2008]. These outcomes depict a severe concern in terms of considerable inconsistency in target validation and safety which emerges as an obstacle in the successful use of siRNA therapeutics. Another phase I clinical trial to treat viral infection of respiratory tract involves a siRNA-based drug ALN-RSV01 that targets the nucleocapsid encoding gene of the respiratory syncytial virus (RSV) to inhibit viral replication in the lung. The trial employed intranasal siRNA delivery (unmodified siRNA formulated in saline) which was well tolerated by adults (doses up to 3 mg/kg) [DeVincenzo et al. 2008, Alnylam Pharmaceuticals Inc.] (http://www.alnylam.com). In early 2008 the Phase II trial ALN-RSV01 was initiated in lung transplant patients naturally infected with RSV and is currently the most advanced program in the global RNAi pipeline. The ALN-RSV program also exemplifies the stability and suitability of RNAi therapeutics with devices used for inhalational delivery. Expression of p53 causes acute kidney injury post cardiac bypass surgery, a systemically administered siRNA targeting p53, QPI-1002 (Quark Pharmaceuticals Inc., Fremont, CA, USA), is also in the phase II development [20, 21]. Studies evaluating the safety of QPI-1002 in diverse patient populations have been accomplished [22] with first part of a phase I/II study in renal transplant recipients for the prevention of delayed graft function. Pachyonychia congenital disorder (PC) is a rare skin disorder which is caused by mutations in one of four keratin (K) genes (K6a, K6b, K16 or K17). Trans Derm, Inc. has developed a siRNA that selectively inhibit a mutant allele of K6a (TD101) which is the most common mutation in keratin causing PC [Leachman et al. 2008]. Trans Derm K6a is in a phase I clinical trial. Recently, another siRNA, REDD14NP, has entered in the phase I trial in patients with blinding choroidal neovascularization secondary to AMD. REDD14NP is a synthetic chemically modified 19-mer siRNA molecule aiming hypoxia-inducible gene RTP801 (or DDIT4) which is implicated in the progression of disease. However, REDD14NP does not activate TLR3 [Feinstein et al. 2009]. In RNAi therapeutics targeting cancer, solid tumors in particular, two siRNA, CALAA-01 and Atu027, arrived for clinical assessments which are designed for intravenous adminis-tration. The first human phase I trial of CALAA-01 was commenced in 2008 in patients having solid tumors refractory to the conventional medication. CALAA-01, siRNA is designed to target the M2 subunit of ribonucleotide reductase for enhancement of cellular specificity this molecule is formulated in cyclodextrin nanoparticles having surface pegylation and conjugation with the transferring ligand [Davis 2009]. The trial results revealed successful delivery of siRNA nanoparticles to intended intracellular localizations and indicated decrease of corresponding mRNA and protein levels in tumor biopsies. So far, this is the first clinical evidence of specific gene inhibition by siRNA in patients after systemic administration [Devis et al. 2010]. The siRNA Atu027 is a lipoplex targeted against the protein kinase N3. The preclinical research indicated that multiple intravenous interventions of Atu027 resulted in RNAi-mediated specific silencing of protein kinase N3 expression in rodents and non-human primates accompanied with considerable inhibition of tumor growth and metastasis [Aleku et al. 2008]. As a next step the phase I trial of Atu027 in patients with advanced solid cancer was recently introduce in 2009. These encouraging clinical results pave the way for transitioning RNAi from a research tool into clinic evaluation and future applications.

RNA Interference: The Interference Within

Having barely over a decade past since its original description, it is quite amazing to witness a plenty of ongoing clinical trials. Ironically, the excitement with which the discovery of siRNAs was perceived as a new straightforward door to gene therapy has not met any FDA approved siRNA drug thus far. Being a cellular process of silencing gene expression, the progress made by RNAi biology to transform into RNAi therapeutics took a swift pace, however, its interference with the other biological processes has emerged as its major unforeseen limitation. The first cautionary note came from studies performed with gene arrays to observe the off-target effects of siRNAs. Their results indicated that ectopically applied siRNAs can modify the expression levels of several of non-targeted transcripts and suggested that even short complementary stretches of siRNAs with non-targeted transcripts can affect their expression levels [Jackson et al. 2003]. Soon the off-targeting by siRNAs was reported to be minimized by incorporation of a 2'-OMe modification on ribose of the siRNA [Jackson et al. 2006], nevertheless, surprisingly this solution did not prove useful for all *in vivo* siRNA applications. Moreover, the fact that in general, cells don't take up naked siRNA and thus a suitable cellular delivery system with great efficacy and least toxicity was the utmost challenge for its migration to the clinic. Advancements on that front in both academia and industry soon have come up as a handful of clinical trials attest. Subsequently another disturbing finding with respect to RNAi therapeutic applications came up from preclinical trials of shRNAs delivered by intraveinous injection into mice, which lead in lethality due to acute liver failure in numerous animals [Grimm et al. 2006]. These studies indicated that the levels of ectopic expression of therapeutic shRNAs have to be precisely controlled as these are processed into siRNAs which can elicit off-target effects and effectively compete with the endogenous microRNAs for the RNAi machinery. At the same time siRNA clinical studies are also being challenged with unanticipated immunological effects and impacts on endogenous RNA processing. Additionally, the type I interferon response produced by certain sequence motifs in siRNAs due to the activation of toll-like receptors (TLRs 7 and 8) further add to the compromise made against the sequence specific knockdown effects of the RNAi pathway [Hornung et al. 2005, Judge et al. 2005, Robbins et al. 2008]. Overcoming the interferon stimulation by TLR3 that recognizes double-stranded RNAs emerges as a more difficult trouble. An unexpected discovery from a murine AMD siRNA treatment study on model [Tracy et al. 2006] described that the inhibition of neo-vascularization was not due to specific knockdown of these targets rather by double-stranded RNA activation of TRL3, triggering interferon-γ and interleukin-12 production with subsequent anti-neo-vascularization effects. And unfavorably, the backbone modifications of the siRNAs did not ameliorate this response [Kleinman et al. 2008]. Despite enthusiasm from a large number of animal model studies, including systemic delivery to non-human primates [Tracy et al. 2006], there are a number of hurdles and concerns that must be overcome before RNAi will be exploited as a new therapeutic modality in the clinic. Like with all biological breakthroughs, understanding of the mechanism is key to effective applications in human disease, we anticipate that RNAi will be a major therapeutic approach in coming years.

The "10–23" DNA Enzyme: Cleaving a Way to Future Genetic Medicine

DNAzymes have been used as gene expression inhibitory agents in a wide array of gene therapeutic and industrial applications in a variety of experimental disease settings, suggesting their possible clinical utility. The 10–23 DNAzyme possess several features that offer a tremendous potential for applications both *in vitro* and *in vivo*. These small molecules are capable of cleaving almost any RNA under simulated physiological conditions where the substrate is strictly recognized on the basis of Watson-Crick base pairing of the binding arms, providing target specificity as well as flexibility. The catalytic efficiency of 10–23 DNAzyme is comparable to and often surpasses, the activity of other antisense molecules [Santoro and Joyce 1997].

The potential of the 10–23 DNAzyme to suppress gene function in biological systems has been explored by a number of groups. Most of the earliest *in vitro* gene suppression studies utilizing DNAzymes are based on viral diseases, however, many other include tuberculosis, Huntington's disease and various types of cancers were also been targeted successfully (Table 2). Considerable reductions of target cellular RNA and protein, by up to 70–90%, has been achieved in these *in vitro* studies. However, the true potential of DNAzymes has been demonstrated in animal models. Specific down-regulation of target genes *in vivo* has been established in the areas of vascular, cardiovascular diseases and cancer models (Table 2). Khachigian and coworkers in particular, have greatly contributed to knowledge in these areas [Khachigian 2000, Khachigian 2002]. The first study of DNAzyme activity in animals was conducted in 1999 with DNAzyme targeting the zinc finger transcription factor Egr-1 where it inhibited restenosis in balloon-injured rat carotid arteries [Santiago et al. 1999]. The DNAzyme was demonstrated to inhibit the smooth muscle cell proliferation and accumulation in the arterial intima. Recently, DNAzyme knockdown studies targeting Egr-1 have also shown its positive regulatory involvement in the induction of transforming growth factor-beta (TGF-β) and inducible nitric oxide synthase (iNOS) expression [Liu et al. 2008]. Alongside, the Egr-1 DNAzymes also demonstrated to block tumor growth in human breast tumor xenografts by reducing the

Table 2. Efficacy of some selected DNAzymes targeting diseases *in vitro* and *in vivo*.

In vitro studies

Disease	Target	Modification	Delivery Method	Activity	Reference
HIV-1	CCR5	none	Lipofectin	Decreased cell membrane fusion	[Goila et al. 2001]
HIV-1	env	none	Lipofectamine	77–81% suppression of viral load; suppression of viral replication	[Zhang et al. 1999]
HIV-1	gag	none	Lipofectin	95% viral inhibition	[Sriram et al. 2000]
HIV-1	TAT/rev	10 G residues at 3'	+/- Lipofectin	Taken up by human macrophage cell line without Lipofectin; inhibition of HIV-1 activity; protection when challenged with HIV-1	[Unwalla et al. 2001]
HIV-1	TAR	none	Lipofectin	10–12 fold reduction in mRNA; >80% reduction of viral antigen; significant inhibition of gene expression in primary and chronically infected cells	[Chakraborti et al. 2003]
Hepatitis B virus	HBx	none	Lipofectamine	Target mRNA and protein levels significantly reduced	[Hou et al. 2006]
Hepatitis B virus	HBx	none	Lipofectin	4–6 fold reduction in intracellular target RNA and protein levels compared to control cells	[Goila et al. 2001]
Hepatitis C virus	HCV	Phosphorothioate 2-base cap	Lipofectamine	32% and 48% reduction in viral target RNA in two human cell lines	[Trepanier et al.2006]
Hepatitis C virus	NS5B, core	5' phosphoro thioate, 3' PG-amine-C7	Lipofectamine	Reduction of HCV viral RNA	[Kumar et al. 2009]

contd. ... Table

Table ... contd.

Disease	Target	Modification	Delivery Method	Activity	Reference
Influenza virus	PB2	N3'-P5' phosphoramidite bonds at 3' and 5'	DOTAP	>99% suppression of viral expression in cells	[Takahashi et al. 2004]
Respiratory syncytial virus	RSV	Phosphorothioate 2-base cap/3' cholesterol	Added to media	Inhibition of target protein expression; 7 log reduction of viral yield; suppression of wild human strains in clinic	[Xie et al. 2006]
Tuberculosis	icl	Phosphorothioate 2-base cap	Added to media	Decreased M. tuberculosis survival in macrophages	[Li et al. 2005]
Vascular disease	c-myc	3'-3' inversion	DOTAP	Blocked target protein expression; 80% suppression of cell growth	[Sun et al. 1999]
Huntington's disease	huntingtin	3'-3' inversion	Lipofectamine	85% protein suppression	[Yen et al. 1999]
Chronic myelogenous leukemia, (CML)	Bcr-abl	2' O-methyl cap	Lipofectin	Apoptotic morphology; 99% suppression of reporter expression	[Warashina et al. 1999]
Cancer (CML)	Bcr-abl	Phosphorothioate 2-base cap	Cytofectin	40% protein suppression; 50% cell growth inhibition; 53–80% growth inhibition	[Wu et al. Year]
Cancer	PKCα	3'-3' inversion	DOTAP	Cell inhibition in four of five cancer cell lines; up to 83% reduction in protein levels	[Sioud et al. 2000]
Pancreatic carcinoma	survivin	phosphorothioate cap	Oligofectamine	Increase in apoptotic cells; inhibition of PANC-1 cell growth	[Liang et al. 2005]

Inflammation	iNOS	5' phosphor thioate, 3'CPG-amine-C7	Lipofectamine	Suppression of target gene, inhibition of LPS induced nitricoxide production.	[Chaudhury et al. 2006]
Inflammation	*TNF-α, receptors*	5' phosphoro thioate, 3'CPG-amine-C7	Lipofectamine	Suppression of autocrine secretion of TNF-α in activated macrophages	[Verma et al. 2009]
Colon cancer	*K-RAS*	None	Lipofectamine	DNAzyme sensitized the K-Ras(G12V) mutant cells to anti-cancer agents such as doxorubicin and radiation	[Yu et al. 2009]
Lung cancer	*MMP-9*	5' and 3' phosphoro thioate	oligofectamine	Inhibition of cell proliferation, migration and invasion	[Yang et al. 2009]
Cancer	c-jun	3'-3' inversion	Lipofectamine	potently induces caspase-2 resulting in apoptosis in a panel of tumor cell lines	[Dass et al. 2010]
***In vivo* studies**					
Squamous cell carcinoma (SCC)	c-Jun	3'-3' inversion	FuGENE6	Inhibition of SCC proliferation and suppression of solid SCC tumour growth in mice	[Zhang et al. 2006]
Breast carcinoma	EGR-1	3'-3' inversion	Intra-tumoral injection	Protein suppression; inhibition of cell proliferation; inhibition of solid tumour growth in mice	[Mitchell et al. 2004]
Tumour angiogenesis	EGR-1	3'-3' inversion	FuGENE6; direct administration	Blocked angiogenesis in mice; inhibited breast carcinoma in mice; repressed neovascularisation of rat cornea	[Fahmy et al. 2003]

contd. ... Table

Table ... contd.

Disease	Target	Modification	Delivery Method	Activity	Reference
Inflammation	c-Jun	3'-3' inversion	FuGENE6; direct administration	Reduction in vascular permeability and inflammation in mice and rats	[Fahmy et al. 2006]
Vascular disease	c-Jun	3'-3' inversion	Intravitreal administration	Inhibition of retinal neovascularisation in mice	[Khachigian et al. 2002]
Cardiovascular disease	EGR-1	3'-3' inversion	Superfect; adventitial/endoluminal delivery	Reduced target protein; inhibition of cell proliferation and wound repair; inhibition of protein expression, cell proliferation and neointima formation after mechanical injury to rat and pig carotid artery wall *in vivo*. Reduced myocardial infarct size in rats.	[Santiago et al. 1999, Lowe et al. 2001, Lowe et al. 2002, Bhindi et al. 2006]
Cardiovascular disease	VDUP1	3'-3' inversion	Superfect; direct intracardiac injection	70% reduction in target cellular mRNA levels; decrease in apoptosis; 2 fold increase in cell survival. Prolonged reduction in cardiomyocyte apoptosis as well as marked reduction in myocardial scar formation in rats.	[Xiang et al. 2005]
Vascular disease	PKC-epsilon	3'-3' inversion	Lipofectin; electroporation	Reduced target protein levels specifically by >60% in mice	[Nunamaker et al. 2003]
Cardiovascular disease	c-Jun	3'-3' inversion	Intramyocardial injection	Decreased pathogenesis of myocardial inflammation and infarction following ischemia-reperfusion injury	[Luo et al. 2009]

Prostate cancer	β1 integrin	phosphorothioate	Intra-tumoral injection	Blocked the adhesion of endothelial cells and abolished their ability to form microcapillary and neovascularization in Matrigel plugs	[Niewiarowska et al. 2009]
Systemic inflammation	iNOS	5'phosphorothioate, 3' CPG-amine-C7	Intraperitoneal direct injection	Reduced target protein, decreased mortality and inflammation	[Verma et al. 2010]

neovascularization and angiogenesis [Fahmy et al. 2003]. Zhang *et al.* accounted the efficacy of vascular endothelial growth factor receptor-2 (VEGF-R2) targeting DNAzymes in inhibiting the tumor angiogenesis [Zhang et al. 2002]. Sel et al. used GATA-3 DNAzymes targeting in allergic bronchial asthma by preventing airway inflammation and hypersensitivity in acute allergic bronchial asthma [Sel et al. 2008].

A few other significant transcription factors such as the leucine zipper transcription factor c-Jun, which is known to be one of the immediate-early genes in tumor proliferation, was knockdown using DNAzymes (Dz13). In various animal models the Dz13 was found to be generating anti-cancerous by inhibiting proliferation, migration, chemo-invasion and tubule formation in endothelial cells, while blocked the VEGF-induced retinal neovascularization and growth of melanoma and squamous cell carcinoma [Zhang et al. 2004, Zhang et al. 2006]. Similar to the Egr-1 DNAzyme, Dz13 was also found to be effective in reducing the intimal thickening in rats, suggests that c-Jun-targeting approaches probably be helpful in controlling tumors [Khachigian et al. 2002]. DNAzymes targeting plasminogen activator inhibitor 1 also shown to prevent the intimal proliferation after balloon injury in animals with diabetes [Witkowski et al. 2007]. Some recent studies demonstrate that Dz13 can also obstruct the key processes involved in inflammation progression like expression of cell adhesion molecules, trans-endothelial leucocyte emigration, neutrophil accumulation and vascular permeability relevant animal models [Fahmy et al. 2006]. Also in murine model of rheumatoid arthritis (RA) which is a chronic systemic inflammatory disease, Dz13 intervention was shown to be successful in reducing the joint swelling, inflammatory cell infiltration and bone erosion [Fahmy et al. 2006], thus imparting an noticeable opportunities for alternative treatments of diseases associated with inflammatory disarrays.

For achieving the best possible efficacy in functional gene silencing the target selection and validation play very critical role. We have shown that the inhibitory potential of DNAzymes is rather a function of target selection within a same mRNA substrate, which is distinct under *in vitro* and physiological conditions. Studies demonstrate that cleavage of murine iNOS mRNA was appreciably higher with DNAzyme targeting at the translation initiation site in the cellular environment as compared to the catalysis carried out in a cell free system [Chaudhury et al. 2006]. Nitric oxide (NO) is one of the major pro-inflammatory mediator implicated form the pathogenesis of various acute and chronic inflammatory diseases. Various chemical inhibitors to iNOS were shown to reduce NO production, however, resulted in deleterious effects such as hypertension, toxicity and even increased mortality in sepsis due to the concurrent non-specific inhibition of vital constitutive isoforms of NOS. Indeed, therapies directed to neutralize NO production have been reported to prevent the development of systemic inflammation, but clinical trials of these therapies in general failed to improve the outcome of patients with sepsis. The DNAzymes designed against the iNOS were found to control the endotoxin induce pathological concentrations of NO in J774 murine macrophage [Chaudhury et al. 2006]. More recently, the therapeutic potential of these DNAzymes was explored in murine model of lethal systemic inflammation as these DNAzymes were specifically targeting the inducible isoform of NOS [Verma et al. 2010]. DNAzyme treatment resulted in significantly reduced NO levels in serum and peritoneal lavage, confirming functional suppression of iNOS gene in LPS-injected mice. These DNAzymes were also able to limit excessive NO production by cytokine and LPS cochallenges in cultured peritoneal macrophages from DNAzyme-treated mice [[Verma et al. 2010]. The iNOS-specific DNAzyme was shown to reduce inflammatory responses and enhanced the survival in murine model of LPS-induced lethal systemic inflammation.

Similarly, recent knockdown studies using specifically designed 10–23 DNAzymes and siRNAs against pro-inflammatory cytokine tumor necrosis factor-α (TNF-α) and its receptors illustrate the involvement of autoctine production of this cytokine through receptor mediated cooperative activation of NF-κB in activated macrophages [Verma et al. 2009]. Understanding the molecular mechanisms by which mediators of inflammation bring about the manifestations of inflammatory diseases is an immediate requirement to design effective therapeutic interventions. Simultaneous silencing of TNF-receptors, R1 and R2 by DNAzyme or siRNA were shown to suppress TNF-α expression more efficiently than silencing them individually in lipopolysaccharides (LPS) stimulated THP-1 macrophages. Co-silencing of TNF-receptors also inhibited TNF-α induced NF-κB activation to a higher extent [[Verma et al. 2009]. In inflammatory disorders with abnormally high TNF-C production the limitation of therapies directed against TNF-α or NF-κB can be further improved by considering the synergistic blockade of TNF-α receptors. These interesting findings have confirmed that DNAzymes are able to act in a specific manner with minimal side-effects *in vivo*, when delivered directly to the experimental animals.

DNAzymes have also been used successfully against viral diseases. For instance, the intranasal application of DNAzyme targeting the conserved genomic RNA sequence of the RSV nucleocapsid protein (named as DZ1133), was shown to inhibit the RSV infection [Zhou et al. 2007]. Knockdown of the Epstein-Barr virus oncoprotein latent membrane protein 1 by DNAzymes augmented apoptosis and rendered nasopharyngeal carcinoma responsive to radiotherapy [Lu et al. 2008]. The hepatitis C virus (HCV) infection has high prevalence all over the globe with significant morbidity which makes it a serious threat on public health. Recent studies conducted using specific 10–23 DNAzymes designed to cleave the HCV RNA at internal ribosome entry site (IRES) and RNA dependent RNA polymerase (RDRP/NS5B) regions were shown to inhibit HCV RNA [Kumar et al. 2009]. Both these targets are important for antiviral drug development, because the 52 UTR having IRES is involved in generation of positive and negative strands and NS5B encodes the viral replicative enzyme, RNA dependent RNA polymerase. Therefore, these DNAzymes can be used as nucleic acid-based antiviral agent in controlling the replication of HCV RNA. Besides therapeutic agents DNAzymes can be effectively exploited to study the effect of one gene expression on that of other. DNAzyme knockdown studies in *Autographa californica* nucleopolyhedrovirus (*Ac*NPV) infected *Sf*9 cells with serine/ threonine kinase (pk1) specific DNAzymes showed the inhibition of protein and mRNA expressions of hyperactive polyhedrin (polh) gene [Mishra et al. 2008], confirming the regulatory role of pk1 in the ranscription of hyperactive polh gene.

DNAzymes perhaps can be developed as promising therapeutic interventions against various disorders. Although, there has been considerable success in these preclinical studies, yet there are certain rigorous parameters over which these DNAzyme molecules have to be polished before being successful in the clinical trials. If clinical studies show assurance and admissibility, these agents could encompass an enormous array of applications.

CONCLUSIONS

The appeal of post-transcriptional gene silencing approaches is that they potentially provide highly specific, non-toxic reagents for safe and effective therapeutics for a wide variety of diseases. RNAi represents a promising new gene therapy approach, however, the development of agents to improve the pharmacokinetics, bio-distribution and its selectivity in systemic delivery will be critical to the future of siRNA therapeutics. Evaluation of the efficacy of siRNA therapeutics requires considerations of off-target effects and innate immune response. The current failures of these molecules at advanced stages of clinical trials enforce us to reconsider several aspect including target selection and validation, delivery technologies and clinical trial design. These challenges are being met with multidisciplinary approaches with the hope that a greater under-standing of each facet of this problem will enable a more optimal utilization of this technology. It would be naive to believe that any one single technology will provide all the solutions, so it is gratifying to see such a broad-based effort across multiple technologies under investigation in academia and industry. As the focus of this research becomes more therapy oriented, with the use of clinically relevant animal models, future impact of catalytic nucleic acid-based therapies should become clearer.

The appeal of these antisense approaches is that they potentially provide highly specific, non-toxic reagents for safe and effective therapeutics for a wide variety of diseases. Perhaps the major obstacle to the further development of these technologies as gene suppressing agents is the difficulty involved in effective cellular delivery and target co-localization. The challenges are being met with a multi-disciplinary approach with the hope that a greater understanding of each facet of this problem will enable a more optimal utilization of this technology. As the focus of this research becomes more therapy oriented, with the use of clinically relevant animal models, future impact of catalytic nucleic acid-based therapies should become clearer.

REFERENCES

Aigner, A. 2006. Gene silencing through RNA interference (RNAi) *in vivo*: Strategies based on the direct application of siRNAs. *Journal of Biotechnology.* **124:**12–25.

Aleku, M., P. Schulz, O. Keil, A. Santel, U. Schaeper and B. Dieckhoff. 2008. Atu027, a liposomal small interfering RNA formulation targeting protein kinase N3, inhibits cancer progression. *Cancer Research.* **68(23):**9788–98.

Alnylam Pharmaceuticals Inc: [http://alnylam.com/Programs-and-Pipeline/index.php].

Anderson, W.F. 1998. Human gene therapy. *Nature.* **392:**25–30.

Bhindi, R., L.M. Khachigian and H.C. Lowe. 2006. DNAzymes targeting the transcription factor Egr-1 reduce myocardial infarct size following ischemia-reperfusion in rats. *Journal of Thrombosis and Haemostasis.* **4(7):**1479–1483.

Blaese, R.M., K.W. Culver, A.D. Miller, C.S. Carter, T. Fleisher and M. Clerici. 1995. T lymphocyte-directed gene transfer for ADA-SCID: Initial trial results after 4 years. *Science.* **270:**475–480.

Bramlage, B., E. Luzi and F. Eckstein. 1998. Designing ribozymes for the inhibition of gene expression. *Trends in Biotechnology.* **16:**434–438. Breaker, R.R. 1997. DNA Enzymes. *Nature Biotechnology.* **15:**427–431.

Chakraborti, S. and A.C. Banerjea. 2003. Inhibition of HIV-1 gene expression by novel DNA enzymes targeted to cleave HIV-1 TAR RNA: potential effectiveness against all HIV-1 isolates. *Molecular Therapy.* **7(6):**817–826.

Chaudhury, I., S.K. Raghav, H.K. Gautam, H.R. Das and R.H. Das. 2006. Suppression of inducible nitric oxide synthase by 10–23 DNAzymes in murine macrophage. *FEBS Letters.* **580**:2046–2052.

Crook, S.T., C.F. Bennett. 1996. Progress in antisense oligonucleotide therapeutics. *Annual Reviews of Pharmacology and Toxicology.* **36**: 107–129.

Dass, C.R., S.J. Galloway and P.F. Choong. 2010. Dz13, a c-jun DNAzyme, is a potent inducer of caspase-2 activation. *Oligonucleotides.* **20(3)**:137–146.

Davis, M.E. 2009. The first targeted delivery of siRNA in humans via a self-assembling, cyclodextrin polymer-based nanoparticle: from concept to clinic. *Molecular Pharmacology.* **6(3)**:659–68.

Davis, M.E., J.E. Zuckerman, C.H. Choi, D. Seligson, A. Tolcher and C.A. Alabi. 2010. Evidence of RNAi in humans from systemically administered siRNA via targeted nanoparticles. *Nature.* **464(7291)**:1067–1070.

DeVincenzo, J., J.E. Cehelsky, R. Alvarez, S. Elbashir, J. Harborth and I. Toudjarska. 2008. Evaluation of the safety, tolerability and pharma-cokinetics of ALN-RSV01, a novel RNAi antiviral therapeutic directed against respiratory syncytial virus (RSV). *Antiviral Reserch.* **77**:225–31.

Dykxhoorn, D.M., C.D. Novina and P.A. Sharp. 2003. Killing the messenger: short RNAs that silence gene expression. *Nature Reviews.* **4**:457– 467.

Fahmy, R.G., A. Waldman, G. Zhang, A. Mitchell, N. Tedla, H. Cai, Geczy, C.N. Chesterman, M. Perry and L.M. Khachigian. 2006. Suppression of permeability and inflammation by targeting of the transcription factor c-Jun. *Biotechnology.* **24(7)**:856-863.

Fahmy, R.G., C.R. Dass, L.Q. Sun, C.N. Chesterman and L.M. Khachigian. 2003. Transcription factor Egr-1 supports FGF-dependent angiogenesis during neovascularization and tumor growth. *Nature Medicine.* **9**:1026–1032.

Feinstein. E., H. Ashush, M.E. Kleinman, M. Nozaki, H. Kalinski and I. Mett. 2009. PF-04523655 (REDD14), a siRNA compound targeting RTP801, penetrates retinal cells producing target gene knockdown and avoiding TLR3 activation. ARVO 2009 Annual Meeting Abstract no. 5693.

Goila, R., A.C. Banerjea. 2001. Inhibition of hepatitis B virus X gene expression by novel DNA enzymes. *Biochemical Journal.* **353**:701–708.

Grimm, D., K.L. Streetz, C.L. Jopling, T.A. Storm, K. Pandey and C.R. Davis. 2006. Fatality in mice due to oversaturation of cellular microRNA/short hairpin RNA pathways. *Nature.* **441(7092)**:537–541.

Hornung, V., M. Guenthner-Biller, C. Bourquin, A. Ablasser, M. Schlee and S. Uematsu. 2005. Sequence-specific potent induction of IFN-α by short interfering RNA in plasmacytoid dendritic cells through TLR7. *Nature Medicine.* **11**:263–270.

Hou, W., Q. Ni, J. Wo, M. Li, K. Liu, L. Chen, Z. Hu, R. Liu and M. Hu. 2006. Inhibition of hepatitis B virus X gene expression by 10–23 DNAzymes. *Antiviral Reserch.* **72(3)**:190–196.

Jackson, A.L., J. Burchard, J. Schelter, B.N. Chau, M. Cleary and L. Lim. 2006. Widespread siRNA "off-target" transcript silencing mediated by seed region sequence complementarity. *RNA.* **12(7)**:1179–1187.

Jackson, A.L., S.R. Bartz, J. Schelter, S.V. Kobayashi, J. Burchard, M. Mao, B. Li, G. Cavet and P.S. Linsley. 2003.Expression profiling reveals offtarget gene regulation by RNAi. *Nature Biotechnology.* **21**:635–637.

Judge, A.D., V. Sood, J.R. Shaw, D. Fang, K. McClintock and Ian MacLachlan. 2005. Sequence-dependent stimulation of the mammalian innate immune response by synthetic siRNA. *Nature Biotechnology.* **23**:457–462.

Khachigian, LM. 2000. Catalytic DNA as therapeutic agents and molecular tools to dissect biological function. *Journal of Clinical Investigations.* **106**:1189–1195.

Khachigian, LM. 2002. DNAzymes: cutting a path to a new class of therapeutics. *Current Opinions in Molecular Therapy.* **4**:119–121.

Khachigian, LM., R.G. Fahmy, G. Zhang, Y.V. Bobryshev and A. Kaniaros. 2002. c-Jun regulates vascular smooth muscle cell growth and neointima formation after arterial injury: inhibition by a novel DNAzyme targeting c-Jun. *Journal of Biology and Chemistry.* **277**:22985–22991.

Kleinman, M.E., K. Yamada, A. Takeda, V. Chandrasekaran, M. Nozaki and J.Z. Baffi. 2008. Sequence- and target-independent angiogenesis suppression by siRNAvia TLR3. *Nature.* **452(7187)**:591–597.

Kumar, D., I. Chaudhury, P. Kar and R.H. Das. 2009. Site-specific cleavage of HCV genomic RNA and its cloned core and NS5B genes by DNAzyme. *Journal of Gastroenterology and Hepatology.* **24(5)**:872–878.

Lavery, K.S., T.H. King. 2003. Antisense and RNAi: powerful tools in drug target discovery and validation. *Current Opinions in Drug Discovery and Development.* **4**:561–569.

Leachman, S.A., Hickerson, P.R. Hull, F.J. Smith, L.M. Milstone and E.B. Lane. 2008. Therapeutic siRNAs for dominant genetic skin disorders including pachyonychia congenita. *Journal of Dermatological Sciences.* **51**:151–157.

Li, J., D. Zhu, Z. Yi, Y. He, Y. Chun, Y. Liu and N. Li. 2005. DNAzymes targeting the icl gene inhibit ICL expression and decrease Mycobacterium tuberculosis survival in macrophages. *Oligonucleotides.* **15(3)**:215–222.

Liang, Z., S. Wei, J. Guan, Y. Luo, J. Gao, H. Zhu, S. Wu and T. Liu. 2005. DNAzyme-mediated cleavage of survivin mRNA and inhibition of the growth of PANC-1 cells. *Journal of Gastroenterology and Hepatology.* **20(10)**:1595–1602.

Liu, G.N., Y.X. Teng and W. Yan. 2008. Transfected synthetic DNA Enzyme Gene specifically inhibits Egr-1 gene expression and reduces neointimal hyperplasia following balloon injury in rats. *International Journal of Cardiology.* **129**:118–24.

Lowe, H.C., C.N. Chesterman and L.M. Khachigian. 2002. Catalytic antisense DNA molecules targeting Egr-1 inhibit neointima formation following permanent ligation of rat common carotid arteries. *Journal of Thrombosis and Haemostasis.* **87(1)**:134–140.

Lowe, H.C., R.G. Fahmy, M.M. Kavurma, A. Baker, C.N. Chesterman and L.M. Khachigian. 2001. Catalytic oligodeoxynucleotides define a key regulatory role for early growth response factor-1 in the porcine model of coronary instent restenosis. *Circulation Research.* **89(8)**:670–677.

Lu, Z.X., X.Q. Ma, L.F. Yang, Z.L. Wang, L. Zeng and Z.J. Li. 2008. DNAzymes targeted to EBV-encoded latent membrane protein-1 induce apoptosis and enhance radiosensitivity in nasopharyngeal carcinoma. *Cancer Letters.* **265:**226–238.

Luo, X., H. Cai, J. Ni, R. Bhindi, H.C. Lowe, C.N. Chesterman and L.M. Khachigian. 2009. c-Jun DNAzymes Inhibit Myocardial Inflammation, ROS Generation, Infarct Size and Improve Cardiac Function After Ischemia-Reperfusion Injury. *Arteriosclerosis, Thrombosis and Vascular Biology.* **29:**1836–1842.

Mann, M.J. 2005. Transcription factor decoys: a new model for disease intervention. *Annals of the New York Academy of Sciences.* **1058:**128–139

Miller, A.D. 1992. Human gene therapy comes of age. *Nature* (Lond). **357:**455–460.

Mishra, G., P. Chaddha, I. Chaudhury and R.H. Das. 2008. Inhibition of Autographa californica nucleopolyhedrovirus (AcNPV) polyhedrin gene expression by DNAzyme of its serine/threonine kinase (pk1) gene. *Virus Ressearch.* **135:**197–201.

Mitchell, A., C.R. Dass, L.Q. Sun and L.M. Khachigian. 2004. Inhibition of human breast carcinoma proliferation, migration, chemoinvasion and solid tumour growth by DNAzymes targeting the zinc finger transcription factor EGR-1. *Nucleic Acids Research.* **32(10):**3065–3069.

Niewiarowska, J., I. Sacewicz, M. Wiktorska, T. Wysocki, O. Stasikowska, M. Wagrowska-Danilewicz and C.S. Cierniewski. 2009. DNAzymes to mouse beta1 integrin mRNA *in vivo*: targeting the tumor vasculature and retarding cancer growth. *Cancer Gene Therapy.* **16(9):**713–722.

Nogoy, N. 2007. Decoy oligonucleotides: silencing the negativity in gene therapy? *Therapy.* **4(2):**171–173.

Nunamaker, E.A., H.Y. Zhang, Y. Shirasawa, J.N. Benoit and D.A. Dean. 2003. Electroporation-mediated delivery of catalytic oligodeoxynucleo-tides for manipulation of vascular gene expression. *American Journal of Physiology-Heart Circulation Physiology.* **285(5):**2240–2247.

Paterson, B.M., B.E. Roberts and E.L. Kuff. 1977. Structural gene identification and mapping by DNA-mRNA hybrid-arrested cell-free translation. *Proceedings of the National Academy of Sciences, USA.* **74:**4370–4374.

Quark Pharma. QPI-1002 for AKI [http://www.quarkpharma.com/qbien/products/QPI-1002/] Rayburn, E.R. and R. Zhang. 2008. Antisense, RNAi and gene silencing strategies for therapy: mission possible or impossible? *Drug Discovery Today.* **13:**513–21.

Registry of Federally and Privately Supported Clinical Trials. A dose escalation and safety study of I5NP to prevent acute kidney injury (AKI) in patients at high risk of AKI undergoing major cardiovascular surgery (QRK.004) [http://clinicaltrials.gov/ct2/show/NCT00683553?term=QPI-1002 andrank=2]

Registry of Federally and Privately Supported Clinical Trials. I5NP for Prophylaxis of Delayed Graft Function in Kidney Transplantation [http://clinicaltrials.gov/ct2/show NCT00802347?term=QPI- 1002andrank=1]

Robbins, M., A. Judge, Ambegia, C. Choi, E. Yaworski, L. Palmer, K. McClintock and I. MacLachlan. 2008. Misinterpreting the therapeutic effects of small interfering RNA caused by immune stimulation. *Human Gene Therapy.* **19:**991–999.

Santiago, F.S., H.C. Lowe, M.M. Kavurma, C.N. Chesterman, A. Baker, D.G. Atkins and L.M. Khachigian. 1999. New DNA enzyme targeting Egr-1 mRNA inhibits vascular smooth muscle proliferation and regrowth factor injury. *Nature Medicine*. **5:**1264–1269.

Santiago, F.S., H.C. Lowe, M.M. Kavurma, C.N. Chesterman, A. Baker, D.G. Atkins and L.M. Khachigian. 1999. New DNA enzyme targeting Egr-1 mRNA inhibits vascular smooth muscle proliferation and regrowth after injury. *Nature Medicine*. **5(11):**1264–1269.

Santoro, S.W. and G.F. Joyce. 1997. A general purpose RNA-cleaving DNA enzyme. *Proceedings of the National Academy of Sciences USA*. **94:**4262–4266.

Sel, S., M. Wegmann, T. Dicke, S. Sel, W. Henke and A.O. Yildirim. 2008. Effective prevention and therapy of experimental allergic asthma using a GATA-3-specific DNAzyme. *Journal of Allergy and Clinical Immunology*. **121:**910–16.

Sioud, M. and M. Leirdal. 2000. Design of nuclease resistant protein kinase calpha DNA enzymes with potential therapeutic application. *Journal of Molecular Biology*. **296(3):**937–947.

Sriram, B. and A.C. Banerjea. 2000. *In vitro*-selected RNA cleaving DNA enzymes from a combinatorial library are potent inhibitors of HIV-1 gene expression. *Biochemical Journal*. **352:**667–673.

Stephenson, M.L. and P.C. Zamecnik. 1978. Inhibition of Rous sarcoma viral RNA translation by a specific oligode oxyribonucleotide. *Proceedings of the National Academy of Sciences, USA*. **75:**285–288.

Sun, L.Q., M.J. Cairns, W.L. Gerlach, C. Witherington, L. Wang and A. King. 1999. Suppression of smooth muscle cell proliferation by a c-myc RNA-cleaving deoxyribozyme. *Journal of Biology and Chemistry*. **274(24):**17236–17241.

Takahashi, H., H. Hamazaki, Y. Habu, M. Hayashi, T. Abe, N. Miyano-Kurosaki and H. Takaku. 2004. A new modified DNA enzyme that targets influenza virus A mRNA inhibits viral infection in cultured cells. *FEBS Letters*. **560(1–3):**69–74.

Tracy, S., A.C. Zimmermann, H. Lee, A. Akinc, B. Bramlage, D. Bumcrot and N. Matthew. 2006. RNAi-mediated gene silencing in non-human primates *Nature*. **441:**111–114.

Trang, P., J.B. Weidhaas and F.J. Slack. 2008. "MicroRNAs as potential cancer therapeutics". *Oncogene 27*. **Suppl 2:**S52–S57.

Trepanier, J., J.E. Tanner, R.L. Momparler, O.N. Le, F. Alvarez and C. Alfieri. 2006. Cleavage of intracellular hepatitis C RNA in the virus core protein coding region by deoxyribozymes. *Journal of Viral Hepatology*. **13(2):**131 318.

Unwalla, H. and A.C. Banerjea. 2001. Inhibition of HIV-1 gene expression by novel macrophage-tropic DNA enzymes targeted to cleave HIV-1 TAT/Rev RNA. *Biochemical Journal*. **357:**147–155.

Vaishnaw, A.K., J. Gollob, C.G. Vitalo, R. Hutabarat, D. Sah, R. Meyers, T. Fougerolles and J. Maraganore. 2010. A status report on RNAi therapeutics. *Silence*. **1:**14.

Verma, N., I. Chaudhury, D. Kumar and R.H. Das. 2009. Silencing of TNF-alpha receptors coordinately suppresses TNF-alpha expression through NF-kappaB activation blockade in THP-1 macrophage. *FEBS Letters.* **583(17):**2968–2974.

Verma, N., S.K. Tripathi, I. Chaudhury, H.R. Das and R.H. Das. 2010. Inducible nitric oxide synthase (iNOS) targeted 10–23 DNAzyme reduces lipopolysaccharide (LPS) induced systemic inflammation and mortality in mice. *Shock.* **33(5):**493–499.

Warashina, M., T. Kuwabara, Y. Nakamatsu and K. Taira. 1999. Extremely high and specific activity of DNA enzymes in cells with a Philadelphia chromosome. *Chemistry and Biology.* **6(4):**237–250.

Witkowski, P., T. Seki, G. Xiang, T. Martens, H. Sondermeijer and F. See. 2007. A DNA enzyme against plasminogen activator inhibitor- type 1 (PAI-1) limits neointima formation after angioplasty in an obese diabetic rodent model. *Journal of Cardiovascular Pharmacology.* **50:**633–640.

Wu, Y., L. Yu, R. McMahon, J.J. Rossi, S.J. Forman and D.S. Snyder. 1999. Inhibition of bcr-abl oncogene expression by novel deoxyribozymes (DNAzymes). *Gene Therapy.* **10(17):**2847–2857.

Xiang, G., T. Seki, M.D. Schuster, P. Witkowski, A.J. Boyle, F. See, P. Martens, A. Kocher, H. Sondermeijer, H. Krum and S. Itescu. 2005. Catalytic degradation of vitamin D up-regulated protein 1 mRNA enhances cardiomyocyte survival prevents left ventricular remodeling after myocardial ischemia. *Journal of Biology and Chemistry.* **280:**39394–39402.

Xie, Y.Y., X.D. Zhao, L.P. Jiang, H.L. Liu, L.J. Wang, P. Fang, K.L. Shen, Z.D. Xie, Y.P. Wu. and X.Q. Yang. 2006. Inhibition of respiratory syncytial virus in cultured cells by nucleocapsid gene targeted deoxyribozyme (DNAzyme). *Antiviral Research.* **71(1):**31–41.

Yang, L., W. Zeng, D. Li and R. Zhou. 2009. Inhibition of cell proliferation, migration and invasion by DNAzyme targeting MMP-9 in A549 cells. *Oncology Reports.* **22(1):**121–126.

Yen, L., S.M. Strittmatter and R.G. Kalb. 1999. Sequence-specific cleavage of Huntingtin mRNA by catalytic DNA. *Annals of Neurology.* **46(3):**366–373.

Yu, S.H., T.H. Wang, and L.C. Au. 2009. Specific repression of mutant K-RAS by 10–23 DNAzyme: Sensitizing cancer cell to anti-cancer therapies. *Biochemical and Biophysical Research Communications.* **378(2):**230–234.

Zhang, G., C.R. Dass, E. Sumithran, N.R. DiGirolimo, L-Q Sun and LM. Khachigian. 2004. Effect of deoxyribozymes targeting c-Jun on solid tumor growth and angiogenesis in rodents. *Journal of the National Cancer Institute.* **96:**683–696.

Zhang, G. and X. Luo, E. Sumithran, V.S. Pua, R.S. Barnetson, G.M. Halliday and L.M Khachigian. 2006. Squamous cell carcinoma growth in mice and in culture is regulated by c-Jun and its control of matrix metallo-proteinase-2 and -9 expression. *Oncogene.* **25:**7260–7266.

Zhang, G., X. Luo, E. Sumithran, V.S.C.Pua, R.S. Barnetson, G.M. Halliday and L.M. Khachigian,. 2006. Squamous cell carcinoma growth in mice and in culture is regulated by c-Jun and its control of matrix metalloproteinase-2 and -9 expression. *Oncogene.* **25:**7260–66.

Zhang, L., W.J. Gasper, S.A. Stass, O.B. Ioffe, M.A. Davis and A.J. Mixson. 2002. Angiogenic inhibition mediated by a DNAzyme that targets vascular endothelial growth factor receptor 2. *Cancer Research.* **62:**5463–5469.

Zhang, X., Y. Xu, H. Ling and T. Hattori. 1999. Inhibition of infection of incoming HIV-1 virus by RNA-cleaving DNA enzyme. *FEBS Letters.* **458(2):**151–516.

Zhou, J., X.Q. Yang, Y.Y. Xie, X.D. Zhao, L.P. Jiang and L.J. Wang. 2007. Inhibition of respiratory syncytial virus of subgroups A and B using deoxyribozyme DZ1133 in mice. *Virus Reserch.* **130:**241–248.

CHAPTER 6

Artificial MicroRNA: A Third Generation RNAi Technology

Pranjal Yadava

ABSTRACT

Gene silencing based on the principles of microRNAs (miRs) is the latest among various RNAi technologies. Since, miR biogenesis is dependent on the secondary structure of its precursor, rather than the sequence, it is possible to design synthetic or artificial miRs (amiRs), to silence any endogenous or exogenous transcript which is originally not targeted by natural miR(s). The designer amiRs utilize cell's own natural miR biogenesis machinery, for their processing and activity. amiRs offer several advantages over conventional haipin based RNAi and this has catapulted amiRs to occupy prime position in functional genomics and provided a potential tool to engineer novel therapeutics for agriculture and medicine. This chapter throws light on basic concept, design rules, new vectors and potential applications of amiRs.

Keywords: microRNA, artificial microRNA, RNAi.

INTRODUCTION

The discovery of phenomenon of RNA interference (RNAi) has brought a paradigm shift in our understanding of the biological world. But it is quite interesting to know that, even before RNAi was discovered in the sense we know it today, its principles were utilized by plant virologists for protecting the plants from deadly viruses. To develop virus resistant transgenic plants, virologists relied on Pathogen-derived Resistance (PDR)—a vaccination like phenomenon where plants containing genes or sequences of a pathogen are protected from cognate or related

---

* *Corresponding author e-mail*: pranjal@icgeb.res.in; pranjal.yadava@gmail.com pathogens. This concept was pioneered in mid-eighties by John C. Sanford of Cornell University (Sanford and Johnston 1985, US Patent 5840481). The PDR approach was itself inspired from an age old agricultural practice of 'cross-protection', in which plants challenged with a mild strain of a virus were protected from super-infection by other related strains. Exact mechanism of PDR/cross-protection mediated resistance was obscure for quite some time (Baulcombe 1996). Initially, the dogma was that this type of resistance was protein-mediated. However, later it was shown that untranslatable transcripts of viral genes alone were sufficient to interfere with virus replication (Lindbo and Dougherty 1992), thus providing an early clue that resistance might be RNA mediated. The phenomena of co-suppression/Post transcriptional gene silencing (PTGS)/RNAi were being unraveled at the same time in various model organisms and plant virologists were rather quick to point out that the PDR mediated virus resistance in plants was related to homology-dependent gene silencing (Mueller et al. 1995). So it can be said that albeit unknowingly, RNAi as a technology was used for the first time even before its discovery and coining of the term 'RNAi'. This phase can be considered as the first generation of RNAi technology.

Later on, the mechanistic details of RNAi were worked out. Double stranded RNA was identified as its trigger and the effector molecule was identified as a 20-25 mer small interfering RNA (siRNA). These insights allowed the scientists to elicit desired RNAi by the use of appropriately designed RNAi constructs. As a matter of fact, intron spilced hairpin constructs emerged as reagents of choice for RNAi. It was this phase which saw the ascent of RNAi from 'science' to 'technology'. Investors smelt big money in this technology and a plethora of biotech ventures cropped up, promising a spectrum of novel therapeutics (Yadava and Mukherjee 2009). This booming phase can be considered as second generation of RNAi and it is still growing strong.

In the meantime, the newly discovered small RNA world grew even bigger, with the discovery of other small RNAs, most notably the microRNAs (miRNAs or miRs). miRs were found to be involved in almost every biological phenomenon. miRs were shown to be not only involved in developmental pathways, but also associated with pathogens and disease. This raised the hope of fine-tuning cellular miR pathway by the use of 'artificial miRs' and sometimes even depleting natural miRs by a novel class of chemically engineered oligonucleotides, termed 'antagomirs. Utilization of miR pathway for gene silencing can be considered as the third generation of RNAi technology. This chapter highlights the current 'state of art' of amiR based gene silencing.

THE NATURAL miR PATHWAY

MiRs are ~19–23 nt long single-stranded RNAs generated from primary transcript having local-hairpin structure (Ambros et al. 2003, Bartel 2004). These transcripts are generated by pol II and therefore possess 5' cap and 3' poly-A tail hallmarks (Cai et al. 2004, Kim 2005). From this transcript arise the 70-100 nt long pre-miR by nuclear miR biogenesis pathway. Mature miRs are processed from the imperfect duplex region of such pre-miR transcripts. The region of the precursor that pairs to the miR is called miR*. Experimental evidence indicates that the structure, rather than the sequence of the pre-miR, directs their correct processing. In animals, the pri-miR is processed in the nucleus into hairpin pre-miR by the double-strand specific ribonuclease III (RNase III) enzyme, DROSHA. The pre-miR is transported to the cytoplasm *via* EXPORTIN-5 (EXP-5) (Yi et al. 2005, Lund et al. 2004). It is then digested by a second, double-strand specific RNAse III enzyme, called DICER. The resulting 19–24 nt miR is bound by a complex known as the RNA-Induced Silencing Complex (RISC) that participates in repressing the target transcript resulting in a reduced expression of the corresponding gene. The plant machinery is different in the sense that both the pri- and pre- miR are processed in the nucleus by the dsRNA specific RNAse III enzyme, DICER LIKE 1 (DCL1) in conjuction with HYPONASTIC LEAVES1 (HYL1) and *Serrate* to release the mature miR (Tang et al. 2003). The HUA ENHANCER1 (HEN1) methylates the terminal sugar residues of miR to prevent 32-end uridylation and increases the stability of miR. DCL1, HYL1 and SE interact with each other and are colocalized in the nuclear bodies, termed Dicing bodies or D-bodies (Fang and Spector 2007). The miR:miR* duplex is then transported to the cytoplasm by EXP-5 homolog, HASTY (Bollman et al. 2003) where it is incorporated into RISC and guides it to repress target mRNA(s). ARGONAUTE (AGO) is the slicer component of RISC. AGO family members contain a PAZ as well as a PIWI domain. PAZ domain may allow interaction with Dicer and other proteins and is also believed to help align miR with its target. The structure of PIWI domain is related to RNaseH and is believed to be involved in target cleavage. Earlier, it was believed that plant miRs lead to cleavage of target while in animals translational repression is the norm. However, now it is known that widespread translational repression also occurs in plants (Brodersen et al. 2008). miR biogenesis is under feedback regulation such that the *DCL1* and *AGO1* genes are themselves regulated by miRs. *DCL1* is itself a target of miR162, which leads to the cleavage of the *DCL1* mRNA (Xie et al. 2003), while AGO1 is targeted by miR 168. Also, over the years few non-canonical miR biogenesis pathways have also been discovered where miRs were found to be processed from introns (termed as mirtrons) (Ruby et al. 2007) or viral tRNA-like sequences (Bogerd et al. 2010). Recently, in a phenomenon sometimes referred to as RNA activation (RNAa), miRs are shown to upregulate translation as well, adding to further complexity (Vasudevan et al. 2007, Huang et al. 2010). As mentioned earlier, miRs are involved in regulation of almost every biological process, viz., development, signaling and stress response including defense against pathogens.

ARTIFICIAL microRNAs

Recently, it has been shown that altering several nucleotides within sense and antisense strands of miR has no bearing on its biogenesis and maturation, as long as secondary structure of its precursor remains unaltered. This fact has been utilized to engineer artificial miRs (amiRs) to silence any transcript of choice. The amiR technology was first used for gene knock down in human cell lines (Zeng et al. 2002) and it was successfully employed to down-regulate gene expression without affecting the expression of other unrelated genes in transgenic plants (Schwab et al. 2006, Alvarez et al. 2006).

AmiR Design Parameters

AmiRs can be defined as single-stranded 21mer RNAs, which are not normally found in plants and which are processed from natural miR precursors. amiRs are designed to mimic natural miRs in all respects. Their sequences are designed according to the determinants of plant miR target selection, such that the 21-mer specifically silences its intended target gene(s). AmiRs are required to resemble natural miRs by three major criteria. Firstly, they start with a U (found in most plant and animal miRs); secondly, they display 5′ instability relative to their amiR* and last but not the least, their 10th nucleotide is either an A or U (Reynolds et al. 2004, Mallory et al. 2004). In the stem region of amiR precursor containing the amiR and amiR* sequences, one end is normally thermodynamically less stable, since it is either mismatched or contains less GC and more AU pairs relative to the other end. The strand with lower thermodynamic stability at its 5′ end (5′ instability) is preferentially incorporated into RISC (Khvorova et al. 2003, Schwarz et al. 2003). This principle of strand asymmetry is followed in amiR design as well. Some important parameters required to fine-tune amiR design are as follows:

1. Pairing of the target to the 5′ portion of the amiR (positions 2 to 12) is most important and this region should not have any mismatch and rarely more than one.

2. Mismatches at the presumptive cleavage site (positions 10 and 11) should be avoided at best.

3. Clusters of more than two mismatches in the 3′ part of the amiR should be avoided.

4. Perfect pairing in the 3′ portion can compensate for the presence of up to two mismatches in the 5′ portion.

5. There should be a low overall free energy of targets paired with their corresponding amiRs (at least 70% compared with a perfect match and a maximum of –30 kcal/mole).

6. AmiRs should have uridine at position 1 and, if possible, adenine at position 10, both of which are overrepresented among natural plant miRNAs.

7. AmiRs should display 5′ instability relative to their amiR*, so that the correct sequence would be incorporated into RISC.

8. To reduce the likelihood that an amiR would act as a primer for RNA-dependent RNA polymerases (RdRp) and thereby trigger secondary RNAi, there should be 1-3 mismatches to the target genes in the 3′ part of the amiRs. This is especially important in plant amiR design. The possibility of transitivity can be investigated by determining potential 21-mer secondary siRNAs for the a**miR** target gene from both strands of a 200–300-bp region, surrounding the initial binding site of the a**miR**. Potential targets of these siRNAs can be identified using **miR** target prediction algorithms.

9. The designed amiRs must not exhibit promiscuity in target selection, i.e., there should not be any off-target effects. The faithfulness of amiRs can be tested by using various miR target prediction algorithms. During this analysis, it is worthwhile to pay special attention to matches between 2-8 nucleotide positions of amiRs, as base pairing to the so-called seed region between **miR** positions 2 and 8 is shown to be often sufficient for target recognition in animal. In contrast to **miRs** in animals, natural plant **miRs** have a very narrow action spectrum and target only mRNAs with few mismatches.

AmiR Transgene Construct and New AmiR Vectors

Once designed, the amiRs are to be incorporated in suitable constructs. First of all, the amiR is brought in the context of a natural pre-miR. For this purpose, the resident miR and miR* of natural miR precursor is replaced by designer amiR and amiR* by site directed mutagenesis. This mutagenesis can be achieved by a series of overlapping PCR reactions using

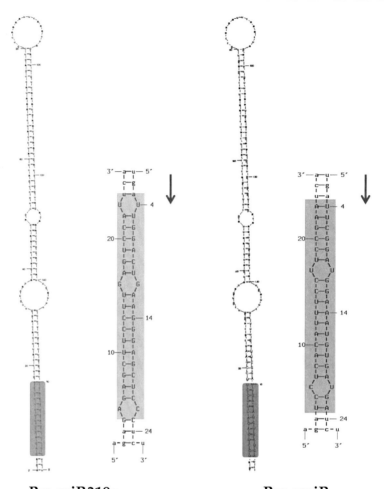

Pre-miR319a   Pre-amiR

Figure 1: Designing artificial miRs on natural pre-miR319a. To design an amiR overexpressing construct, the resident miR319a and miR*319a of natural pre-miR319a are replaced by amiR and amiR* through site-directed mutagenesis by overlapping PCR. The secondary structures of natural pre-miR319a and designed pre-amiRs as depicted by m-fold software are shown here. The overall secondary structures of pre-amiRs, including the terminal loop, other loops, and their relative distance from the stem exactly mimics the natural pre-miR319a. The shaded blow-ups of the stem region harboring mature miR/amiR show the subtle differences as a result of designing. Recent evidences demonstrate a non-canonical loop to base processing of pre-miR319a, where the position of loops is critical. Hence maintaining the natural loop conformation for amiR precursors ensures their correct and efficient processing.

(Color image of this figure appears in the color plate section at the end of the book.)

appropriately designed oligonucleotides, wherein the amiR precursor is synthesized in parts and later fused together to get the full length amiR precursor. Alternatively, the full length pre-amiR gene itself can be commercially synthesized.

So far, most of the amiRs in plants are based on the natural precursor structures of ath-miR159a, ath-miR164b, ath-miR172a, ath-miR319a and osa-miR528. Recently, a simple amiR vector (pAmiR169d) based on the structure of *Arabidopsis* miR169d precursor (pre-miR169d) has been constructed (Liu et al. 2010). Novel amiR vectors for simultaneous targeting of multiple genes in mammalian cells have also been developed (Hu et al. 2010). The recommended number of concatenated amiRs in a multi-amiR expression vector should not be more than four and the relative position of an amiR in the multi-amiR expression vector has no apparent influence on its RNAi activity. Regulatable RNA polymerase II promoters can also be used to overexpress authentic miRs in cell culture (Zeng et al. 2005). A modified *Cabbage leaf-curl virus* vector has been designed to express artificial and endogenous miRs in plants. Using this viral miR expression system, it was demonstrated that VIGS using amiRs, dubbed as "MIR VIGS," was effective to silence the expression of endogenous genes, including PDS, Su, CLA1 and SGT1, in *Nicotiana benthamiana* (Tang et al. 2010). Highly Efficient gene Silencing Compatible vector' (HESC vector) is a new amiR vector for rice and is deemed suitable for use in a systems biology approach for functional genomic research (Wang et al. 2010).

APPLICATIONS OF AmiRs

AmiRs are generally seen as third generation RNAi technology and have found tremendous applications in agriculture and medicine. amiR screens are becoming reagents of choice for large scale functional genomics studies for both animals and plants. The *Arabidopsis* 2010 project, which aims at finding functions of all the genes in *A. thaliana* by the year 2010, has used amiRs to achieve this stupendous goal (Small 2007). Each of the estimated 22,000 *Arabidopsis thaliana* genes is targeted with 3 amiRs, making this the first genome wide resource for plant RNAi. These constructs use miR319a backbone and CaMV35S promoter. In plants, amiR technology has been validated in rice (Warthmann et al. 2008), moss (Khraiwesh et al. 2008) and *Chlamydomonas,* apart from *Arabidopsis*, tobacco and tomato (Alvarez et al. 2006).

In animal cell culture based system, amiR libraries have been produced to target mammalian genes like mouse p53 ORF (Xue et al. 2009). Such an enzymatically prepared library has the potential to construct an amiR library that targets a whole transcriptome for genome-wide RNAi screening, or a randomized amiR library to search for functional amiRs. In medicine, amiRs are finding application towards developing therapy against cancer (Liang et al. 2007), hepatitis B (Gao et al. 2008), rabies (Israsena et al. 2009) etc. amiR based therapy is more desirable over small hairpin RNAs (shRNAs) as amiRs can mitigate shRNA-mediated toxicity in the brain (McBride et al. 2008)

Application of this technology to achieve plant virus resistance has been proved successful for few viruses and in model plants (Niu et al. 2006, Qu et al. 2007). However, till a year ago, the potential of this approach was not been tested on geminiviruses or in any crop plant (Shepherd et al. 2009). Now, amiRs have been shown to be effective in engineering geminivirus resistance in tomato crop (Yadava et al. 2010).

UNIQUE ADVANTAGES OF AmiRs

AmiRs have several unique advantages which have catapulted them to occupy prime position in functional genomics. As with natural miRNAs, amiRs are likely to be particularly useful for targeting groups of closely related genes, including tandemly arrayed genes. Approximately 4000 genes in *Arabidopsis* are found in tandem arrays and no convenient tool exists as yet for their knockout. Because of their exquisite specificity, amiRs can possibly be adapted for allele-specific knockouts. There is a substantial level of alternative splicing and amiRs have the potential to target only specific splice forms. In contrast with conventional hairpin mediated RNAi, in which small RNAs are generated from both strands, amiRs have the advantage of being strand specific. In addition, amiR sequences can be optimized for high efficiency, since they are always produced from the same locus in their precursors. Most importantly, amiR induced mutants can be complemented by amiR resistant targets, where silent mutations can be introduced in the region of amiR binding, which disrupt amiR mediated degradation.

In the hairpin RNAi approach, the hairpin transgene is transcribed into long fold-back RNA and multiple siRNAs are formed from one precursor. Further, as the positions of Dicer cleavage are not known, the 5' ends of siRNAs cannot be accurately determined. In addition, it remains elusive as to which parameters are used by the siRNAs to select their target transcripts. Prediction of small RNA targets other then the perfectly complementary intended targets is generally considered to be quite difficult. AmiRs, in contrast, are produced from amiR precursors, which are normally processed such that only one single stable amiR is preferentially generated. Since the determinants of amiR target selection have been fairly worked out, the complete spectrum of amiR targets is easily predictable. siRNAs are perfectly complementary to their targets and thus their binding to the target can lead to transitivity, i.e., RdRp dependent amplification and generation of secondary siRNAs, which in turn would have large number of targets. This increases the chances of off-target effects in RNAi based transgenic plants. On the other hand, amiRs can be consciously designed to contain few mismatches with respect to their target, thereby altogether avoiding complications of transitivity. Mismatches will not compromise with silencing as; even imperfectly matched miRs are known to efficiently repress the targets. One of the major concerns of RNAi in developing transgenic crops is the potential pleotropic effects caused by off-targeting. This concern is addressed most satisfactorily by amiRs, which paves the way for their utilization in developing novel traits in GM crops.

Unlike conventional siRNA based RNAi, amiRs can function in tissue-specific and inducible manner, as they have only limited cell-autonomous effect. siRNAs on the other hand are considered non-cell autonomous and systemic in action. This is especially important when, silencing is required to be restricted to specific organs only. Also, siRNA based silencing is known to be compromised in extremes of temperature. On the other hand miR biogenesis machinery is more robust as natural miRs are produced by the organism in all the conditions, including extremes of temperatures. Thus, amiR based silencing is expected to be more resilient and amiR transgenics can possibly be widely adapted to temperate and tropical agro-ecosystems.

CONCLUSION

The advent of amiR technology has widened the available options of gene silencing. Despite the availability of several knock-down tools for last many years, functions of all the genes of even model plant *Arabidopsis* remained a holy grail of functional genomics. Ultimately, this extraordinary feat is being accomplished by amiR technology (The Arabidopsis 2010 project). This vouches for the exquisite efficiency of amiRs and in coming years many more conundrums of functional genomics would be solved by using amiRs. The outstanding straightforwardness of design, efficacy of action and specificity of targeting would make amiRs an attractive tool to develop novel therapeutics for agriculture and human health, in coming years. Like conventional RNAi, an obvious repercussion of this technology would be to use it against pathogenic viral ORFs and also target practically every human/plant disorder with a gain-of-function genetic lesion. amiRs could potentially be used to fine-tune cellular metabolic pathways for use in molecular farming and other industrial applications.

The ongoing research in amiRs focuses on use of novel amiR precursors, like chimeric and polycistronic precursors for multiple targeting. These developments would further strengthen the amiRs as a third generation platform technology for gene silencing.

ACKNOWLEDGEMENT

The author is thankful to Dr. Sunil Kumar Mukherjee for valuable discussions and to Council of Scientific and Industrial Research, Government of India for conferring Shyama Prasad Mukherjee Fellowship.

REFERENCES

Alvarez, J.P., I. Pekker, A. Goldshmidt, E.Blum, Z. Amsellem and Y. Eshed. 2006. Endogenous and synthetic microRNAs stimulate simultaneous, efficient and localized regulation of multiple targets in diverse species. *Plant Cell.* **18:**1134–1151.

Ambros, V., R.C. Lee, A. Lavanway, P.T. Williams and D. Jewell. 2003. MicroRNAs and other tiny endogenous RNAs in C. elegans. *Curr. Biol.* **13:**807–818.

Bartel, D.P. 2004. MicroRNAs: genomics, biogenesis, mechanism and function. *Cell.* **116:**281–297.

Baulcombe, D. 1996. Mechanisms of Pathogen-Derived resistance to viruses in transgenic plants. *The Plant Cell.* **8:**1833–1844.

Bogerd, H.P., H.W. Karnowski, X. Cai, J. Shin, M. Pohlers and B.R. Cullen. 2010. A mammalian herpesvirus uses noncanonical expression and processing mechanisms to generate viral MicroRNAs. *Mol. Cell.* **37:**135–142.

Bollman, K.M., M.J. Aukerman, M. Park, C. Hunter, T.Z. Berardini and R.S. Poethig. 2003. HASTY, the Arabidopsis ortholog of exportin 5/MSN5, regulates phase change and morphogenesis. *Development.* **130:**1493–1504.

Brodersen, P., L. Sakvarelidze-Achard, M. Bruun-Rasmussen, P. Dunoyer, Y.V. Yamamoto, L. Sieburth and O. Voinnet. 2008. Widespread translational inhibition by plant miRNAs and siRNAs. *Science.* **320:**1185–1190.

Cai, X., C.H. Hagedorn and B.R. Cullen. 2004. Human microRNAs are processed from capped, polyadenylated transcripts that can also function as mRNAs. *RNA.* **10:**1957–1966.

Fang, Y. and D.L. Spector. 2007. Identification of nuclear dicing bodies containing proteins for MicroRNA biogenesis in living Arabidopsis plants. *Curr. Biol.* **17:**818–823.

Gao, Y., L. Yu, W. Wei, J. Li, Q. Luo and J. Shen. 2008. Inhibition of hepatitis B virus gene expression and replication by artificial microRNA. *World J. Gastroenterol.* **14:**4684–4689.

Hu, T., P. Chen, Q. Fu, Y. Liu, M. Ishaq, J. Li, L. Ma and D. Guo. 2010. Comparative studies of various artificial microRNA expression vectors for RNAi in mammalian cells. *Mol. Biotechnol.* **46:**34–40.

Huang, V., Y. Qin, J. Wang, X. Wang, R.F. Place, G. Lin, T.F. Lue and L. Li. 2010. RNAa is conserved in mammalian cells. *PLoS One.* **5:**e8848.

Israsena, N., P. Supavonwong, N. Ratanasetyuth, P. Khawplod and T. Hemachudha. 2009. Inhibition of rabies virus replication by multiple artificial microRNAs. *Antiviral Res.* **84:**76–83.

Khraiwesh, B., S. Ossowski, D. Weigel, R. Reski and W. Frank. 2008. Specific gene silencing by artificial microRNAs in Physcomitrella patens: An alternative to targeted gene knockouts. *Plant Physiol.* **148:**684–693.

Khvorova, A., A. Reynolds and S.D. Jayasena. 2003. Functional siRNAs and miRNAs exhibit strand bias. *Cell.* **115:**209–216. Kim, V.N. 2005. MicroRNA biogenesis: coordinated cropping and dicing. *Nature.* Reviews. *Molecular Cell. Biol.* **6:**376–385.

Liang, Z., H. Wu, S. Reddy, A. Zhu, S. Wang, D. Blevins, Y. Yoon, Y. Zhang and H. Shim. 2007. Blockade of invasion and metastasis of breast cancer cells via targeting CXCR4 with an artificial microRNA. *Biochem. Biophys. Res. Commun.* **363:**542–546.

Lindbo, J.A. and W.G. Dougherty. 1992. Untranslatable transcripts of the tobacco etch virus coat protein gene sequence can interfere with tobacco etch virus replication in transgenic plants and protoplasts. *Virology.* **189:**725–733.

Liu, C., L. Zhang, J. Sun, Y. Luo, M. Wang, Y. Fan and L. Wang. 2010. A simple artificial microRNA vector based on ath-miR169d precursor from Arabidopsis. *Mol. Biol. Rep.* **37:**903–909.

Lund, E., S. Güttinger, A. Calado, J.E. Dahlberg and U. Kutay. 2004. Nuclear export of microRNA precursors. *Science.* **303:**95–98.

Mallory, A.C., B.J. Reinhart, M.W. Jones-Rhoades, G. Tang, P.D. Zamore, M.K. Barton and D.P. Bartel. 2004. MicroRNA control of PHABULOSA in leaf development: importance of pairing to the microRNA 5' region. *EMBO J.* **23:**3356-3364.

McBride, J.L., R.L. Boudreau, S.Q. Harper, P.D. Staber, A.M. Monteys, I. Martins, B.L. Gilmore, H. Burstein, R.W. Peluso, B. Polisky, B.J. Carter and B.L. Davidson. 2008. Artificial miRNAs mitigate shRNA-mediated toxicity in the brain: Implications for the therapeutic development of RNAi. *Proc. Natl. Acad. Sci. USA.* **105:**5868–5873.

Mueller, E., J. Gilbert, G. Davenport, G. Brigneti and D.C. Baulcombe. 1995. Homology-dependent resistance: transgenic virus resistance in plants related to homology-dependent gene silencing. *Plant J.* **7:**1001–1013.

Niu, Q.W., S.S. Lin, J.L. Reyes, K.C. Chen, H.W. Wu, S.D. Yeh and N.H. Chua. 2006. Expression of artificial microRNAs in transgenic Arabidopsis thaliana confers virus resistance. *Nature Biotechnol.* **24:**1420–1428.

Qu, J., J. Ye and R. Fang. 2007. Artificial microRNA-mediated virus resistance in plants. *J. Virol.* 81:6690. Reynolds, A., D. Leake, Q. Boese, S. Scaringe, W.S. Marshall and A. Khvorova. 2004. Rational siRNA design for RNA interference. *Nature Biotechnol.* **22:**326–330.

Ruby, J.G., C.H. Jan and D.P. Bartel. 2007. Intronic microRNA precursors that bypass Drosha processing. *Nature.* **448:**83–86.

Sanford, J. and S. Johnston. 1985. The concept of parasite-derived resistance—Deriving resistance genes from the parasite's own genome. *J. Theoretic. Biol.* **113**:395–405.

Schwab, R., S. Ossowski, M. Riester, N. Warthmann and D. Weigel. 2006. Highly specific gene silencing by artificial microRNAs in Arabidopsis. *Plant Cell.* **18**:1121.

Schwarz, D.S., G. Hutvágner, T. Du, Z. Xu, N. Aronin and P.D. Zamore. 2003. Asymmetry in the assembly of the RNAi enzyme complex. *Cell.* **115**:199–208.

Shepherd, D.N., D.P. Martin and J.A. Thomson. 2009. Transgenic strategies for developing crops resistant to geminiviruses. *Plant Sci.* **176**:1–11.

Small, I. 2007. RNAi for revealing and engineering plant gene functions. *Curr. Opinion Biotechnol.* **18**:148–153.

Tang, G., B.J. Reinhart, D.P. Bartel and P.D. Zamore. 2003. A biochemical framework for RNA silencing in plants. *Genes Dev.* **17**:49–63.

Tang, Y., F. Wang, J. Zhao, K. Xie, Y. Hong and Y. Liu. 2010. Virus-based MicroRNA expression for gene functional analysis in plants. *Plant Physiol.* **153**:632–641.

Vasudevan, S., Tong Y. and Steitz J.A. 2007. Switching from repression to activation: MicroRNAs can up-regulate translation. *Science.* **318**:1931–1934.

Wang, X., Y. Yang, C. Yu, J. Zhou, Y. Cheng, C. Yan and J. Chen. 2010. A highly efficient method for construction of rice artificial MicroRNA vectors. *Mol. Biotechnol.* 22 May (advance online).

Warthmann, N., H. Chen, S. Ossowski, D. Weigel and P. Hervé. 2008. Highly specific gene silencing by artificial miRNAs in rice. *PLoS One* **3**:e1829.

Xie, Z., K.D. Kasschau and J.C. Carrington. 2003. Negative feedback regulation of Dicer-Like1 in Arabidopsis by microRNA-guided mRNA degradation. *Curr. Biol.* **13**:784–789.

Xue, L., Q. Yuan, Y. Yang and J. Wu. 2009. Enzymatic preparation of an artificial microRNA library. *Biochem. Biophys. Res. Commun.* 390:791–796. Yadava, P. and S.K. Mukherjee. 2009. RNAi technology: Strands of promise. *Biotech. News.* **4**:10–14

Yadava, P. and S.K. Mukherjee. 2010. Engineering geminivirus resistance in tomatoes using artificial microRNAs. Keystone Symposium on RNA Silencing Mechanisms in Plants, Santa Fe, NM, USA, 21–26 February 2010.

Yi, R., B.P. Doehle, Y. Qin, I.G. Macara and B.R. Cullen. 2005. Over-expression of Exportin 5 enhances RNA interference mediated by short hairpin RNAs and microRNAs. *RNA.* **11**:220–226.

Zeng, Y., E.J. Wagner and B.R. Cullen. 2002. Both natural and designed micro RNAs can inhibit the expression of cognate mRNAs when expressed in human cells. *Mol. Cell.* **9**:1327–1333.

Zeng, Y., X. Cai and B.R. Cullen. 2005. Use of RNA polymerase II to transcribe artificial microRNAs. *Methods Enzymol.* **392**:371.

CHAPTER 7

Pectin Methylesterase Enhances
Tomato Bushy Stunt Virus P19 Rna
Silencing Suppressor Effects

Tatiana V. Komarova, Vitaly Citovsky
and Yuri L. Dorokhov[*]

ABSTRACT

An ubiquitous cell wall enzyme, pectin methylesterase (PME), a multifunctional protein which catalyzes pectin deesterification, producing atmospheric methanol, participates in cell wall modulation during general plant growth and pollen tube growth. The tobacco PME interacts with the movement protein of *Tobacco mosaic virus* (TMV) suggesting that PME may be involved in cell-to-cell movement of plant viruses, whereas methanol produced by PME-catalyzed reactions may act as a signal molecule in plant-plant and plant-pathogen interactions. Furthermore, PME also is an efficient enhancer of virus- and transgene-induced gene silencing (VIGS and TIGS, respectively) via activation of siRNA and miRNA production. Enhancement of VIGS and TIGS is accompanied with relocation of the DCL1 protein from the cell nucleus to the cytoplasm. This chapter shows that PME can prevent nuclear transport of reporter proteins, GFP and GFP-tagged TMV movement protein (MP) and MS2 phage coat protein fused with a nuclear localization signal (NLS). Furthermore, PME reduced nuclear transport of a plant protein ALY. Interestingly, this PME activity enhanced the effect of a plant virus RNA silencing suppressor, the *Tomato bushy stunt virus* (TBSV) P19 protein, which is known to relocate the host ALY to the cell cytoplasm. PME production is elevated under stress conditions and the invading virus may exploit the ability of PME to impair ALY nuclear transport for augmenting the activity of its P19 suppressor. Indeed, coexpression of PME and P19 substantially enhanced TMV infection. The crucifer-infecting TMV (crTMV) RNA accumulation was increased 10 times at 3 dpi in leaf areas co-agroinjected with P19 and PME. The GFP production in leaves infiltrated with crTMV:GFP together with binary vectors expressing PME

* *Corresponding author e-mail*: YLD: dorokhov@genebee.msu.su

and P19 was doubled in comparison with P19 alone. Because only enzymatically active PME, which is secreted into cell wall, was able to suppress nuclear import, we hypothesize that the products of pectin deesterification, demethylesterified homogalacturonan or methanol, may immediate the effects of PME on nuclear import.

Keywords: ALY protein, nuclear transport, pectin methylesterase, *tobacco mosaic virus, tomato bushy stunt virus* P19 protein, silencing suppressor.

INTRODUCTION

All animals and plants utilize small RNA molecules to control protein expression during different developmental stages and in response to viral infection. The multiple plant proteins involved in RNAi may be subdivided into the main and auxiliary proteins. The main proteins involved in the plant siRNA pathways, including RNA-dependent RNA polymerases (RDRs), dsRNA-binding proteins (DRBs), DCLs and AGOs, are members of large protein families (Vazquez et al. 2010, Hiraguri et al. 2005, Wassenegger and Krezal 2006). Auxiliary proteins such as a JmjC domain-containing protein (Searle et al. 2010), GW repeat proteins (Jin and Zhu 2010), silencing defective 5 (SDE5) (Hernandez-Pinzon et al. 20007), SNF2 domain-containing proteins (Smith et al. 2007), Aly or Hin19 protein (Uhrig et al. 2004, Park et al. 2004, Canto et al. 2006, Hsieh et al. 2009) modify VIGS and TIGS development. Identification of novel additional host proteins associated with RNAi would be helpful for understanding their contribution to gene silencing as a part of host defense mechanism.

An ubiquitous cell wall enzyme, PME, a secreted protein which catalyzes pectin deesterification, is shown to be involved in binding to the movement protein (MP) of TMV (Dorokhov et al. 1999, Chen et al. 2000), suggesting that PME is engaged in the cell-to-cell movement of plant viruses (Chen and Citovsky 2003). A novel function of PME as an efficient enhancer of RNA silencing was shown recently (Dorokhov et al. 2006). Co-agroinjection of *Nicotiana benthamiana* leaves with the proPME gene and the TMV:GFP vector resulted in a stimulation of VIGS manifested by inhibition of GFP production, viral RNA degradation and stimulation of siRNAs production. The expression of proPME enhanced the GFP transgene-induced gene silencing accompanied by relocation of the DCL1 protein from nucleus to the cytoplasm and activation of siRNAs and miRNAs production. Moreover, the suppression of TMV short- and long-distance movement was observed in PME transgenic plants (Gasanova et al. 2008 Uhrig et al. 2004). It was hypothesized that relocated to the cytoplasm DCL1 may use both miRNA precursor and viral RNA as substrates.

This chapter shows that PME as an efficient enhancer of VIGS and TIGS can prevent nuclear traffic of the plant protein ALY which relocates the TBSV silencing suppressor P19 protein into the cell nucleus. PME induced inhibition of ALY:GFP nuclear transport and enhanced P19 silencing suppression effects, resulting in substantial increase in TMV reproduction.

PME BLOCKS NUCLEAR TRANSPORT OF NLS-CONTAINING PROTEINS

To understand whether PME-mediated DCL1 relocation is a specific event or PME interferes with nuclear proteins in general, we first examined intracellular localization of an NLS-containing reporter, GFP:NLS, expressed in the absence or presence of PME. Figure 1A shows that free GFP expressed in *N. benthamiana* leaves partitions between of the nucleus and the cytoplasm, due to its small size, which is below diffusion limit of the nuclear pore. Also as expected, GFP fused to the prothymosin α NLS accumulated exclusively in the cell nucleus (Figure 1B).

Nuclear localization of GFP:NLS was largely impaired by coexpression of a full-length PME, resulting in appearance of a significant pool of cytoplasmic GFP:NLS (Figure 1C). Epifluorescent microscopy analysis showed that coincubation of the GFP:NLS-expressing leaf with a PME-expressing leaf resulted in significant accumulation of percentage of cells with GFP fluorescence in cytoplasm as compared to coincubation with an vector control (93.8±0.60 vs. 15.5±1.34). This effect of PME expression was not specific only for GFP:NLS. We tested two other unrelated proteins tagged with GFP and fused to SV40 NLS: MS2 bacteriophage CP (GFP:NLS:CP$_{MS2}$) (Zhang and Simon 2004) and TMV MP (MP:GFP:NLS). Figure 1D shows that GFP:NLS:CP$_{MS2}$, indeed, accumulated in the cell nucleus and PME coexpression impeded this accumulation (Figure 1E) in more than 90% cells (vs. 10% in control). As GFP:NLS:CP$_{MS2}$ is larger than free GFP, it did not diffuse into the nucleus (compare panels C and E in Figure 1) and remained exclusively cytoplasmic. When wild-type TMV MP tagged with GFP was expressed in leaf tissue, we observed clearly distinguishable irregular cortical bodies and cell wall-associated puncta, characteristic of the well-known ER and plasmodesmata-specific localization of this protein (Figure 1F). Fusion with NLS directed virtually all MP:GFP into the cell nucleus (Figure 1G), indicating that this signal overcomes the putative plasmodesmata-targeting activity of MP. Importantly, coexpression of PME partially restored the characteristic wild-type pattern of MP localization (Figure 1H), suggesting that the nuclear import of MP:GFP:NLS was significantly impaired. In control experiments, coexpression of another cell wall protein NtGUT1 (Iwai et al. 2002) did not alter nuclear accumulation of all three NLS-containing reporters (data not shown). Furthermore, PME mutants,

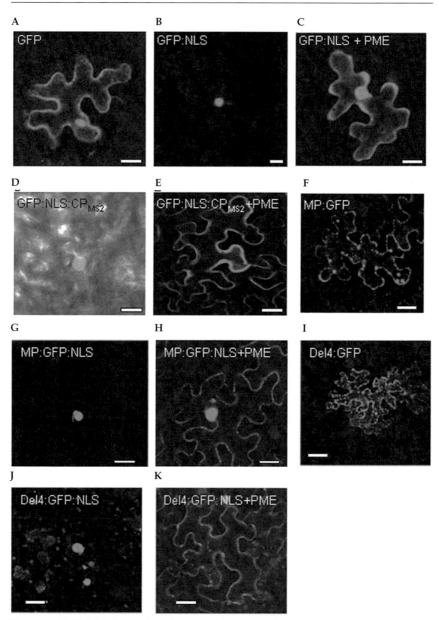

Figure 1. PME inhibits nuclear import of NLS-containing proteins: GFP, GFP:CP$_{MS2}$ and TMV MP:GFP. (A) GFP. (B) GFP:NLS. (C) GFP:NLS coexpressed with PME (D) GFP:NLS:CP$_{MS2}$. (E) GFP:NLS:CP$_{MS2}$ coexpressed with PME. (F) MP:GFP. (G) MP:GFP:NLS. (H) MP:GFP:NLS coexpressed with PME. (I) Del4:GFP. (J) Del4:GFP:NLS. (K) Del4:GFP:NLS coexpressed with PME. Images are projections of several confocal sections. Panel D represents a confocal image superimposed on a bright field image of the same cell. Bars = 20 μm.

(Color image of this figure appears in the color plate section at the end of the book.)

PME (396A397A) and A2-PME, which lack the enzymatic activity (Dorokhov et al. 2006), also showed no effect on nuclear import of the reporters (data not shown). We then used an MP mutant, Del4, which lacks the plasmodesmata-dilating domain (Waigmann et al. 1994), to demonstrate that inhibition of MP:GFP:NLS nuclear import by PME is not dependent on MP interaction with plasmodesmata. Figure 1 shows that Del4 tagged with GFP exhibited predominantly cytoplasmic and ER-associated localization (panel I) and relocalized to the cell nucleus upon fusion to SV40 NLS (panel J). Coexpression of PME largely blocked nuclear import of Del4:GFP:NLS (Figure 1K). Similarly to MP:GFP:NLS, nuclear import of Del4:GFP:NLS was not affected by inactive mutants of PME or by NtGUT1 (data not shown). Next, we used western blot analyses to address the possibility that the reporter protein may have undergone proteolytic cleavage which may result in cytoplasmic accumulation of the cleavage products. These experiments (data not shown) detected no cleavage products of intracellular MP:GFP:NLS or Del4:GFP:NLS following expression alone or coexpression with PME, its mutant derivatives, or NtGUT1.

PME RELOCATES THE PLANT NUCLEAR PROTEIN ALY AND ENHANCES THE RNA SILENCING SUPPRESSOR ACTIVITY OF TBSV P19

As PME interfered with the nuclear import of reporters containing a model NLS, most likely it also can affect import of at least some cellular nuclear proteins. Indeed, PME expression impaired nuclear import of the plant mRNA nuclear export factor ALY (Uhrig et al. 2001). Nuclear import of ALY, in turn, is known to correlate with the biological activity of the TBSV silencing suppressor, P19, with which it interacts; if ALY succeeds to direct associated P19 into the cell nucleus, then P19 is rendered inactive, but if P19 sequesters ALY in the cell cytoplasm, it remains capable to suppress silencing (Uhrig et al. 2004, Canto et al. 2006). Consistent with this model, PME-induced inhibition of ALY nuclear import enhanced silencing suppression by P19, resulting in substantial increase in viral reproduction. Figure 2A shows that GFP-tagged ALY expressed in plant leaves accumulated in the cell nucleus. Coexpression of PME altered this pattern of ALY:GFP subcellular localization, redirecting most of it to the cell cytoplasm (Figure 2B) with significant accumulation of percentage of cells with GFP fluorescence in cytoplasm as compared to coexpression with vector control (Uhrig et al. 2004) 7.3±0.51 vs. 11.5±1.43). Within the time course of the experiment, negative controls, an anti-sense PME (asPME) construct and the construct expressing enzymatically inactive PME (396A397A), had no effect on the

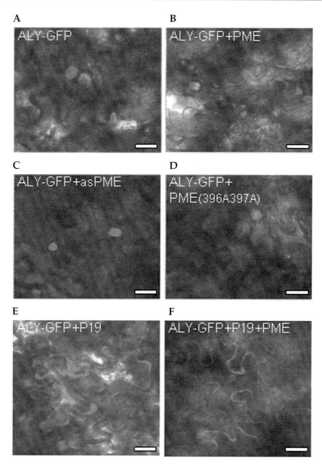

Figure 2. PME relocates ALY from the cell nucleus to the cytoplasm. (A) ALY:GFP. (B) ALY:GFP coexpressed with PME. (C) ALY:GFP coexpressed with asPME. (D) ALY:GFP coexpressed with PME(396A397A). (E) ALY:GFP coexpressed with P19. (F) ALY:GFP coexpressed with P19 and PME. Images are projections of several confocal sections superimposed on a bright field image of the same cell. Bars = 20 µm.

(Color image of this figure appears in the color plate section at the end of the book.)

ALY:GFP nuclear accumulation (Figure 2C, D, respectively). This change in ALY subcellular localization promoted by PME expression resembled that induced by coexpression of TBSV P19 (Uhrig et al. 2004) (Figure 2E). PME or P19 relocated virtually all detectible ALY:GFP, thus coexpression of both PME and P19 with ALY:GFP had no further discernable effect on ALY:GFP localization (Figure 2F).

As PME effect on ALY subcellular localization mimics that of P19 and P19 function in plant cells may correlate with its ability to interact with ALY and change its intracellular localization (Uhrig et al. 2004), we examined whether exogenous PME may also enhance silencing suppressor activity of P19. To this end, we agroinfiltrated leaves with a GFP-tagged, movement-deficient mutant of crucifer-infecting TMV (crTMV), crTMV:GFP-MP(fs); coinfiltration of a tested protein with crTMV:GFP-MP(fs) allows to estimate the effects on silencing by counting the number of cells displaying GFP fluorescence. Confirming our previous observations that PME is able to promote gene silencing (Dorokhov et al. 2006), coinfiltration of crTMV:GFP-MP(fs) and PME reduced the number of transformed cells exhibiting the GFP signal (Figure 3A) twofold, as compared to crTMV:GFP-MP(fs) expressed alone (Figure 3B). Compared to the same control, coinfiltration with P19 increased the number of the GFP-expressing cells also twofold, indicating silencing suppression (Figure 3B). Joint expression of crTMV:GFP-MP(fs) with both P19 and exogenous PME increased the level of GFP accumulation by a factor of four (Figure 3B), suggesting that PME enhances the silencing suppressor effect of P19.

Figure 3C shows that joint injection of P19 and PME drastically increased accumulation of another marker RNA, crTMV:GFP with wild-type MP. Our analysis of the GFP amounts in the agroinfiltrated tissues confirmed that PME indeed suppressed GFP expression as compared to crTMV:GFP agroinfiltrated alone, but it substantially increased this expression in the presence of both P19 and PME. As expected this enhanced GFP expression was not observed when P19 was coinfiltrated with the asPME construct (Figure 3D).

Furthermore, PME had no such effects on an unrelated viral silencing suppressor HC-Pro, the activity of which does not involve nuclear import (not shown). It is tempting to speculate that TBSV has evolved to take the advantage of plant stress and defense response which includes induction of PME expression. In this scenario, the host plant responds to the viral invasion via mechanical inoculation, for example, by inducing PME synthesis whereas the virus uses the resulting decrease in nuclear import to circumvent ALY nuclear uptake to inactivate the viral P19 suppressor.

The mechanism by which PME affects nuclear import is unknown. PME itself is a cell wall-associated protein and only enzymatically-active PME capable of targeting to the cell wall (Dorokhov et al. 2006a, b) could impair nuclear import. Thus, it is likely that PME does not interfere with the nuclear import machinery directly; instead, the products of PME-mediated pectin deesterification, such as methanol (Nemecek-Marshall et al. 1995) and other metabolites, may serve mediate the PME effect on nuclear import.

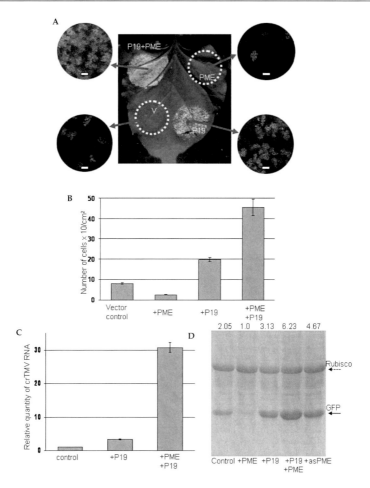

Figure 3. PME enhances the RNA silencing suppressor activity of TBSV P19. (A) Agroinfiltration experiments with a GFP-tagged, movement-deficient mutant crTMV:GFP-MP(fs). Visualization of GFP expression in leaf areas coinfiltrated with crTMV:GFP-MP(fs) and an empty, control binary vector (V) or binary vectors expressing PME, P19, or both PME and P19. GFP signal is in green and plastid autofluorescence is in magenta. Bars = 200 μm. (B) Numbers of GFP-expressing cells within the leaf areas shown in (A). The data represent mean values with standard error for 5–8 independent experiments. (C) Relative quantity of crTMV RNA at 3 dpi as determined by quantitative real-time PCR in leaf areas agroinjected either by the GFP-expressing vector alone (control) or together with P19 or P19 with PME (co-injection). (D) GFP expression at 10 dpi in leaves infiltrated with crTMV:GFP alone (control) or together with binary vectors expressing the indicated the expressing PME, P19, both PME and P19, or asPME. Protein was detected by polyacrylamide gel electrophoresis and Coomassie blue staining. Equal loading of all protein samples is indicated by the intensity of the Rubisco protein band. Numbers above gel wells are relative GFP band density units.

(Color image of this figure appears in the color plate section at the end of the book.)

ACKNOWLEDGEMENTS

We thank members of the MSU Department of Virology for helpful discussions and technical assistance. This work was partly supported by grants from RFBR, Moscow Government and Icon Genetic GmbH. The work in the VC laboratory is supported by grants from NIH, NSF, NIFA, BARD and BSF.

REFERENCES

Canto, T., J.F. Uhrig, M. Swanson, K.M. Wright and S.A. MacFarlane. 2006. Translocation of *Tomato bushy stunt virus* P19 protein into the nucleus by ALY proteins compromises its silencing suppressor activity. *J. Virol.* **80**:9064–9072.

Chen, M.H., V. Citovsky. 2003. Systemic movement of a tobamovirus requires host cell pectin methylesterase. *Plant J.* **35**:386–392.

Chen, M.H., J. Sheng, G. Hind, A.K. Handa and V. Citovsky. 2000. Interaction between the tobacco mosaic virus movement protein and host cell pectin methylesterases is required for viral cell-to-cell movement. *EMBO J.* **19**:913–920.

Dorokhov, Y.L., K.M. Makinen, O.Y. Frolova, A. Merits, N. Kalkkinen, J. Saarinen, J.G. Atabekov and M. Saarma. 1999. A novel function for a ubiquitous plant enzyme pectin methylesterase: the host-cell receptor for the tobacco mosaic virus movement protein. *FEBS Lett.* **461**:223–228.

Dorokhov, Y.L., E.V. Skurat, O.Y. Frolova, T.V. Gasanova, P.A. Ivanov, N.V. Ravin, K.G. Skryabin, K. Mäkinen, V. Klimyuk, Y.Y. Gleba and J.G. Atabekov. 2006a. Role of the leader sequences in tobacco pectin methylesterase secretion. *FEBS Lett.* **260**:3329–3334.

Dorokhov, Y.L., O.Y. Frolova, E.V. Skurat, P.A. Ivanov, T.V. Gasanova, A.S. Sheveleva, N.V. Ravin, K. Mäkinen, V.I. Klimyuk, K.G. Skryabin, Y.Y. Gleba and J.G. Atabekov. 2006. A novel function for a ubiquitous plant enzyme pectin methylesterase: the enhancer of RNA silencing. *FEBS Lett.* **580**:3872–3878.

Gasanova, T.V., Skurat, E.V., Frolova, O.Y., Semashko, M.A. and Dorokhov, Y.L. Uhrig et al. 2004. Pectin methylesterase as a factor for plant transcriptome stability. *Mol. Biol. (Moscow)*. **42**:421–429.

Hernandez-Pinzon, I., N.E. Yeli na, F. Schwach, D.J. Studholme, D. Baulcombe and T. Dalmay. 2007. SDE5, the putative homologue of a human mRNA export factor, is required for transgene silencing and accumulation of trans-acting endogenous siRNA. *Plant J.* **50**:140–148.

Hiraguri, A., R., Itoh, N. Kondo, Y. Nomura, D. Aizawa, Y. Murai, H. Koiwa, M. Seki, K. Shinozaki and T. Fukuhara. 2005. Specific interactions between Dicer-like proteins and HYL1/DRB-family dsRNA-binding proteins in Arabidopsis thaliana. *Plant Mol. Biol.* **57**:173–188.

Hsieh, Y.C., R.T. Omarov and H.B. Scholthof. 2009. Diverse and newly recognized effects associated with short interfering RNA binding site modifications on the Tomato bushy stunt virus p19 silencing suppressor. *J. Virol.* **83**:2188–2200.

Iwai, H., N. Masaoka, T. Ishii and S. Satoh. 2002. A pectin glucuronyl-transferase gene is essential for intercellular attachment in the plant meristem. *Proc. Natl. Acad. Sci. USA.* **99:**16319–16324.

Jin, H. and J.K. Zhu. 2010. A viral suppressor protein inhibits host RNA silencing by hooking up with Argonautes. *Genes Dev.* **24:**853–856.

Nemeèek-Marshall, M., R.C. MacDonald, J.J. Franzen, C.L. Wojciechowski and R. Fall. 1995. Methanol emission from leaves. *Plant Physiol.* **108:**1359–1368.

Park, J.W., S. Faure-Rabasse, M.A. Robinson, B. Desvoyes and H.B. Scholthof. 2004. The multifunctional plant viral suppressor of gene silencing P19 interacts with itself and an RNA binding host protein. *Virology.* **323:**49–58.

Searle, I.R., O. Pontes, C.W. Melnyk, L.M. Smith and D.C. Baulcombe. 2010. JMJ14, a JmjC domain protein, is required for RNA silencing and cell-to-cell movement of an RNA silencing signal in Arabidopsis. *Genes Dev.* **24:**986–991.

Smith, L.M., O. Pontes, I. Searle, N. Yelina, F.K. Yousafzai, A.J. Herr, C.S. Pikaard and D.C. Baulcombe. 2007. An SNF2 protein associated with nuclear RNA silencing and the spread of a silencing signal between cells in Arabidopsis. *Plant Cell.* **19:**1507–1521.

Uhrig, J., T. Canto, D. Marshall and S.A. MacFarlane. 2004. Relocalization of nuclear ALY proteins to the cytoplasm by the *Tomato bushy stunt virus* P19 pathogenicity protein. *Plant Physiol.* **135:**2411–2423.

Vazquez, F., S. Legrand and D. Windels. 2010. The biosynthetic pathways and biological scopes of plant small RNAs. *Trends Plant Sci.* **15:**337–345.

Waigmann, E., W.J. Lucas, V. Citovsky and P.C. Zambryski. 1994. Direct functional assay for tobacco mosaic virus cell-to-cell movement protein and identification of a domain involved in increasing plasmodesmal permeability. *Proc. Natl. Acad. Sci. USA.* **91:**1433–1437.

Wassenegger, M. and G. Krczal. 2006. Nomenclature and functions of RNA-directed RNA polymerases. *Trends Plant Sci.* **11:**142–151.

Zhang, F. and A. Simon. 2003. A novel procedure for the localization of viral RNAs in protoplasts and whole plants. *Plant J.* **35:**665–673.

CHAPTER 8

MicroRNAs: A New Paradigm for Gene Regulation

Luciana dos Reis Vasques,* Cláudia Gasque Schoof and Eder Leite da Silva Botelho

ABSTRACT

MicroRNAs (miRNAs) are small non-coding RNAs that act in the regulation of gene expression, targeting messenger RNAs and causing their degradation or translation repression. These small RNAs are involved in important biological processes and alteration in their expression patterns has been associated to several diseases. This chapter will focus on mammalian miRNAs pathway and review its processing mechanisms and genomic organization. In addition, the chapter will provide a brief overview of normal development function and maintenance of differentiated state, possible causes and consequences of miRNAs misregulation in cancer and other diseases, as well as advances on miRNAs-based therapy.

Keywords: miRNA, biogenesis, RISC, hESC, mESC, microRNA-based therapy, microRNA misregulation.

INTRODUCTION

The first findings on gene expression regulation date from 1960s, when two researchers, François Jacob and Jacques Monod, uncovered the Lac operon system in *Escherichia coli*, providing a new perspective for the mechanism of gene regulation (Jacob et al. 1961). Since then, different mechanisms of gene expression modulation have been discovered, such as specific transcription factors, epigenetic modifications, DNA regulatory elements, among others. The latest paradigmatical breakthrough in this field was the identification of a new class of RNAs: microRNAs (miRNAs). These are small non-coding RNAs, with approximately 22nt, which

* *Corresponding author e-mail*: lrvasques@gmail.com

are involved in gene expression regulation related to a variety of biological processes, such as cell proliferation and apoptosis. However, its main function is to establish and maintain the differentiated state of several cellular types of an organism (reviewed in Bartel, 2004, Bushati and Cohen, 2007). MiRNAs are well conserved and are of fundamental importance for the development of a wide range of eukaryotic organisms such as plants, nematodes, insects and mammals (Pasquinelli et al. 2000, Lagos-Quintana et al. 2001, Reinhart et al. 2002).

These small RNAs were observed for the first time in *C. elegans*, in 1993, by two independent groups (Lee et al. 1993, Wightman et al. 1993). They found that the control of early larval development is related to the suppression of translation of a gene, *lin-14*. The major discovery lies in the fact that this translation inhibition is accomplished by the complementary binding of endogenous small RNA, lin-4, to 3'-end UTR of the mRNA of lin-14. Although another target of lin-4, the mRNA of lin-28, was identified four years later (Moss et al. 1997), this specific type of translation blockage seemed to be an isolated case, until the year 2000, when another small RNA, let-7, which inhibits translation of lin-41 was identified (Reinhart et al. 2000, Pasquinelli et al. 2000). Since then, many others were subsequently identified, not only in *C. elegans*, but also in several other organisms and then, the term "microRNAs" was coined to classified them (Lau et al. 2001, Lagos-Quintana et al. 2001, Lee et al. 2001). Currently, more than 1000 miRNAs have already been identified in humans, (miRBase-http://www.mirbase.org/cgi-bin/browse.pl), representing 1 to 3% of genes present in our genome (Zhao et al. 2007). A recent study indicates that miRNAs may be responsible for regulating at least 60% of human protein-coding genes (Friedman et al. 2009). As new miRNAs are identified, their expression profile in several tissues, in normal and altered states, as well as the identification of their targets, provide the first data about their biological functions in different pathways. This chapter will review the biogenesis and the mechanism of action of miRNAs, as well as provide an overview of recent publication in the field.

BIOGENESIS AND MECHANISM OF ACTION OF miRAs

The miRNAs pathway occurs in 3 main steps (Figure 1): The first consists in transcription of primary miRNA (pri-miRNA), which has one or several loops of 60 to 80 nucleotides (nt), showing imperfect pairings (reviewed in Bartel, 2004). In the second step, each loop is recognized by Drosha, an RNase III endoribonuclease and its partner, Pasha/ DGCR-8 (*Drosophila* and *C. elegans*/ mammals orthologs respectively), that has affinity for double-stranded RNA (Landthaler et al. 2004, Han et al. 2004). Drosha

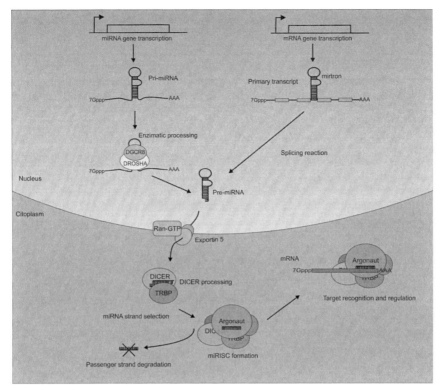

Figure 1: Schematic illustration of microRNAs biogenesis. For further explanation, see text.

(Color image of this figure appears in the color plate section at the end of the book.)

cleaves the pri-miRNAs in miRNAs precursors, pre-miRNAs, which have approximately 70 nt with 2-nt 3′ overhang (Lee et al. 2003, Basyuk et al. 2003). Once processed, the pre-miRNAs are exported by Exportin 5 core and its co-factor Ran-GTP (Yi et al. 2003, Lund et al. 2004, Bohnsack et al. 2004). The hairpin structure of pre-miRNA and 2 nt at 3′-end left by Drosha processing are recognized by Exportin 5 in sequence-independent manner, facilitating its transport to the cytoplasm (Zeng et al. 2004). In the third step, the pre-miRNAs are processed in miRNAs duplexes of 21–25 nt by Dicer, another RNase III endoribonuclease and its partner TRBP (Trans Activator RNA Binding Protein) in humans (Loquacious in *Drosophila*) (Bernstein et al. 2001, Zhang et al. 2002, Chendrimada et al. 2005, Förstemann et al. 2005, Saito et al. 2005).

As will be seen below, most miRNAs are hosted in intronic regions of other genes. In these cases, the beginning of processing occurs simultaneously to the transcript, prior to splicing (Kim et al. 2007, Morlando et al., 2008). Alternatively, they can also be generated directly as a product of splicing, known as Mirtrons (intronic miRNAs), avoiding processing by the complex Drosha/DGCR8(Pasha) (Figure 1) (Ruby et al. 2007, Berezikov et al., 2007, Okamura et al. 2007). Once mature, miRNAs duplexes are loaded through Dicer into RISC complex (RNA-Induced Silencing Complex), which also recognizes the 2 nt 3' overhang and then trigger the post-transcriptional silencing. Usually the duplex strand with less stability at 5'-end is selected to be part of miRNA/RISC complex (miRISC), while the other strand, called passenger, is degraded (Schwarz et al. 2003, Khvorova et al. 2003). The protagonist of RISC protein complex belongs to the Argonaut family (AGO), which in humans is composed of eight members, four of them (AGO 1 to 4) are associated with miRNAs as well as RISC activity. Among them, only AGO2 has a catalytic activity (Liu et al. 2003, Meister et al. 2004 and reviewed in Peters and Meister, 2007).

Endogenous genes may be regulated by miRNAs in two different manners determined by the degree of miRNAs complementarity with their targets. In one, miRNAs have full or nearly full complementarity and the targeted RNA are sliced by RISC containing Ago1 (in plants) or Ago2 (in *Drosophila* and mammals) and, subsequently, degraded by RNases (Llave et al. 2002, Meister et al. 2004, Okamura et al. 2004). This mechanism is commonly found in plants and the target region is usually in the ORF (Rhoades et al. 2002). In another, miRNAs present partial complementarity, repressing target RNA translation without their degradation (Doench et al. 2003, Doench et al. 2004). In this case, besides partial complementarity with their RNA targets, each miRNA has a region, located between the nucleotides 2 to 8, with total complementarity. This sequence is known as seed region and is responsible for the miRNA specificity (Lewis et al. 2003, Doench et al. 2004, Lewis et al. 2005). The repression, without further target RNA degradation, is generally found in animals, in which the seed region is located most frequently in the 3' UTR of the target RNA (Lai. 2002). Rare cases of pairing regions other than 3' UTR are also found; examples are: the miR-148, which represses the translation binding to the coding region of DNMT3B transcript (Duursma et al. 2008) and the miR-122, which binds to 5' UTR of *Hepatitis C virus* (HCV) RNA (Díaz-Toledano et al. 2009).

However, even in circumstances where the complementarity is partial, the target RNA degradation may occur. This phenomenon has been observed in some cytoplasmatic foci, called P-bodies (processing bodies), that, alternatively, can group targetRNAs/miRISC complexes to an enzyme, Dcp1. This enzyme cleaves 5' Cap of the target RNA, which is subjected to degradation by RNases (Liu et al. 2005).

In addition to the canonical pathways, recent studies have reported some new functions for protein Ago2. In 2007, Diederichs and Haber showed that

Ago2 can, in some cases, cleave the pre-miRNA hairpin 12 nucleotides from its 3' end, prior to Dicer processing. Such event generates an intermediate product, named ac-pre-miRNA (Ago2-cleaved pre-miRNA), which is then processed by Dicer and loaded into RISC complex. Through this process, the nicked passenger strand cleaved by Ago2 may facilitate removal from RISC, after maturation. Another study has identified a murine miRNA (miR-451) that is not processed by Dicer. In this case, the pre-miRNA is directly loaded into and cleaved by Ago2 at the 3'-end (Cheloufi et al. 2010).

The cellular mechanisms described above are multifaceted genetic regulatory processes and a fine orchestration of expression/repression of several genes occur through differential expression of miRNAs. It seems that more than one miRNA must bind to the same target RNA to achieve an efficient gene silencing (Doench et al. 2004). Another aspect is that the same miRNA may target different RNAs (Yuan et al. 2009). Other forms of miRNAs action continue to be described. One is that miRNAs, as molecules of si- and pi-RNA (small interfering- and Piwi-interacting-RNAs), can guide chromatin remodeling. Recently, Gonzales et al. (2008) demonstrate this possibility, studying miR17-5p and miR20a from an oncomiR cluster, miR17-92. Promoters that undergo overlapping transcription and are complementary to miR17-5p and miR20a seed regions can induce heterochromatinization by a mechanism still to be fully elucidated. Another surprising discovery in the last few years, is that some miRNAs were found to stimulate translation, instead of suppress, by different mechanisms. Examples are: miR-369-3 in humans, upon cell cycle arrest, can induce translation through association of miRISC complex to ARE (AU-rich element), present in 3' UTR of the TNFα mRNA (Vasudevan et al. 2007); and miR-10b that binds to 5' UTR of mRNAs that encode ribosomal proteins, facilitating their translation (Ørom et al. 2008).

ORGANIZATION OF miRNA GENES

In general, miRNAs are subjected to RNA polymerase II transcriptional control (Lee et al. 2004), although some miRNAs are transcribed by RNA polymerase III, (Borchert et al. 2006) and possess poli-A tail and 5'-Cap (Cai et al. 2004). In humans, about 30% of miRNAs are located in intergenic regions and 70% inserted into other genes. In the latter case, the miRNA and the host gene usually are in the same orientation and are synergistically expressed, both of which subjected to the same promoter region (Rodriguez et al. 2004). Intragenic miRNA have been found in introns of coding and non-coding RNAs (ncRNAs) and even in exons of mRNAs that undergo alternative splicing (Lagos-Quintana et al. 2001, Lau et al. 2001, Lagos-Quintana et al. 2003, Rodriguez et al. 2004, Maselli et al.

2008). In addition, approximately 40% of miRNAs can be found in clusters, forming polycistronic transcripts that have coordinated expression (Lee et al. 2004, Rodriguez et al. 2004, Griffiths-Jones et al. 2008). Interestingly, it has been proposed that a miRNA cluster may regulate a group of mRNAs that encode proteins related to the same pathway (Yuan et al. 2009), which highlights the complexity of these regulatory molecules. Few miRNAs can still be found within 3'-UTR region of coding transcripts, as miR-198 (Cullen. 2004). However, in vertebrates, intronic miRNAs are the most abundant, constituting 40 to 70% of all miRNAs (Baskerville et al. 2005, Griffiths-Jones et al. 2008, reviewed by Olena and Patton, 2009).

miRNA FUNCTION

The characteristics described above show how difficult miRNAs are to be studied. The newest works in mammals are focused on the identification and validation of new miRNAs and their RNA targets, location in the genome and organization. Thus, miRNA expression panels from different tissues and organisms (Baskerville et al. 2005, He et al. 2005), from human and mouse embryonic stem cells (hESC and mESC) (Houbaviy et al. 2003, Suh et al. 2004), as well as tumor tissues (Lu et al. 2005), are being obtained by Northern Blot or Microarray. In addition, prediction programs identifying possible target regions, using miRNAs known sequences against different RNA database from several organisms, are being developed and are available on the Internet (MiRanda/PicTar/Targetscan/RNAHybrid/Diana-microT/among others). Therefore, the identification of each miRNA function is the most difficult task and it is being done slowly. Here, we will describe some of the recent advances in this field in different cellular pathways.

As previously mentioned, the first miRNAs findings were related to early stages of development. Thus, an increasing number of studies in development of different species has been performed. In 2003, the Sharp Group was the first to publish a work in which the expression profile of miRNAs in undifferentiated and differentiated mESC was obtained. They have found a cluster miR-290-295, which encodes for eight mature miRNAs and is highly expressed in undifferentiated cells. Interestingly, its expression decreases upon differentiation (Houbaviy et al. 2003). Thus, this cluster was credited to be involved in maintenance of mESC undifferentiated state. However, later studies observed that this cluster acts by an indirect manner, controlling epigenetic modifications, since its miRNAs regulate the expression of DNA metyltransferases through inhibition of its repressor, Rbl2 (Retinoblastoma-like 2 protein), (Benetti et al. 2008, Sinkkonen et al. 2008). In a similar study conducted on hESC, the exclusive group of miRNAs expressed in non-differentiated cells was different from mESC counterparts, except for one homologous cluster (Suh et al. 2004). Additionally, it was observed by Kanellopoulou et al. (2005) that Dicer-deficient mESCs are viable, but lack the ability to differentiate, providing further evidence of the importance of miRNAs in the early stages of development. In agreement with these findings, at the beginning of organogenesis, several studies have reported the key role of miRNAs in determining the cell destination and differentiation of hematopoietic stem cells (Chen et al. 2004, Felli et al. 2005, Georgantas et al. 2007, Zhou et al. 2007, Wang et al. 2008); in myogenesis (Kim et al. 2006, Wong et al. 2008), in differentiation of neural stem cells (Smirnova et al. 2005), in osteogenesis (Mizuno et al. 2008); in differentiation of epithelial stem cells (Yi et al. 2008, Lena et al. 2008); in differentiation of mature adipocytes with ubiquitous localization (Esau et al. 2004); among others.

Despite the importance of miRNAs in the normal development of many organisms, as seen above, deregulation of their expression has been increasingly related to several pathologies, being cancer the most studied one. The first deregulated miRNAs described in cancer cells were miR-15a and miR-16-1, which are located in a chromosomal region deleted in more than half of B-cell Chronic Lymphocytic Leukemia (B-CLL) cases (Calin et al. 2002). In later works, these miRNAs were found to control the expression of BCL2, an antiapoptotic proto-oncogene (Cimmino et al. 2005). Over the years, this field has experienced a great expansion and currently, several miRNAs aberrantly expressed in cancer have been described. MiRNAs that suppress expression of proto-oncogenes preventing tumorigenesis and tumor progression, named "tumor suppressor miRNAs", are downregulated in tumor cells. Among these, one of the most thoroughly studied examples is the let-7 family, which members are located in fragile chromosomal sites and are often deleted in lung cancer (Takamizawa et al. 2004, Calin et al. 2004, Johnson et al. 2005). On the other hand, miRNAs that regulate the expression of tumor suppressor genes, contributing to tumor development, are known as "OncomiRs" and are overexpressed in tumor cells. One of the best-known examples is the cluster miR-17-92, which is located in a chromosomal region often amplified in several tumor types and, consequently, is highly expressed (Ota et al. 2004, Hayashita et al. 2005, Matsubara et al. 2007). In addition, an increasing number of studies has associated aberrant expression profile of miRNAs with different stages of tumor progression, degree of malignancy, subtypes, association with prognosis and resistance to chemotherapy agents (reviewed in Grady and Tewari, 2010, Fabbri. 2010, Nana-Sinkam et al. 2010).

As shown above, there are a number of causes for deregulation of miRNAs. More than half of miRNAs is located at fragile sites of the genome, subject to amplifications, deletions, breaks and rearrangements (Calin et al. 2004, Sevignani et al. 2007). Thus, changes in copy number of miRNA genes play an important role in tumor formation (Keating et al. 2002, Bottoni et al. 2005, He et al. 2005, Zhang et al. 2006). Moreover, alterations in miRNAs expression can be caused by transcriptional deregulation (O'Donnell et al. 2005, Bommer et al. 2007, Chang et al. 2008), epigenetic modifications (Saito et al. 2006, Zhang et al. 2008, Yang et al. 2008), defect or disruption in the expression of proteins that participate in its biogenesis (Karube et al. 2005, Chiosea et al. 2006, Thomson et al. 2006, Muralidhar et al. 2007, Chiosea et al. 2007, Kumar et al. 2007) and mutations or polymorphisms (Calin et al. 2005, Diederichs et al. 2006, Raveche et al. 2007, Wu et al. 2008). The latter can create new illegitimate interactions, since the mutation or polymorphism at miRNAs or miRNA target sites, may affect the complementarity, preventing binding or creating a new target site (Saunders et al. 2007, reviewed by Fabbri et al. 2009, Ryan et al. 2010). As an example, a polymorphism next to the miR-24 binding site at 3' UTR of Human Dihydrofolate Reductase gene (DHFR), prevents its inhibition and, as a result, DHFR gene is found overexpressed, conferring resistance to the chemotherapic agent Methotrexate (Mishra et al. 2007). Finally, miRNAs can also target an exon of a RNA that undergoes alternative splicing by exon skipping. Thus, different products from the same gene may be sensitive or not to a particular miRNA, depending on the presence or absence of a target site in its mature RNA (Laneve et al. 2007, Duursma et al. 2008).

Besides cancer, other pathologies are also being increasingly associated with miRNAs deregulation; among them, several neuro-degenerative diseases (Wu. 2006, Kim et al. 2007, Lukiw. 2007, Hébert et al. 2008, Cogswell et al. 2008, Johnson et al. 2008, Stark et al. 2008, Otaegui et al. 2009), cardiovascular (Thum et al. 2007, Ikeda et al. 2007, Sucharov et al. 2008, Naga et al. 2009) autoimmune (Dai et al. 2007, Pauley et al. 2008, Nakamachi et al. 2009, Stanczyk et al. 2009, Padgett et al. 2009, Fulci et al. 2010), Schizophrenia (Hansen et al. 2007, Perkins et al. 2007, Beveridge et al. 2008), Tourette Syndrome (Abelson et al. 2005) and others.

Even viral genomes encode miRNAs (Pfeffer et al. 2004, Pfeffer et al. 2005). They can inactivate the innate defense mechanisms of the host cell, as well as its own endogenous-viral RNAs (reviewed in He et al. 2008). The opposite was also reported, where miRNAs from host cells may control viral infection by suppressing viral replication (Zhang et al. 2010).

The understanding of the proper miRNAs processing and functioning is essential for identifying alterations in different pathways and their related pathologies. Thus, in recent years a growing number of studies of miRNA-based therapy have been performed. The introduction of exogenous miRNAs, recovering therefore the correct miRNAs expression in a cell, would indirectly downregulate the expression of its targets. The opposite can also be done, by inhibiting the expression of specific miRNA aberrantly upregulated through an anti-miRNA oligonucleotides (AMOs) and which would retrieve the expression of their RNA targets (reviewed in Wahid et al. 2010, Garofalo et al. 2010, Croce. 2010). The miRNA-based therapy is goal to pharmaceutical companies. In 2008, the company Santaris first announced the beginning of a clinical trial with an anti-miR-122 molecule to treat Hepatitis C, since miR-122 is specifically overexpressed in patients with this pathology. Currently, several other clinical trials based on miRNA therapy for a number of diseases such as sepsis, asthma, leukemia and other cancers are in progress (source: http://www.clinicaltrials.gov).

CONCLUSION

When the term "RNA World" was coined in the 1980s, scientists were beginning to understand that, in fact, the RNAs have a myriad of functions than what was initially proposed by the Biology Central Dogma. The identification of a class of small RNA molecules provoked a colossal and paradigmatical shift in gene regulation knowledge. Moreover, these molecules are well conserved and present in most multicellular organisms and are involved in regulating many cellular biochemical pathways. In humans they may control the expression of two thirds of the coding genes. The regulation of most genes already studied, might have to be revisited through another angle. Thus, the identification of new miRNAs, their targets, organization in genomes, as well as their role is essential for moving toward the understanding of the complexity of eukaryotic cells.

REFERENCES

Abelson, J.F., K.Y. Kwan, B.J. O'Roak, D.Y. Baek, A.A. Stillman, T.M. Morgan, C.A. Mathews, D.L. Pauls, M.R. Rasin and M. Gunel. 2005. Sequence variants in SLITRK1 are associated with Tourette's syndrome. *Science.* **310:**317–320.

Bartel, D.P. 2004. MicroRNAs: Genomics, Biogenesis, Mechanism and Function. *Cell.* **116:**281–297.

Baskerville S. and D.P. Bartel 2005. Microarray profiling of microRNAs reveals frequent coexpression with neighboring miRNAs and host genes. *RNA.* **11(3):**241–7.

Basyuk, E., F. Suavet, A. Doglio, R. Bordonné and E. Bertrand. 2003. Human let-7 stem-loop precursors harbor features of RNase III cleavage products. *Nucleic Acids Res.* **31:**6593–6597.

Benetti, R., S. Gonzalo, I. Jaco, P. Muñoz, S. Gonzalez, S. Schoeftner, E. Murchison, T. Andl, T. Chen, P. Klatt, E. Li, M. Serrano, S. Millar, G. Hannon and M.A. Blasco. 2008. A mammalian microRNA cluster controls DNA methylation and telomere recombination via Rbl2-dependent regulation of DNA methyltransferases. *Nature. Struct. Biol.* **15**:268–279.

Berezikov, E., W.J. Chung, J. Willis, E. Cuppen and E.C. Lai. 2007. Mammalian mirtron genes. *Mol. Cell.* **28**:328–336.

Bernstein, E., A.A. Caudy, S.M. Hammond and G.J. Hannon. 2001. Role for a bidentate ribonuclease in the initiation step of RNA interference. *Nature.* **409**:363–366.

Beveridge, N.J., P.A. Tooney, A.P. Carroll, E. Gardiner, N. Bowden, R.J. Scott, N. Tran, I. Dedova and M.J. Cairns. 2008. Dysregulation of miRNA 181b in the temporal cortex in schizophrenia. *Hum. Mol. Genet.* **17(8)**:1156–68.

Bohnsack, M.T., K. Czaplinski and D. Gorlich. 2004. Exportin 5 is a RanGTP-dependent dsRNA-binding protein that mediates nuclear export of pre-miRNAs. *RNA.* **10**:185–191.

Bommer, G.T., I. Gerin, Y. Feng, A.J. Kaczorowski, A.J. Kuick, R.E. Love, Y. Zhai, T.J. Giordano, Z.S. Qin, B.B. Moore, O.A. Macdougald, K.R. Cho and E.R. Fearon. 2007. p53-Mediated Activation of miRNA34 Candidate Tumor-Suppressor Genes. *Curr. Biol.* **17(15)**:1298–307.

Borchert, G.M., W. Lanier and B.L. Davidson. 2006. RNA polymerase III transcribes human microRNAs. *Nat. Struct. Mol. Biol.* **13**:1097–1101.

Bottoni, A., D. Piccin, F. Tagliati, A. Luchin, M.C. Zatelli and E.C. degli Uberti. 2005. miR-15a and miR-16-1 down-regulation in pituitary adenomas. *J. Cell. Physiol.* **204(1)**:280–5.

Bushati, N. and S.M. Cohen. 2007. microRNA Functions. *Annu. Rev. Cell. Dev. Biol.* **23**:175–205.

Cai, X., C.H. Hagedorn and B.R. Cullen. 2004. Human microRNAs are processed from capped, polyadenylated transcripts that can also function as mRNAs. *RNA.* **10(12)**:1957–66.

Calin, G.A., C.D. Dumitru, M. Shimizu, R. Bichi, S. Zupo, E. Noch, H. Aldler, S. Rattan, M. Keating, K. Rai, L. Rassenti, T. Kipps, M. Negrini, F. Bullrich and C.M. Croce. 2002. Frequent deletions and down-regulation of micro- RNA genes miR15 and miR16 at 13q14 in chronic lymphocytic leukemia. *Proc. Natl. Acad. Sci. USA.* **99**:15524–15529.

Calin, G.A., C. Sevignani, C.D. Dumitru, T. Hyslop, E. Noch, S. Yendamuri, M. Shimizu, S. Rattan, F. Bullrich, M. Negrini and C.M. Croce. 2004. Human microRNA genes are frequently located at fragile sites and genomic regions involved in cancer. *Proc. Natl. Acad. Sci. USA.* **101**:2999–3004.

Calin, G.A., M. Ferracin, A. Cimmino, G. Di Leva, M. Shimizu, S.E. Wojcik, M.V. Iorio, R. Visone, N.I. Sever, M. Fabbri, R. Iuliano, T. Palumbo, F. .Pichiorri, C. Roldo, R. Garzon, C. Sevignani, L. Rassenti, H. Alder, S. Volinia, C.G. Liu, T.J. Kipps, M. Negrini and C.M. Croce. 2005. A MicroRNA sig*Nature.* associated with prognosis and progression in chronic lymphocytic leukemia. *N. Engl. J. Med.* **353**:1793–801.

Chang, T.C., D. Yu, Y.S. Lee, E.A. Wentzel, D.E. Arking, K.M. West, C.V. Dang, A. Thomas-Tikhonenko and J.T. Mendell. 2008. Widespread microRNA repression by Myc contributes to tumorigenesis. *Nat. Genet.* **40**:43–50.

Cheloufi, S., C.O. Dos Santos, M.M. Chong and G.J. Hannon. 2010. A dicer-independent miRNA biogenesis pathway that requires Ago catalysis. *Nature.* **465(7298)**:584–9.

Chen, C.Z., L. Li, H.F. Lodish and D.P. Bartel. 2004. MicroRNAs modulate hematopoietic lineage differentiation. *Science.* **303**:83–86.

Chendrimada, T.P., R.I. Gregory, E. Kumaraswamy, J. Norman, N. Cooch, K. Nishikura and R. Shiekhattar. 2005. TRBP recruits the Dicer complex to Ago2 for microRNA processing and gene silencing. *Nature.* **436**: 740–744.

Chiosea, S., E. Jelezcova, U. Chandran, M. Acquafondata, T. McHale, R.W. Sobol and R. Dhir. 2006. Up-Regulation of DICER, a Component of MicroRNA machinery, in Prostate Adenocarcinoma. *Am. J. Pathol.* **169**:1812–1820.

Chiosea, S., E. Jelezcova, U. Chandran, J. Luo, G. Mantha, R.W. Sobol and S. Dacic. 2007. Overexpression of DICER in Precursor lesions of Lung Adenocarcinoma. *Cancer Res.* **67**:2345–2350.

Cimmino, A., G.A. Calin, M. Fabbri, M.V. Iorio, M. Ferracin, M. Shimizu, S.E. Wojcik, R.I., C.M. Aqeilan, S. Zupo, M. Dono, L. Rassenti, H. Alder, S. Volinia, C.G. Liu, T.J. Kipps, M. Negrini and Croce. 2005. miR-15 and miR-16 induce apoptosis by targeting BCL2. *Proc. Natl. Acad. Sci. USA.* **102**:13944–13949.

Cogswell, J.P., J. Ward, I.A. Taylor, M. Waters, Y. Shi, B. Cannon, K. Kelnar, J. Kemppainen, D. Brown, C. Chen, R.K. Prinjha, J.C. Richardson, A.M. Saunders, A.D. Roses and C.A. Richards. 2008. Identification of miRNA changes in Alzheimer's disease brain and CSF yields putative biomarkers and insights into disease pathways. *J. Alzheimers Dis.* **14**:27–41.

Cullen, B.R. 2004. Transcription and processing of human microRNA precursors. *Mol Cell.* **16**: 861-865.

Dai, Y., Y.S. Huang, M. Tang, T.Y. Lv, C.X. Hu, Y.H. Tan, Z.M. Xu and Y.B. Yin. 2007. Microarray analysis of microRNA expression in peripheral blood cells of systemic lupus erythematosus patients, *Lupus.* **16**:939–946.

Díaz-Toledano, R., A. Ariza-Mateos, A. Birk, B. Martínez-García and J. Gómez. 2009. *In vitro* characterization of a miR-122-sensitive double-helical switch element in the 5' region of hepatitis C virus RNA. *Nucleic Acids Res.* **37(16)**:5498–510.

Diederichs, S. and D.A. Haber. 2006. Sequence variations of microRNAs in human cancer: alterations in predicted secondary structure do not affect processing. *Cancer Res.* **66(12)**:6097–6104.

Diederichs, S. and D.A. Haber. 2007. Dual role for Argonautes in microRNA processing and posttranscriptional regulation of microRNA expression. *Cell.* **131**:1097–1108.

Doench, J.G., C.P. Petersen and P.A. Sharp. 2003. siRNAs can function as miRNAs. *Genes Dev.* **17(4)**:438–42.

Doench, J.G. and P.A. Sharp. 2004. Specificity of microRNA target selection in translational repression. *Genes Dev.* **18**:504–511.

Duursma, A.M., M. Kedde, M. Schrier, C. le Sage and R. Agami. 2008. miR-148 targets human DNMT3b protein coding region. *RNA*. **14(5):**872–7.

Esau, C., X. Kang, E. Peralta, E. Hanson, E.G. Marcusson, L.V. Ravichandran, Y. Sun, S. Koo, R.J. Perera, R. Jain, N.M. Dean, S.M. Freier, C.F. Bennett, B. Lollo and R. Griffey. 2004. MicroRNA-143 regulates adipocyte differentiation. *J. Biol. Chem.* **279(50):**52361–5.

Fabbri, M., N. Valeri and G.A. Calin. 2009. MicroRNAs and genomic variations: from Proteus tricks to Prometheus gift. *Carcinogenesis*. **30(6):**912–7.

Fabbri, M. 2010. miRNAs as molecular biomarkers of cancer. *Expert Rev. Mol. Diagn.* **10(4):**435–44.

Felli, N., L. Fontana, E. Pelosi, R. Botta, D. Bonci, F. Facchiano, F. Liuzzi, V. Lulli, O. Morsilli, S. Santoro, M. Valtieri, G.A. Calin, C.G. Liu, A. Sorrentino, C.M. Croce and C.M. Peschle. 2005. MicroRNAs 221 and 222 inhibit normal erythropoiesis and erythroleukemic cell growth via kit receptor down-modulation. *Proc. Natl. Acad. Sci. USA*. **102:**18081–18086.

Förstemann, K., Y. Tomari, T. Du, V.V. Vagin, A.M. Denli, D.P. Bratu, C. Klattenhoff, W.E. Theurkauf and P.D. Zamore. 2005. Normal microRNA maturation and germ-line stem cell maintenance requires Loquacious, a double-stranded RNA-binding domain protein. *PloS Biol*. **3(7):**e236.

Friedman, R.C., K.K. Farh, C.B. Burge and D.P. Bartel. 2009. Most mammalian mRNAs are conserved targets of microRNAs. *Genome Res*. **19(1):**92–105.

Fulci, V., G. Scappucci, G.D. Sebastiani, C. Giannitti, D. Franceschini, F. Meloni, T. Colombo, F. Citarella, V. Barnaba, G. Minisola, M. Galeazzi and G. Macino. 2010. miR-223 is overexpressed in T-lymphocytes of patients affected by rheumatoid arthritis. *Hum. Immunol*. **71(2):**206–211.

Garofalo, M., C.M. Croce. 2010. microRNAs: Master Regulators as Potential Therapeutics in Cancer. *Annu. Rev. Pharmacol. Toxicol*. **51:**25–43.

Georgantas, R.W., R. Hildreth, S. Morisot, J. Alder, C.G. Liu, S. Heimfeld, G.A. Calin, C.M. Croce and C.I. Civin. 2007. CD34+ hematopoietic stem-progenitor cell microRNA expression and function: a circuit diagram of differentiation control. *Proc. Natl. Acad. Sci. USA*. **104:**2750–2755.

Gonzalez, S., D.G. Pisano and M. Serrano . 2008. Mechanistic principles of chromatin remodeling guided by siRNAs and miRNAs. *Cell Cycle*. **7(16):**2601–8.

Grady, W.M. and M. Tewari. 2010. The next thing in prognostic molecular markers: microRNA signatures of cancer. *Gut*. **59(6):**706–8.

Griffiths-Jones, S., H.K. Saini, S. van Dongen and A.J. Enright. 2008. miRBase: tools for microRNA genomics. *Nucleic Acids Res*. **36:**D154–8.

Han, J., Y. Lee, K.H. Yeom, Y.K. Kim, H. Jin and V.N. Kim. 2004. The Drosha-DGCR8 complex in primary microRNA processing. *Genes Dev*. **18:**3016–3027.

Hansen, T., L. Olsen, M. Lindow, K.D. Jakobsen, H. Ullum, E. Jonsson, O.A. Andreassen, S. Djurovic, I. Melle and I. Agartz. 2007. Expressed microRNAs Implicated in Schizophrenia Etiology. *PLoS ONE*. **2:**e873.

Hayashita, Y., H. Osada, Y. Tatematsu, H. Yamada, K. Yanagisawa, S. Tomida, Y. Yatabe, K. Kawahara, Y. Sekido and T. Takahashi. 2005. A polycistronic microRNA cluster, miR-17-92, is overexpressed in human lung cancers and enhances cell proliferation. *Cancer Res.* **65:**9628–9632.

He, L., J.M. Thomson, M.T. Hemann, E. Hernando-Monge, D. Mu, S. Goodson, S. Powers, C. Cordon-Cardo, S.W. Lowe, G.J. Hannon and S.M. Hammond. 2005. A microRNA polycistron as a potential human oncogene. *Nature.* **435:**828–833.

He, S., Y., Zang, G. Skogerbo, F. Ren, H. Cui, H. Zhao, R. Chen and Y. Zhao. 2008. The properties and functions of virus encoded microRNA, siRNA and other small noncoding RNAs. *Crit. Rev. Microbiol.* 34(3-4):175–88.

Hébert, S.S., K. Horré, L. Nicolaï, A.S. Papadopoulou, W. Mandemakers, A.N. Silahtaroglu, S. Kauppinen, A. Delacourte and B. De Strooper. 2008. Loss of microRNA cluster miR-29a/b-1 in sporadic Alzheimer's disease correlates with increased BACE1/beta-secretase expression. *Proc. Natl. Acad. Sci. USA.* **105(17):**6415–20.

Houbaviy, H.B., M.F. Murray and P.A. Sharp. 2003. Embryonic stem cell-specific MicroRNAs. *Dev. Cell. Aug.* **5(2):**351–8.

Ikeda, S. and S.W. Kong, J. Lu, E. Bisping, H. Zhang, P.D. Allen, R.D. Golub, B. Pieske, W.T. Pu. 2007. Altered microRNA expression in human heart disease. *Physiol. Genomics.* **31:**367–373.

Jacob, F. and J. Monod. 1961 Genetic regulatory mechanisms in the synthesis of proteins. *J. Mol. Biol.* **3:**318-56.

Johnson, S.M., H. Grosshans, J. Shingara, M. Byrom, R. Jarvis, A. Cheng, E. Labourier, K.L. Reinert, D. Brown and F.J. Slack. 2005. RAS is regulated by the let-7 microRNA family. *Cell.* **120:**635–647.

Johnson, R., C. Zuccato, N.D. Belyaev, D.J. Guest, E. Cattaneo and N.J. Buckley. 2008. A microRNA-based gene dysregulation pathway in Huntington's disease. *Neurobiol. Dis.* **29:**438–445.

Kanellopoulou, C., S.A. Muljo, A.L. Kung, S. Ganesan, R. Drapkin, T. Jenuwein,

D.M. Livingston and K. Rajewsky. 2005. Dicer-deficient mouse embryonic stem cells are defective in differentiation and centromeric silencing. *Genes Dev.* **19(4):**489–501.

Karube, Y., H. Tanaka, H. Osada, S. Tomida, Y. Tatematsu, K. Yanagisawa, Y. Yatabe, J. Takamizawa, S. Miyoshi, T. Mitsudomi and T. Takahashi. 2005. Reduced expression of Dicer associated with poor prognosis in lung cancer patients. *Cancer Sci.* **96:**111–5.

Keating, M., K. Rai, L. Rassenti, T. Kipps, M. Negrini, F. Bullrich and C.M. Croce. 2002. Frequent deletions and downregulation of micro-RNA genes miR15 and miR16 at 13q14 in chronic lymphocytic leukemia. *Proc. Natl. Acad. Sci. USA.* **99:**15524–9.

Khvorova, A., A. Reynolds and S.D. Jayasena. 2003. Functional siRNAs and miRNAs Exhibit Strand Bias. *Cell.* **115:**209–216.

Kim, H.K., Y.S. Lee, U. Sivaprasad, A. Malhotra and A. Dutta. 2006. Muscle-specific microRNA miR-206 promotes muscle differentiation. *J. Cell. Biol.* **174(5):**677–87.

Kim, J., K. Inoue, J. Ishii, W.B. Vanti, S.V. Voronov, E. Murchison, G. Hannon and A. Abeliovich. 2007. A MicroRNA feedback circuit in midbrain dopamine neurons. *Science* **317:**1220–1224.

Kim, Y.K. and V.N. Kim. 2007. Processing of intronic microRNAs. *EMBO J.* **26:**775–783.

Kumar, M.S., J. Lu, K.L. Mercer, T.R. Golub and T. Jacks. 2007. Impaired microRNA processing enhances cellular transformation and tumorigenesis. *Nat. Genet.* **39:**673–7.

Lagos-Quintana, M., R. Rauhut, W. Lendeckel and T. Tuschl. 2001. Identification of novel genes coding for small expressed RNAs. *Science.* **294:**853–858.

Lagos-Quintana, M., R. Rauhut, J. Meyer, A. Borkhardt and T. Tuschl. 2003. New microRNAs from mouse and human. *RNA.* **9:**175–179.

Lai, E.C. 2002. MicroRNAs are complementary to 3'UTR motifs that mediate negative post-transcriptional regulation. *Nat Genet* **30:**363–364.

Landthaler, M., A. Yalcin and T. Tuschl. 2004. The human DiGeorge syndrome critical region gene 8 and Its *D. melanogaster* homolog are required for miRNA biogenesis. *Curr. Biol.* **14:**2162–2167.

Laneve, P., Di L. Marcotullio, U. Gioia, M.E. Fiori, E. Ferretti, A. Gulino, I. Bozzoni and E. Caffarelli. 2007. The interplay between microRNAs and the neurotrophin receptor tropomyosin-related kinase C controls proliferation of human neuroblastoma cells. *Proc. Natl. Acad. Sci. USA.* **104(19):**7957–62.

Lau, N.C., L.P. Lim, E.G. Weinstein and D.P. Bartel. 2001. An abundant class of tiny RNAs with probable regulatory roles in *Caenorhabditis elegans. Science.* **294:**858–862.

Lee, R.C., R.L. Feinbaum and V. Ambros. 1993. The *C. elegans* Heterochronic Gene lin-4 Encodes Small RNAs with Antisense Complementarity to lin-14. *Cell.* **75:**843–854.

Lee, R.C. and V. Ambros. 2001. An extensive class of small RNAs in *Caenorhabditis elegans. Science.* **294(5543):**862–4.

Lee, Y., K. Jeon, J.T. Lee, S. Kim and V.N. Kim. 2002. MicroRNA maturation: Stepwise processing and subcellular localization. *EMBO J.* **21:**4663–4670.

Lee, Y., C. Ahn, J. Han, H. Choi, J. Kim, J. Yim, J. Lee, P. Provost, O. Rådmark, S. Kim and V.N. Kim. 2003. The nuclear RNase III Drosha initiates microRNA processing. *Nature.* **425:**415–419.

Lee, Y., M. Kim, J. Han, K.H. Yeom, S. Lee, S.H. Baek and V.N. Kim. 2004. MicroRNA genes are transcribed by RNA polymerase II. *EMBO J.* **23:**4051–4060.

Lena, A.M., R. Shalom-Feuerstein, P. Rivetti di Val Cervo, D. Aberdam, R.A. Knight, G. Melino and E. Candi . 2008. miR-203 represses 'stemness' by repressing DeltaNp63. *Cell. Death Differ.* **15(7):**1187–95.

Lewis, B.P., I.H. Shih, M.W. Jones-Rhoades, D.P. Bartel and C.B. Burge. 2003. Prediction of mammalian microRNA targets. *Cell.* **115:**787–798

Lewis, B.P., C.B. Burge and D.P. Bartel. 2005. Conserved seed pairing, often flanked by adenosines, indicates that thousands of human genes are microRNA targets. *Cell.* **120:**15–20.

Liu, J., M.A. Carmell, F.V. Rivas, C.G. Marsden, J.M. Thomson, J.J. Song, S.M. Hammond, L. Joshua-Tor and G.J. Hannon. 2003. Argonaute2 is the catalytic engine of mammalian RNAi. *Science.* **305:**1437–1441.

Liu, J., M.A. Valencia-Sanchez, G.J. Hannon and R. Parker. 2005. MicroRNA-dependent localization of targeted mRNAs to mammalian P-bodies. *Nat. Cell. Biol.* **7(7):**719–23.

Llave, C., Z. Xie, K.D. Kasschau and J.C. Carrington. 2002. Cleavage of Scarecrow-like mRNA targets directed by a class of Arabidopsis miRNA. *Science.* **297(5589):**2053–6.

Lu, J., G. Getz, E.A. Miska, E. Alvarez-Saavedra, J. Lamb, D. Peck, A. Sweet-Cordero, B.L. Ebert, R.H. Mak, A.A. Ferrando, J.R. Downing, T. Jacks, H.R. Horvitz and T.R. Golub. 2005. MicroRNA expression profiles classify human cancers. *Nature.* **435:**834–838.

Lukiw, W.J. 2007. Micro-RNA speciation in fetal, adult and Alzheimer's disease hippocampus. *NeuroReport.* **18:**297–300.

Lund, E., S. Güttinger, A. Calado, J.E. Dahlberg and U. Kutay. 2004. Nuclear export of microRNA precursors. *Science.* **303(5654):**95-8.

Maselli, V., D. Di Bernardo and S. Banfi. 2008. CoGemiR: A comparative genomics microRNA database. *BMC Genomics.* **9:**457.

Matsubara, H., T. Takeuchi, E. Nishikawa, K. Yanagisawa, Y. Hayashita, H. Ebi, H. Yamada, M. Suzuki, M. Nagino, Y. Nimura, H. Osada and T. Takahashi. 2007. Apoptosis induction by antisense oligonucleotides against miR-17-5p and miR-20a in lung cancers overexpressing miR-17-92. *Oncogene.* **26(41):**6099–105.

Meister, G., M. Landthaler, A. Patkaniowska, Y. Dorsett, G. Teng and and T. Tuschl. 2004. Human Argonaute2 mediates RNA cleavage targeted by miRNAs and siRNAs. *Mol. Cell.* **15:**185–197.

Mishra, P.J., R. Humeniuk, P.J. Mishra, G.S. Longo-Sorbello, D. Banerjee and J.R. Bertino. 2007. A miR-24 microRNA binding-site polymorphism in dihydrofolate reductase gene leads to methotrexate resistance. *Proc. Natl. Acad. Sci. USA.* **104(33):**13513–8.

Mizuno, Y., K. Yagi, Y. Tokuzawa, Y. Kanesaki-Yatsuka, T. Suda, T. Katagiri, T. Fukuda, M. Maruyama, A. Okuda, T. Amemiya, Y. Kondoh, H. Tashiro and Y. Okazaki. 2008. miR-125b inhibits osteoblastic differentiation by down-regulation of cell proliferation. *Biochem Biophys Res Commun. Apr 4.* **368(2):**267–72.

Morlando, M., M. Ballarino, N. Gromak, F. Pagano, I. Bozzoni and N.J. Proudfoot. 2008. Primary microRNA transcripts are processed co-transcriptionally. *Nat. Struct. Mol. Biol.* **15:**902–909.

Moss, E.G., R.C. Lee and V. Ambros. 1997. The Cold Shock Domain Protein LIN-28 Controls Developmental Timing in *C. elegans* and Is Regulated by the lin-4 RNA. *Cell.* **88:**637–648.

Muralidhar, B., L.D. Goldstein, G. Ng, D.M. Winder, R.D. Palmer, E.L. Gooding, N.L. Barbosa- Morais, G. Mukherjee, N.P. Thorne, I. Roberts, M.R. Pett and N. Coleman. 2007. Global microRNA profiles in cervical squamous cell carcinoma depend on Drosha expression levels. *J. Pathol.* **212:**368–77.

Naga Prasad, S.V., Z.H. Duan, M.K. Gupta, V.S. Surampudi, S. Volinia, G.A. Calin, C.G. Liu, A. Kotwal, C.S. Moravec, R.C. Starling, D.M. Perez, S. Sen, Q. Wu, E.F. Plow, C.M. Croce and S. Karnik. 2009. Unique microRNA profile in end-stage heart failure indicates alterations in specific cardiovascular signaling networks. *J. Biol. Chem.* **284:**27487–27499.

Nana-Sinkam, P. and C.M. Croce. 2010. MicroRNAs in diagnosis and prognosis in cancer: what does the future hold? *Pharmacogenomics.* **11(5):**667–9.

Nakamachi, Y., S. Kawano, M. Takenokuchi, K. Nishimura, Y. Sakai, T. Chin, R. Saura, M. Kurosaka and S. Kumagai. 2009. MicroRNA 124a is a key regulator of proliferation and monocyte chemoattractant protein 1 secretion in fibroblast-like synoviocytes from patients with rheumatoid arthritis. *Arthritis Rheum.* **60:**1294–1304.

Okamura, K., A. Ishizuka, H. Siomi and M.C. Siomi. 2004. Distinct roles for Argonaute proteins in small RNA-directed RNA cleavage pathways. *Genes Dev.* **18(14):**1655–66.

Okamura, K., J.W. Hagen, H. Duan, D.M. Tyler and E. Lai. 2007. The mirtron pathway generates microRNA-class regulatory RNAs in Drosophila. *Cell.* **130:**89–100.

Olena, A.F. and J.G. Patton. 2010. Genomic organization of microRNAs. *J. Cell. Physiol.* **222(3):**540–5.

Ørom, U.A., F.C. Nielsen and A.H. Lund. 2008. MicroRNA-10a binds the 5'UTR of ribosomal protein mRNAs and enhances their translation. *Mol. Cell.* **30(4):**460–71.

Ota, A., H. Tagawa, S. Karnan, S. Tsuzuki, A. Karpas, K.A. O'Donnell, E.A. Wentzel, K.I. Zeller, C.V. Dang and J.T. Mendell. 2005. c-Myc-regulated microRNAs modulate E2F1 expression. *Nature.* 435:839–43, as a target for 13q31-q32 amplification in malignant lymphoma. *Cancer Res.* **64:**3087–3095.

Otaegui, D., S.E. Baranzini, R. Armañanzas, B. Calvo, M. Muñoz-Culla, P. Khankhanian, I. Inza, J.A. Lozano, T. Castillo-Triviño, A. Asensio, J. Olaskoaga and A. López de Munain. 2009. Differential micro RNA expression in PBMC from multiple sclerosis patients, *PLoS One.* **4(7):**e6309.

O'Donnell, K.A., E.A. Wentzel, K.I. Zeller, C.V. Dang and J.T. Mendell. 2005. c-Myc-regulated microRNAs modulate E2F1 expression. *Nature;* **435:**839–43.

Padgett, K.A., R.Y. Lan, P.C. Leung, A. Lleo, K. Dawson, J. Pfeiff, T.K. Mao, R.L. Coppel, A.A. Ansari and M.E. Gershwin. 2009. Primary biliary cirrhosis is associated with altered hepatic microRNA expression, *J. Autoimmun.* **32:**246–253.

Pasquinelli, A.E., B.J. Reinhart, F. Slack, M.Q. Martindale, M.I. Kuroda, B. Maller, D.C. Hayward, E.E. Ball, B. Degnan, P. Müller, J. Spring, A. Srinivasan, M. Fishman and J. Finnerty, J. Corbo, M. Levine, P. Leahy, E. Davidson and G. Ruvkun. 2000. Conservation of the sequence and temporal expression of let-7 heterochronic regulatory RNA. *Nature.* **408:**86–89.

Pauley, K.M., M. Satoh, A.L. Chan, M.R. Bubb, W.H. Reeves and E.K. Chan. 2008. Upregulated miR-146a expression in peripheral blood mononuclear cells from rheumatoid arthritis patients, *Arthritis Res. Ther.* **10(4):**R101.

Perkins, D.O., C.D. Jeffries, L.F. Jarskog, J.M. Thomson, K. Woods, M.A. Newman, J.S. Parker, J. Jin and S.M. Hammond. 2007. microRNA expression in the prefrontal cortex of individuals with schizophrenia and schizoaffective disorder. *Genome Biol.* **8:**R27.

Peters, L. and G. Meister. 2007. Argonaute proteins: mediators of RNA silencing. *Mol. Cell.* **26:**611–623.

Pfeffer, S., M. Zavolan, F.A. Grässer, M. Chien, J.J. Russo, J. Ju, B. John, A.J. Enright, D. Marks, C. Sander and T. Tuschl. 2004. Identification of virus-encoded microRNAs. *Science.* **304:**734–736.

Pfeffer, S., A. Sewer, M. Lagos-Quintana, R. Sheridan, C. Sander, F.A. Grässer, L.F. van Dyk, C.K. Ho, S. Shuman, M. Chien, J.J. Russo, J. Ju, G. Randall, B.D. Lindenbach, C.M. Rice, V. Simon, D.D. Ho, M. Zavolan and T. Tuschl. 2005. Identification of microRNAs of the herpesvirus family. *Nat. Methods.* **2:**269–276.

Raveche, E.S., E. Salerno, B.J. Scaglione, V. Manohar, F. Abbasi, Y.C. Lin, T. Fredrickson, P. Landgraf, S. Ramachandra, K. Huppi, J.R. Toro, V.E. Zenger, R.A. Metcalf and G.E. Marti. 2007. Abnormal microRNA-16 locus with synteny to human 13q14 linked to CLL in NZB mice. *Blood.* **109:**5079–86.

Reinhart, B.J., F.J. Slack, M. Basson, A.E. PasquinellI, J.C. Bettinger, A.E. Rougvie, H.R. Horvitz and G. Ruvkun. 2000. The 21-nucleotide let-7 RNA regulates developmental timing in *Caenorhabditis elegans. Nature.* **403:**901–906.

Reinhart, B.J., E.G. Weinstein, M.W. Rhoades, B. Bartel and D.P. Bartel. 2002. MicroRNAs in plants. *Genes Dev.* **16(13):**1616–26.

Rhoades, M.W., B.J. Reinhart, L.P. Lim, C.B. Burge, B. Bartel and D.P. Bartel. 2002. Prediction of plant microRNA targets. *Cell.* **110(4):**513–20.

Rodriguez, A., S. Griffiths-Jones, J.L. Ashurst and A. Bradley. 2004. Identification of mammalian microRNA host genes and transcription units. *Genome Res.* **14:**1902–1910.

Ruby, J.G., C.H. Jan and D.P. Bartel. 2007. Intronic microRNA precursors that bypass Drosha processing. *Nature.* **448:**83–86.

Ryan, B.M., A.I. Robles and C.C. Harris. 2010. Genetic variation in microRNA networks: the implications for cancer research. *Nat. Rev. Cancer.* **10(6):**389–402.

Saito, K., A. Ishizuka, H. Siomi and M.C. Siomi. 2005. Processing of pre-microRNAs by the Dicer-1-Loquacious complex in Drosophila cells. *PLoS Biol.* **3(7):**e235.

Saito, Y., G. Liang, G. Egger, J.M. Friedman and J.C. Chuang, G.A. Coetzee and P.A. Jones. 2006. Specific activation of microRNA-127 with downregu-lation of the proto-oncogene BCL6 by chromatin-modifying drugs in human cancer cells. *Cancer Cell.* **9:**435–443.

Saunders, M.A., H. Liang and W.H. Li. 2007. Human polymorphism at microRNAs and microRNA target sites. *Proc. Natl. Acad. Sci. USA.* **104(9):**3300–5.

Schwarz, D.S., G. Hutvágner, T. Du, Z. Xu, N. Aronin and P.D. Zamore. 2003. Asymmetry in the assembly of the RNAi enzyme complex. *Cell.* **115:**199–208.

Sevignani, C., G.A. Calin, S.C. Nnadi, M. Shimizu, R.V. Davuluri, T. Hyslop, P. Demant, C.M. Croce and L.D. Siracusa. 2007. MicroRNA genes are frequently located near mouse cancer susceptibility loci. *Proc. Natl. Acad. Sci. USA.* **104(19):**8017–22.

Sinkkonen, L., T. Hugenschmidt, P. Berninger, D. Gaidatzis, F. Mohn, C.G. Artus-Revel, M. Zavolan, P. Svoboda and W. Filipowicz. 2008. MicroRNAs control de novo DNA methylation through regulation of transcriptional repressors in mouse embryonic stem cells. *Nature Struct. Biol.* **15:**259–267.

Smirnova, L., A. Gräfe, A. Seiler, S. Schumacher, R. Nitsch and F.G. Wulczyn. 2005. Regulation of miRNA expression during neural cell specification. *Eur. J. Neurosci.* **21(6):**1469–77.

Stanczyk, J., E. Karouzakis, A. Jungel, C. Ospelt, C. Kolling, B. Michel, R. Gay, S. Gay and D. Kyburz. 2009. Mir-203 regulates the expression of IL-6 and matrix metalloproteinase(MMP)-1 in RA synovial fibroblasts. ACR/ARHP Scientific Meeting.

Stark, K.L., B. Xu, A. Bagchi, W.S. Lai, H. Liu, R. Hsu, X. Wan, P. Pavlidis, A.A. Mills, M. Karayiorgou and J.A. Gogos. 2008. Altered brain microRNA biogenesis contributes to phenotypic deficits in a 22q11-deletion mouse model. *Nat. Genet.* **40:**751–760.

Sucharov, C., M.R. Bristow and J.D. Port. 2008. miRNA expression in the failing human heart: functional correlates. *J. Mol. Cell. Cardiol.* **45:**185–192.

Suh, M.R., Y. Lee, J.Y. Kim, S.K. Kim, S.H. Moon, J.Y. Lee, K.Y. Cha, H.M. Chung, H.S. Yoon, S.Y. Moon, V.N. Kim and K.S. Kim. 2004. Human embryonic stem cells express a unique set of microRNAs. *Dev. Biol.* **270(2):**488–98.

Takamizawa, J., H. Konishi, K. Yanagisawa, S. Tomida, H. Osada, H. Endoh, T. Harano, Y. Yatabe, M. Nagino, Y. Nimura, T. Mitsudomi, T. Takahashi. 2004. Reduced expression of the let-7 microRNAs in human lung cancers in association with shortened postoperative survival. *Cancer Res.* **64:**3753–3756.

Thomson, J.M., M. Newman, J.S. Parker, E.M. Morin-Kensicki, T. Wright and S.M. Hammond. 2006. Extensive post-transcriptional regulation of microRNAs and its implications for cancer. *Genes Dev.* **20:**2202–7.

Thum, T., P. Galuppo, C. Wolf, J. Fiedler, S. Kneitz, L.W. vanLaake, P.A. Doevendans, C.L. Mummery, J. Borlak, A. Haverich, C. Gross, S. Engelhardt, G. Ertl and J. Bauersachs. 2007. MicroRNAs in the human heart: a clue to fetal gene reprogramming in heart failure. *Circulation.* **116:**258–267.

Vasudevan, S., Y. Tong and J.A. Steitz. 2007. Switching from repression to activation: microRNAs can up-regulate translation. *Science.* **318(5858):**1931–4.

Yang, N., G. Coukos and L. Zhang. 2008. MicroRNA epigenetic alterations in human cancer: one step forward in diagnosis and treatment. *Int. J. Cancer.* **122:**963–8.

Yi, R., Y. Qin, I.G. Macara and B.R. Cullen. 2003. Exportin-5 mediates the nuclear export of pre-microRNAs and short hairpin RNAs. *Genes Dev.* **17(24):**3011–6.

Yi, R., M.N. Poy, M. Stoffel and E. Fuchs. 2008. A skin microRNA promotes differentiation by repressing 'stemness'. *Nature. Mar 13.* **452(7184):**225–9.

Yuan, X., C. Liu, P. Yang, S. He, Q. Liao, S. Kang and Y. Zhao. 2009. Clustered microRNAs' coordination in regulating protein-protein interaction network. *BMC Syst. Biol.* **3:**65.

Wahid, F., A. Shehzad, T. Khan and Y.Y. Kim. 2010. MicroRNAs: Synthesis, mechanism, function and recent clinical trials. *Biochim. Biophys. Acta.* **1803(11):**1231–43.

Wang, Q., Z. Huang, H. Xue, C. Jin, X.L. Ju, J.D. Han and Y.G. Chen. 2008. MicroRNA miR-24 inhibits erythropoiesis by targeting activin type I receptor ALK4. *Blood.* **111:**588–595.

Wightman, B., I. Ha and G. Ruvkun. 1993. Posttranscriptional Regulation of the Heterochronic Gene lin-14 by lin-4 Mediates Temporal Pattern Formation in *C. elegans. Cell.* **75:**855–862.

Wong, C.F. and R.L. Tellam. 2008. MicroRNA-26a targets the histone methyltransferase Enhancer of Zeste homolog 2 during myogenesis. *J. Biol. Chem.* **283(15):**9836–43.

Wu, J. and X. Xie. 2006. Comparative sequence analysis reveals an intricate network among REST, CREB and miRNA in mediating neuronal gene expression. *Genome Biol.* **7(9):**R85.

Wu, M., N. Jolicoeur, Z. Li, L. Zhang, Y. Fortin, L.A. Denis, Z. Yu and S.H. Shen. 2008. Genetic variations of microRNAs in human cancer and their effects on the expression of miRNAs. *Carcinogenesis.* **29(9):**1710–6.

Zeng, Y. and B.R. Cullen. 2004. Structural requirements for pre-microRNA binding and nuclear export by Exportin 5. *Nucleic Acids Res.* **32:**4776–4785.

Zhang, H., F.A. Kolb, V. Brondani, E. Billy and W. Filipowicz. 2002. Human Dicer preferentially cleaves dsRNAs at their termini without a requirement for ATP. *EMBO J.* **21:**5875–5885.

Zhang, L., Huang, N. Yang, J. Greshock, M.S. Megraw, A. Giannakakis, S. Liang, T.L. Naylor, A. Barchetti, M.R. Ward, G. Yao, A. Medina, A. O'brien Jenkins, D. Katsaros, A. Hatzigeorgiou, P.A. Gimotty, Weber B.L. and G. Coukos. 2006. microRNAs exhibit high frequency genomic alterations in human cancer. *Proc. Natl. Acad. Sci. USA.* **103:**9136–9141.

Zhang, L., S. Volinia, T. Bonome, G.A. Calin, J. Greshock, N. Yang, C.G. Liu, A. Giannakakis, P. Alexiou, K. Hasegawa, C.N. Johnstone, M.S. Megraw, S. Adams, H. Lassus, J. Huang, S. Kaur, S. Liang, P. Sethupathy, A. Leminen, V.A. Simossis, R. Sandaltzopoulos, Y. Naomoto, D. Katsaros, P.A. Gimotty, A. DeMichele, Q. Huang, R. Bützow, A.K. Rustgi, B.L. Weber, M.J. Birrer, A.G. Hatzigeorgiou, C.M. Croce and G. Coukos. 2008. Genomic and epigenetic alterations deregulate microRNA expression in human epithelial ovarian cancer. *Proc. Natl. Acad. Sci. USA.* **105:**7004–7009.

Zhang, G.L., Y.X. Li, S.Q. Zheng, M. Liu, X. Li and H. Tang. 2010. Suppression of hepatitis B virus replication by microRNA-199a-3p and microRNA-210. *Antiviral Res.* [Epub ahead of print].

Zhao, Y. and D. Srivastava. 2007. A developmental view of microRNA function. *Trends Biochem Sci.* **32(4):**189–97.

Zhou B., S. Wang, C. Mayr, D.P. Bartel, and H.F. Lodish. 2007. miR-150, a microRNA expressed in mature B and T cells, blocks early B cell development when expressed prematurely. *Proc. Natl. Acad. Sci. USA.* **104:**7080–7085.

Chapter 9

A RNAi-based Genome-wide Screen to Discover Genes Involved in Resistance to *Tomato Yellow Leaf Curl Virus* (TYLCV) in Tomato

Henryk Czosnek*, Dagan Sade, Rena Gorovits, Favi Vidavski, Hila Beeri, Iris Sobol and Assaf Eybishtz

ABSTRACT

The cultivated tomato *Solanum lycopersicum* is under the threat of diseases caused by the *Tomato yellow leaf curl virus* (TYLCV). Several wild tomato species are resistant to TYLCV. Five loci, coined *Ty-1* to *Ty-5*, have been shown to be associated with resistance. Breeding for resistance consisted in introgressing resistance from the wild tomato species into *S. lycopersicum*. The genes conferring TYLCV resistance have not been identified so far.

We have proposed to use a TRV-VIGS RNAi-based genome screen to uncover the genes and gene networks underlying TYLCV resistance in tomato. We have postulated that genes preferentially expressed in resistant plants (compared to susceptible) and upregulated upon TYLCV infection, are part of resistance gene network(s). Moreover if a given gene has a critical role in the resistance network, silencing this gene will lead to the collapse of resistance. To decipher the networks we have used two inbred tomato lines issued from the same breeding program that used *S. habrochaites* as a source of resistance: one was resistant (R), the other was susceptible (S) to the virus. Sixty nine genes preferentially expressed in R tomatoes were identified by differential screening of cDNA libraries from infected and uninfected R and S tomato plants. From the twenty genes silenced so far, six answered to this criterion. Hence, not all the genes that have been found

* *Corresponding author e-mail*: czosnek@agri.huji.ac.il

to be preferentially expressed in R plants have the same cardinal role in the establishment of resistance to TYLCV.

We are presenting the results obtained with three genes: a permease, a sucrose transporter and a lipocalin-like protein. If we define a pathogen resistance gene as a gene that if silenced the host becomes susceptible, then it is obvious that TYLCV resistance is controlled by more than one gene. At this time there is no apparent biochemical and physiological link between these genes and the way each one affects the expression of the others is not known. Future investigations will focus on the known resistance pathways in plants and in the hierarchy of the genes we have discovered. Since, in our case, resistance to TYLCV has been introgressed from the wild tomato species *S. habrochaites*, it will be of interest to find out whether the genes preferentially expressed in R tomato plants have been introgressed from this wild tomato species.

Keywords: Gene network, resistance genes, tomato, virus resistance.

INTRODUCTION

Plants have evolved efficient defense mechanisms against biotic and abiotic stresses (Dangl et al. 2001, Takken et al. 2010, Walling 2000) regulated by different cross-communicating signaling pathways (Baker et al. 1997, Champigny et al.2009, Dong 2001, Gilliland et al. 2006, Kunkel et al. 2002, McCarty et al. 2000). Plants can mobilize panoply of responses to cope with virus invasion. Innate responses include triggering resistance gene products, local cell death and systemic acquired resistance (Whitham et al. 2006). RNA silencing is a universal defense mechanism. The first discovered natural function of RNA silencing was anti-viral response in plants (Carrington et al. 2001) wherein replication of RNA and DNA viruses is associated with the accumulation of virus-derived small RNAs that help cleave viral messengers in a sequence specific manner (Hamilton et al. 1999). This mode of RNA silencing was referred as posttranscriptional gene silencing (PTGS). The molecular basis of RNA silencing is being quickly clarified (Pratt et al. 2009). RNA silencing has helped explain the phenomena of recovery from virus infection (Lindbo et al. 1993) and of cross protection (Ratcliff et al. 1999). To evade gene silencing, viruses utilize various proteins, including proteases, coat proteins, replicative enhancers, and, most commonly, transport proteins and pathogenicity factors. These viral suppressors prevent the accumulation of short RNAs and abolish both local and systemic silencing (Dunoyer et al. 2004, Roth et al. 2004).

RNA silencing of viruses led to the development of an outstanding reverse genetic tool now widely used in plant biology, known as virus-induced gene silencing (VIGS). In plants infected with unmodified viruses the mechanism is specifically targeted against the viral genome. However, with virus vectors carrying inserts derived from host genes the process can be additionally targeted against the corresponding mRNAs (Lu et al.

2003). For instance, the endogenous *Phytoene desaturase* (PDS) mRNA could be silenced upon replication of the *Tobacco mosaic virus* (TMV) harboring a stretch of PDS (Kumagai et al. 1995). It was subsequently shown that a GUS transgene with homology to *Cauliflower mosaic virus* (CaMV) was silenced upon CaMV infection (Al-Kaff et al. 1998).

VIGS has emerged as one of the most reliable means to study gene silencing in plants (Burch-Smith et al. 2004, Purkayastha et al. 2009). One of the most common vectors currently used are based on the *Tobacco rattle virus* (TRV) (Burch-Smith et al. 2004). This technique is quite rapid and inexpensive and loss-of-function effects can be easily observed. Among other features, VIGS has been used to dissect the genetics of floral development (Dong et al. 2007), water deficit stress tolerance (Senthil-Kumar et al. 2008), embryogenesis, chlorophyll biosynthesis and disease resistance (Burch-Smith et al. 2004) and scent production (Spitzer et al. 2007).

TOMATO YELLOW LEAF CURL VIRUS (TYLCV); NATURAL RESISTANCE TO THE VIRUS

The cultivated tomato (*Solanum lycopersicum*) is widely grown in the world. Tomatoes are under the constant threat of diseases caused by the *Tomato yellow leaf curl virus* (TYLCV) and related viruses (Czosnek 2007). TYLCV has a single 2781 bp circular ssDNA genome encapsidated in a geminate particle (genus *Begomovirus*, family *Geminiviridae*) (Czosnek et al. 1988, Navot et al. 1991). TYLCV is solely transmitted by the whitefly *Bemisia tabaci* (Zeidan et al. 1991). It replicates in a rolling-circle fashion using a double stranded DNA intermediate replicative form as a template (Gutierrez 2000, Hanley-Bowdoin et al. 2000). The TYLCV genome encodes two open reading frames ORFs (V1 and V2); the complementary viral strand encodes four ORFs: C1 to C4 (Díaz-Pendón et al. 2010). A 300 nucleotide-long intergenic region IR contains the origin of replication and bidirectional promoters. V1 encodes the coat protein (CP); V2 encodes a movement protein (MP) and may also function as a silencing suppressor (Zrachya et al. 2007). C1 encodes a protein (Rep) necessary for replication protein, C2 a transcription activator protein (TrAP), C3 a replication enhancer protein (REn) and C4 a small protein embedded within the Rep (C2, C3 and C4 have also role in movement, transcription and replication (Gronenborn 2007).

S. lycopersicum is susceptible to the virus. Several accessions of wild tomato species such as *S. pimpinellifolium, S. chilense, S. peruvianum* and *S. habrochaites*, have been found to be resistant to TYLCV (Zakay et al. 1991). Domestication of wild tomato and selection for high yield and fruit quality led to the loss of alleles conferring resistance to biotic, including TYLCV and abiotic stresses (Bai et al. 2007, Sim et al. 2009). Breeding for resistance to TYLCV, which started about 40 years ago, consisted in re-introducing the resistance genes from the wild tomato species into *S. lycopersicum* lines selected for TYLCV resistance as well as for fruit quality and yield. The genetics of the resistance trait depends on the wild tomato genitor. *S. chilense* provides resistance determined by a semi dominant locus and several minor loci (Agrama et al. 2006, Zamir et al.1994). Resistance from *S. peruvianum* is controlled by 3-5 recessive loci (Friedmann et al. 1998). Resistance from *S. habrochaites* is under the control of a major dominant locus and several minor loci (Vidavski et al. 1998).

The TYLCV-resistant tomato genotypes obtained by intensive breeding possess chromosomal fragments from wild species on the background of the domesticated tomato karyotype (Bai et al. 2007). The chromosomal fragments from the wild, embedding the resistance-conferring loci, could be identified thanks to DNA polymorphism (Labate et al. 2009). Up to now, five major loci have been located on the tomato chromosome. The first identified locus, *Ty-1*, originating from *S. chilense* was mapped to chromosome 6; two modifiers were mapped to chromosomes 3 and 7 (Zamir et al. 1994). *Ty-2*, originating from *S. habrochaites*, was mapped to the long arm of chromosome 11 (Hanson et al. 2000). *Ty-3*, originating from *S. chilense*, was also mapped to chromosome 6 (Ji et al. 2006, Ji et al. 2007). *Ty-4*, also originating from *S. chilense*, was mapped to chromosome 3 (Ji et al. 2008, Ji et al. 2009). *Ty-5*, which originates from *S. peruvianum*, was mapped to chromosome 4; four additional loci were mapped to chromosomes 1, 7, 9 and 11 (Anbinder et al. 2009). None of these loci accounted for 100% of the resistance trait. The genes conferring TYLCV resistance at these, or other, loci have not been identified so far.

AN RNAi-BASED REVERSE GENETICS APPROACH TO DISCOVER GENES INVOLVED IN RESISTANCE TO *TOMATO YELLOW LEAF CURL VIRUS*

The discovery of multiple loci associated with resistance to TYLCV suggests that TYLCV resistance is sustained by a multigene interacting network. Plant resistance to viruses may be described as the outcome of interconnecting gene networks and signaling pathways leading to inhibition of virus replication and/or movement (Culver et al. 2007). We have used TRV-VIGS to decipherer the gene network underlying resistance to TYLCV. To identify genes involved in resistance to TYLCV, we hypothesized that these genes are expressed at higher levels in resistance than in susceptible lines. Further we assumed that if these genes are located at important nodes of the resistance network, silencing them would lead to the collapse of resistance. This approach proved fertile.

To discover genes involved in resistance to TYLCV, we took advantage of two inbred tomato lines issued from the same breeding program that used *S. habrochaites* as a source of resistance: the TYLCV-susceptible line 906–4 and the TYLCV-resistant line 902, respectively designated S and R hereafter (Vidavski et al. 1998). Upon whitefly-mediated inoculation of TYLCV, the R line remains symptomless and yielded, while the S line presents the typical disease symptoms of stunting, leaf yellowing and leaf curling and did not yield. The virus accumulated in the R plants with a velocity far lower than in the S plants; four weeks after inoculation, the S plants contained three orders of magnitude more virus than the R plants (Eybishtz et al. 2009). None of the polymorphic DNA markers tagging resistance to TYLCV (*Ty-1* to *Ty-5*) have been found to be linked with resistance of line R (unpublished). The presence or absence of a SNP in the *Mnu*I site of an *Hsp70* intron identified the tomato plants as R or S (Eybishtz et al. 2009).

In order to discover genes preferentially expressed in R plants (versus S plants), we prepared several cDNA libraries from S and R plants, before and after TYLCV inoculation. The libraries were made with RNA collected 1, 3, 5 and 7 days post-inoculation to represent the genes expressed at the early stages of infection, avoiding intensive cell damage due to infection (Michelson et al. 1997). From the ca. 300,000 clones assayed, 69 transcripts were found to represent genes preferentially expressed in R plants (Eybishtz et al. 2009). They were identified by sequence comparison with public databases. As a first approach we assumed that those genes which are known to be responsive to various pathogens and genes that encode membrane proteins such as receptors may be involved in the establishment of TYLCV resistance (Lucas 2006, Stange 2006).

We used the TRV-VIGS reverse genetics approach to find out whether the genes preferentially expressed in R plants are involved in resistance to TYLCV (Eybishtz et al. 2009, Eybishtz et al. 2010). This method uses a bipartite vector system designed between left and right borders of the Agrobacterium Ti -plasmid. *TRVI* contains the RNA-dependant RNA polymerase (RdRp) and the MP components of the virus whereas *TRVII* contains multiple cloning site (MCS) and the CP sequences. The bipartite plasmids are equipped with the 35S *Cauliflower Mosaic Virus* (CaMV) promoter and a *Nopaline synthase* (*NOS*) gene terminator. The multiple cloning sites (MCS) in *TRVII* allow ligation of DNA target sequences that will induce PTGS in the plant upon delivery by agroinoculation (Liu et al. 2002). The multiplication of the vector in the plant tissue triggers the cleavage of target sequence resulting in loss of expression (Baulcombe

1999, Burch-Smith et al. 2004, Ratcliff et al. 2001). VIGS has proven multiple advantages as a high through-put functional genomics tool (Benedito et al. 2004) in the discovery of gene networks in plants (Briggs et al. 2005, Robertson 2004, Wellmer et al. 2005).

At present, we have silenced 20 out of the 69 genes preferentially expressed in R plants. Fragments of 150 to 200 bp of the target genes were cloned in the *TRVII* vector. The *TRVI* and recombinant *TRVII* vectors were delivered to R and S tomato plants by agroinoculation (20 days after sowing). Seven days later, the plants were caged with viruliferous whiteflies for a three-day-long inoculation period; at this time, the RNAi signal was conspicuous in the plant leaves and its amount remained significant thereafter (40). The effect of silencing was appraised during the next 40 days. Assuming that if these genes are sustaining resistance to TYLCV in R plants, silencing them would lead to the collapse of resistance upon TYLCV infection. Six genes answered to this premise. Hence, it seems that many genes are involved in the establishment of natural resistance to TYLCV. We present here the behavior of four genes preferentially expressed in R plants upon silencing and TYLCV inoculation.

TRV-VIGS SILENCING OF A SINGLE GENE IN TOMATO PLANTS RESISTANT TO TYLCV RENDERS THEM SUSCEPTIBLE TO THE VIRUS

Permease I-like protein

The first gene we choose to silence was a *Permease I-like protein* (Figure 1A). This gene has been known to be involved in lipid transfer from endoplasmic reticulum (ER) to thylakoid in *Arabidopsis* (Xu et al. 2003). *Permease-I like protein* was preferentially expressed in non-inoculated R plants (compared to S plants) and was strongly upregulated upon TYLCV inoculation (Eybishtz et al. 2009). *Permease I-like protein* was silenced in R plants. The levels of *Permease-I* transcripts, as measured by real-time quantitative PCR (qPCR), were much depleted in the silenced R plants. Therefore, at the time the R plants were inoculated with TYLCV, about one week after TRV-VIGS treatment, the expression of *Permease I like protein* was inhibited. Silencing of *Permease I-like protein* led to the collapse of the resistant phenotype and the treated R plants presented the typical characteristic of infected S plants. Four weeks after TYLCV inoculation, the silenced *Permease I-like protein* R plants ceased growing, they developed typical yellowing and curling of leaves and they contained amounts of virus similar to those measured in infected S plants. Hence, shown for the first time, silencing a single gene can lead to the loss of TYLCV resistance in tomato plants (Eybishtz et al. 2009). It is difficult

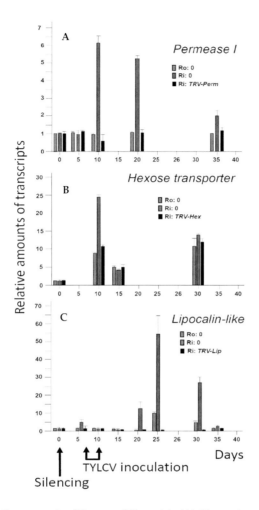

Figure 1. Relative amounts of *Permease I-like protein* (A), Hexose transporter *LeHT1* (B) and lipocalin-like protein (C) gene transcripts in TYLCV-resistant tomato plants R as determined by quantitative real-time PCR. Tubulin RNA was used as a reference gene transcript for each of the plants analyzed. Twenty days after sowing plants were silenced (day 0); plants were inoculated with viruliferous whiteflies for a three-day-long period (day 7 to 10). Ro:0, untreated and not infected plants; Ri:0 infected untreated plants; Ri. *TRV-Perm* infected silenced plants. The amount of transcript immediately before silencing (at day 0) is taken as 1. Bars represent the average of triplicate measures of three different plants; the standard error is indicated.

(Color image of this figure appears in the color plate section at the end of the book.)

to provide a simple model describing the role of *Permease I-like protein* in the gene network and signaling pathways involved in resistance to TYLCV. We may assume that the permease-I like protein is involved in the movement of macromolecular or metabolites that can be a part of the signaling pathway leading to resistance.

Pectin methylesterase

Another gene preferentially expressed in R plants was a *Pectin methylesterase*. An homologue of this gene has been found to be involved in plant-virus interactions, in virus movement and in resistance of *Nicotiana tabacum* to *Tobacco mosaic virus* (TMV) in (Chen et al. 2000, Chen et al. 2003). This gene also plays a role in silencing the plant host defense against foreign nucleic acid entering the cell (Dorokhov et al. 2006). Contrary to the *Permease I-like protein* gene, silencing the *Pectin methylesterase* gene had no influence on the resistance abilities of the R plant against TYLCV. The plants remained symptomless and the amounts of virus were as low as in the non-silenced infected R plants. Hence *Pectin methylesterase* maybe part of the resistance pathway but this gene is not located at a bottleneck of the resistance network. Thus, not all the genes that are preferentially expressed in R plants play a cardinal role in the establishment of resistance to TYLCV.

Hexose transporter LeHT1

In tomato, the interest in hexose transporter was related mainly to sugar accumulation in the tomato fruit as desirable commercial trait (Dibley et al. 2005). However, it is well known that sugars mediate transduction signals in plants (Price et al. 2004, Sinha et al. 2002, Smeekens 1998, Smeekens et al. 1997). One of the genes preferentially expressed in R plants was the Hexose transporter *LeHT1*, one of the three known tomato hexose transporter genes *LeHT1, LeHT2, LeHT3* (Gear et al. 2000). Plant hexose transporters are plasma membrane carriers, which function as proton/hexose symporters (Bush 1993) and are involved in life cycle processes (Büttner et al. 2000) as well as in pathogenesis (Wahl et al. 2010). Long-distance and inter-cellular transport of sugars is mediated by proton-coupled sucrose transporters (belonging to the disaccharide transporter family) and hexose and polyol transporters (belonging to the monosaccharide transporter family) (Lecourieux et al. 2010). Sugar signaling cascades are also involved in pathogen defense and programmed cell death (PCD) (Fotopoulos et al. 2003, Nørholm et al. 2006). Hexokinases can sense soluble hexoses and regulate PCD (Kim et al. 2006), providing a direct link between carbohydrate metabolism and the plant defense response (Bolton 2009).

LeHT1 expression was in R plants was developmentally regulated in R plant leaves. The gene was upregulated about 2.5 folds upon inoculation of TYLCV, compared to non-inoculated plants of the same age (Figure 1B). Silencing *LeHT1* abolished the virus-induced upregulation. *LeHT1* silencing caused growth inhibition. As a mirror image, over-expression of the hexose transporter gene *STP13* in *Arabidopsis* improved plant growth (Schofield et al. 2009). The amount of TYLCV in the infected *LeHT1*-silenced R plants was similar to that of the infected S plants. *In situ* localization of TYLCV confirmed that silencing of *LeHTI* dramatically increased virus spread (Eybishtz et al. 2010), confirming that *LeHT1* has a role in confining the virus in the infected cells and emphasizing the role of the hexose transporters as part of a defense mechanism limiting virus movement. *LeHT1* silencing affected the concentration of sucrose (but not that of glucose and fructose), which was significantly lower in silenced than in non-silenced R tomatoes. Therefore, alteration of sugar transport as a result of *LeHT1* silencing may lead to an increase in phloem loading and permeability of plasmodesmata, enhancing cell-to-cell and long-distance spread of TYLCV.

Necrosis started to appear on the stem and petioles of the *LeHT1*-silenced R plants about three weeks after TYLCV inoculation. Necrosis has not been reported before to accompany infection of tomato (susceptible or resistant) by any of the tomato-infecting begomoviruses of the species TYLCV or *Tomato leaf curl virus* (Abhary et al. 2007). Necrosis occurred in *LeHT1*-silenced R plants only after TYLCV infection; it did not appear in non-infected *LeHT1*-silenced R plants and in infected and non-infected *LeHT1*-silenced S tomatoes. Furthermore, necrosis also appeared when the *LeHT1*-silenced R plants were infected with different RNA and DNA viruses (*Bean dwarf mosaic virus* BDMV, *Cucumber mosaic virus* CMV and TMV) (Eybishtz et al. 2010). Therefore, once resistance to TYLCV has been impaired by *LeHT1* silencing, the necrotic response developed by R plants was no more specific of TYLCV.

Necrosis is often associated with PCD, a phenomenon associated with chromatin condensation and nuclear fragmentation (Gunawardena et al. 2004). DNA extracted from stem necrotic tissues of TYLCV-infected *LeHT1*-silenced R plants presented a typical laddering pattern, which was absent in the DNA from non-silenced infected R plants. Among the MAPKs, the Jun N-terminal kinases (JNKs) have been involved in cell proliferation and apoptosis of various mammalian cells (Behrens et al. 2001, Davis 2000). Using an antibody against human JNK, we found that the amount of JNK-like protein was higher in the stem of R plants than in that of S plants. The amount of JNK increased upon TYLCV infection of *LeHT1*-silenced R plants suggesting that JNK is part of the signaling cascade leading to stem necrosis (Zhang et al. 2001). Apoptosis was confirmed by detecting H_2O_2, a molecule formed when reactive oxygen species (ROS) are released during oxidative bursts accompanying the induction of the plant anti-oxidative defenses (Levine et al. 1994, Sasabe et al. 2000). H_2O_2 production can be visualized by using the DAB stain; the intensity of the brownish color correlates with the hypersensitive response (HR) and PCD (Alvarez et al. 1998, Rivas et al. 2000). The necrotic tissues of *LeHT1*-silenced R plants presented a dark brown color in the vascular system, which was enhanced following TYLCV infection. Hence, *LeHT1* silencing increased the anti-oxidative defense response, which was exacerbated by TYLCV infection.

In most cases, PCD minimizes the pathogen spread by a combination of a layer of dead cells and induction of systemic acquired resistance (Chisholm et al. 2006, Greenberg et al. 2004, Hofius et al. 2007). In the case of infection of *LeHT1*-silenced R plants, the plant defense mechanisms were unable to confine virus infection and the resistance collapsed. This study has uncovered two layers of plant defense against TYLCV infection: inhibition of virus spread and necrosis.

Lipocalin-like protein

A cDNAs representing a gene preferentially expressed in R plants had a sequence similar to a tomato clone from GenBank (accession number AK323915.1) that had no homology with any known functional gene or protein (Aoki et al. 2010). However, modeling the three-dimensional structure of the putative translated protein pointed to a lipocalin with its typical barrel-shaped architecture (Kelley et al. 2009). The lipocalin-like gene from the T and S tomato lines was coined *SlVSRLip*. Lipocalins are ancient proteins found in bacteria, invertebrates, vertebrates, green algae and plants (Åkerstrom et al. 2000, Charron et al. 2005). Members of this protein family are very diverse and share low amino acid sequence identity. The biological functions of lipocalins are closely related to the ligands they bind (Flower et al. 2000). The first true plant lipocalins were identified in wheat *Triticum aestivum* (*TaTIL-1*) and *Arabidopsis thaliana* (*AtTIL*) (Charron et al. 2002). Their expression was induced upon exposure to low temperature; high salt and wounding treatments had no measurable effect (Charron et al. 2008, Chi et al. 2009). It has been suggested that these lipocalins act as scavengers of harmful molecules induced by temperature stress, preventing lipid peroxidation of membranes (Charron et al. 2008, Levesque-Tremblay et al. 2009). Tomato lipocalins have been identified initially by sequence homology with *TaTIL-1* (Charron et al. 2009). Proteomic studies have shown that tomato lipocalins are associated with cold (84) and salt (Chen et al. 2009) stress responses. Some tomato lipocalins are developmentally regulated during fruit development and ripening (Faurobert et al. 2007).

SlVSRLip behaved differently than the tomato lipocalins described so far. *SlVSRLip* was developmentally regulated in the leaves of S and R tomato genotypes. In both genotypes, the gene was significantly expressed during a time window lasting about two weeks, starting about 40 days after sowing, peaking approximately one week later and decreasing thereafter to reach basal levels 55 days after sowing (Figure 1C). In our experimental conditions, *SlVSRLip* was not significantly regulated by heat or cold stress or by salt, either in R or S tomato plants. It has to be noted that the stress applied to the R and S plants was rather mild and limited in time compared to the treatments inflicted in other investigations (Chen et al. 2009, Page et al. 2010).

SlVSRLip of R plants was upregulated following whitefly feeding on leaves of R plants. Insects did not change the developmental pattern of *SlVSRLip* expression; however the levels of expression were approximately three times higher than in control R plants. It is interesting to note that although whiteflies fed on the R plants for three days, starting 28 days after sowing, the expression of *SlVSRLip* started to increase 10 days later only. Inoculating R tomatoes with TYLCV by feeding R plants with viruliferous whiteflies also did not change the pattern of *SlVSRLip* expression, but the levels of expression were 5-6 times higher than that of non-infected R plants. Strikingly, the expression of *SlVSRLip* in S tomato plants was affected neither by non-viruliferous nor by viruliferous whiteflies.

This study was the first showing that a lipocalin-like gene is upregulated in plants as a response to insect feeding, whether carrying viruses or not. The changes in *Arabidopsis* transcriptome in response to whitefly (B biotype) nymph and to aphid feeding have been studied using microarrays (Kempema et al. 2007). Lipocalin-like transcripts were not found to be up- or down-regulated by whitefly nymphs, perhaps because lipocalins are not significantly regulated in *Arabidopsis* or because the expression of this gene is triggered only by adult feeding.

The association of lipocalins with virus infection has not been reported before. To study the role of *SlVSRLip* in R tomato plants, the gene was silenced immediately after planting. *SlVSRLip* expression was totally inhibited for the next 45 days, at least (Figure 1C). Silencing of *SlVSRLip* of R plants had dramatic effects. Following TYLCV inoculation, infected *SlVSRLip*-silenced R plants ceased to grow and within 4 weeks after inoculation presented symptoms similar to those of infected S plants. The amount of virus in the infected silenced R plants increased by five orders of magnitude compared to infected non-silenced R tomatoes. Hence silencing of the SlVSRLip gene in R plants changed the plant phenotype from TYLCV-resistant to TYLCV-susceptible. As in the case of the sucrose transporter *LeHT1*, *SlVSRLip*-silenced R plants presented a necrotic response along the stems and petioles, an apoptotic response accompanied by ROS emission (Eybishtz et al. 2010). These results indicated that the R lipocalin-like gene product may have a protective effect against damages induced by TYLCV. Directly or indirectly, over-expression of lipocalin induced by viruliferous whitefly feeding keeps the virus titer in leaves of R plants low enough to avoid damages to tomato tissues and cells, preventing stunting and symptom development. In contrast, *SlVSRLip*, which is barely expressed in the leaves of S tomato plants, is unable to prevent virus accumulation and disease in S plants.

CONCLUSION

Pathogen infection affect many metabolic pathways and cellular activities include membrane disorganization, increase in levels of toxic metabolites, inhibition of nutrient uptake and photosynthesis, generation of ROS and ultimately cell and plant death. Several tens of genes conferring resistance to plant pathogens have been identified (Hulbert et al. 2001, McDowell et al. 2003). The study of plant pathogen interactions have been accelerated by high-throughput gene expression platforms (Restrepo et al. 2005).

Reverse genetics is a powerful tool that establishes a direct link between the biochemical function of a gene product and its role *in vivo*. Several approaches have been developed for plants, some of which are applicable to many species and each of which has advantages and limitations. T-DNA and transposon insertion lines make those resources attractive (Alonso et al. 2003, Gilchrist et al. 2010). In addition "mutation libraries" derived from ethyl methane sulfonate (EMS) and fast-neutron bombardment have been generated, facilitating reverse genetics analyses (Menda et al. 2004). All have their advantages and disadvantages depending on the species and researcher and the questions being asked. The low cost of VIGS and RNAi makes it attractive for studying genes of unknown function. Here, the viral transcripts harboring the targeted gene sequence spreads in plants allowing the systemic silencing of the targeted gene in the plant tissues (Benedito et al. 2004, Liu et al. 2002).

TYLCV is a geminivirus species complex whose members cause severe crop losses worldwide. Domestication of tomato from the wild and selection for high yield led to the loss of many biotic, including TYLCV and abiotic resistant alleles, which might have been linked with poor fruit quality. It has been postulated that resistance to viruses is the result of gene networks that respond to biochemical triggers induced by virus inoculation (Genoud et al. 1999, Goldbach et al. 2003, Murphy et al. 2001). Hence, the genetic erosion accompanying domestication has disrupted the gene networks sustaining resistance to viruses. Breeding tomato varieties for resistance to TYLCV has consisted in identifying resistance traits in the wild tomato species and introgressing the resistance into the susceptible cultivated tomato. As a result of the crossing and selection processes, the networks sustaining virus resistance have been (partially?) reconstituted. Five major genetic loci conferring resistance to TYLCV in tomato have been tagged by polymorphic DNA markers. The genes conferring TYLCV resistance at these loci have not been identified so far.

We have proposed an original approach to uncover the genes and gene networks underlying TYLCV resistance in tomato. We have suggested that the differential gene expression patterns in susceptible and resistant tomato plants may provide hints on the involvement of certain in genes in the establishment of resistance. We have postulated that genes preferentially expressed in resistant plants (compared to susceptible) and upregulated upon TYLCV infection, may be part of resistance gene network(s). To decipher the networks we have used two inbred tomato lines issued from the same breeding program for TYLCV-resistance that used *S. habrochaites* (accessions LA1777 and LA386) as a source of resistance: one was resistant to whitefly-mediated inoculation of TYLCV (R), the other was susceptible (S) to the virus (Vidavski et al. 1998).

In order to discover genes preferentially expressed in R plants (versus S plants), we have prepared four cDNA libraries: from uninfected R and S plants and from infected R and S tomatoes. Sixty-nine deferentially expressed genes were identified (Eybishtz et al. 2009). In order to identify the genes involved in TYLCV resistance in R plants and subsequently to peel the resistance layers, we have used the TRV-VIGS approach. The rationale was that if a given gene has a critical role in the resistance network and has a dominant effect in R tomatoes, silencing this gene will lead to the collapse of resistance and to the appearance of a susceptible phenotype. From the twenty genes tested, six answered to this criterion. Hence, not all the genes that have been found to be preferentially expressed in R plants have the same cardinal role in the establishment of resistance to TYLCV. We have presented the results obtained with three genes that do not seem to be directly interconnected: a permease, a sucrose transporter and lipocalin-like protein. All three are more expressed in R than in S plants and are strongly upregulated in R tomatoes upon whitefly-mediated inoculation of TYLCV. The permease and the sucrose transporter genes are immediately upregulated, while lipocalin-like transcripts start to accumulate significantly only 10 days after TYLCV inoculation. The magnitude of gene expression is also different: x6 for the permease, x25 for the sucrose transporter and x55 for the lipocalin. When these genes are silenced, resistance to TYLCV collapses upon virus inoculation. Infection of R plants where the sucrose transporter and lipocalin-like protein genes have been silenced induces an apoptotic response, but not in permease-silenced plants. Hence silencing some of the genes involved in TYLCV resistance in R plants uncovers a second line of defense, which at the end, is ineffective to confine the viral disease. In general, plants express natural resistance pathways against pathogen by inducing a multi layer response that starts from the basal response and production of general pathogen-associated molecular pattern molecules (PAMPs) followed by activation of MAPK signaling cascades and production of anti-microbial compounds. The next layer of resistance usually involves gene expression related to the plant immune response; at this point, the defense system shows more specificity against a particular pathogen and may involve a PCD response (van Doorn et al. 2005).

If we define a pathogen resistance gene as a gene that if silenced the host becomes susceptible, then it is obvious that TYLCV resistance is controlled by more than one gene. At this time there is no obvious biochemical and physiological link between these genes and the way each one affects the expression of the others is not known. Future investigations will focus on the known resistance pathways in plants and in the hierarchy of the genes we have discovered. It is feasible to silence one of the genes and measure the effect of the expression of other. It is also desirable to examine the expression of the genes that network with them, as can be found in the Kyoto Encyclopedia of Genes and genomes—KEGG (and other softwares for system biology such as ariadnegenomics and atted).

Since in our case resistance to TYLCV has been introgressed from wild tomato species, *S. habrochaites*, it will be of interest to find out whether the genes preferentially expressed in R tomato plants have been introgressed from *S. habrochaites*. It might be possible to clone the gene from the resistant wild tomato genitor and from the susceptible domesticated tomato using genomic libraries. Of particular interest is the sequence of the promoters and of the putative introns. Sequence comparisons of gene homologues from *S. habrochaites* and *S. lycopersicum* may allow discovering differences accounting for the resistance to TYLCV. Finding that the genes preferentially expressed in R plants originate from the wild tomato genitor will confirm our hypothesis that breeding for resistance has allowed reconstituting the TYLCV resistance network. Sequence comparisons may also be instrumental in developing polymorphic DNA markers that tag resistance. Such markers may be useful in breeding for resistance.

ACKNOWLEDGEMENTS

This research was supported by a grant from the U.S. Agency for International Development, Middle East Research and Cooperation (MERC) program to H.C. (GEG-G-00-02-00003-00), Project M21037.

The authors thank Prof. David Baulcombe and the Gatsby Charitable Foundation, The Sainsbury Laboratory, for providing the TRV vectors.

REFERENCES

Abhary, M., B.L. Pati and C.M. Fauquet. 2007. Molecular biodiversity, taxonomy and nomenclature of Tomato yellow leaf curl-like viruses. In: Czosnek H, editor, *Tomato Yellow Leaf Curl Virus Disease: Management, Molecular Biology, Breeding for Resistance*. Dordrecht, The Netherlands: Springer; pp. 85–118.

Agrama, H.A. and J.W. Scott. 2006. Quantitative trait loci for *Tomato yellow leaf curl virus* and *Tomato mottle virus* resistance in tomato. *Journal of the American Society for Horticultural Sciences*. **131**:267–272.

Åkerstrom, B.D., R. Flower and J.P. Salier. 2000. Lipocalins: unity in diversity. *Biochimica et Biophysica Acta*. **1482**:1–8.

Al-Kaff, N.S., S.N. Covey, M.M. Kreike, A.M. Page, R. Pinder and P.J. Dale. 1998. Transcriptional and posttranscriptional plant gene silencing in response to a pathogen. *Science*. **279**:2113–2115.

Alonso, J.M., A.N. Stepanova and T.J. Leisse. 2003. Genome-wide insertional mutagenesis of *Arabidopsis thaliana*. *Science*. **301**:653–657.

Alvarez, M.E., R.I. Pennell, P.J. Meijer, A. Ishikawa, R.A. Dixon and C. Lamb. 1998. Reactive oxygen intermediates mediate a systemic signal network in the establishment of plant immunity. *Cell*. **92**:773–84.

Anbinder, I., M. Reuveni, R. Azari, I. Paran, S. Nahon, H. Shlomo, L. Chen, M. Lapidot and I. Levin. 2009. Molecular dissection of *Tomato yellow leaf curl virus* (TYLCV) resistance in tomato line TY172 derived from *Solanum peruvianum*. *Theoretical and Applied Genetics*. **119**:519–530.

Aoki, K., K. Yano and A. Suzuki. 2010. Large-scale analysis of full-length cDNAs from the tomato (*Solanum lycopersicum*) cultivar Micro-Tom, a reference system for the Solanaceae genomics. *BMC Genomics*. **11**:210.

Bai, Y. and P. Linghout. 2007. Domestication and breeding of tomatoes: what have we gained and what can we gain in the future? *Annals of Botany*. **100**:1085–1094.

Baker, B., P. Zambryski, B. Staskawicz and S.P. Dinesh-Kumar. 1997. Signaling in plant-microbe interactions. *Science*. **276**:726–733.

Baulcombe, D.C. 1999. Fast forward genetics based on virus induced gene silencing. *Current Opinion in Plant Biology*. **2**:109–13.

Behrens, A., K. Sabapathy, I. Graef, M. Cleary, G.R. Crabtree and E.F. Wagner. 2001. Jun N-terminal kinase 2 modulates thymocyte apoptosis and T cell activation through c-Jun and nuclear factor of activated T cell (NF-AT). *Proceedings of the National Academy of Sciences, USA*. **98**:1769–1774.

Benedito, V.A., P.B. Visser, G.C. Angenent and F.A. Krens. 2004. The potential of virus-induced gene silencing for speeding up functional characterization of plant genes. *Genetic and Molecular Research*. **3**:323–341.

Bolton, M.D. 2009. Primary metabolism and plant defense-fuel for the fire. *Molecular Plant-Microbe Interactions.* **22:**487–497.

Briggs, S.P. and T. Singer. 2005. Genetic networks. *Plant Physiology.* **138:**542–544.

Burch-Smith, T., M. Schiff, Y. Liu and S.P. Dinesh-Kumar. 2004. Efficient virus-induced silencing in Arabidopsis. *Plant Physiology.* **142:**21–27.

Bush, D.R. 1993. Proton-coupled sugar and amino acid transporters in plants. *Annual Review of Plant Physiology and Plant Molecular Biology.* **44:**513–542.

Büttner, M. and N. Sauer. 2000. Monosaccharide transporters in plants: structure, function and physiology. *Biochimica et Biophysica Acta– Biomembranes.* **1465:**263–274.

Carrington, J.C., K.D. Kasschau and L.K. Johansen. 2001. Activation and suppression of RNA silencing by plant viruses. *Virology.* **281:**1–5.

Champigny, M. and R. Cameron. 2009. Action at a distance: Long-distance signals in induced resistance. *Advances in Botanical Research.* **51:**124–156.

Charron, J.B.F., G. Breton, M. Badawi and F. Sarhan. 2002. Molecular and structural analyses of a novel temperature stress-induced lipocalin from wheat and Arabidopsis. *FEBS Letters.* **517:**129–132.

Charron, J.B.F., F. Ouellet, M. Pelletier, J. Danyluk, C. Chauve and F. Sarhan. 2005. Identification, expression and evolutionary analyses of plant lipocalins. *Plant Physiology.* **139:**2017–2028.

Charron, J.B.F., F. Ouellet, M. Houde and F. Sarhan. 2008. The plant Apolipoprotein D ortholog protects Arabidopsis against oxidative stress. *BMC Plant Biology.* **8:**86.

Chen, M., J. Sheng, G. Hind, A. Handa and V. Citovsky. 2000. Interaction between the tobacco mosaic virus movement protein and host cell pectin methylesterases is required for viral cell-to-cell movement. *The EMBO Journal.* **19:**913–920.

Chen, M. and V. Citovsky. 2003. Systemic movement of a tobamovirus requires host cell pectin methylesterase. *The Plant Journal.* **35:**386–392.

Chen, S., N. Gollop and B. Heuer. 2009. Proteomic analysis of salt-stressed tomato (*Solanum lycposersicum*) seedlings: effect of genotype and exogenous application of glycinebetaine. *Journal of Experimental Botany.* **60:**2005–2019.

Chi, W.T., R. W.M. Fung, H.S. Liu, C.C. Hsu and Y.Y. Charng. 2009. Temperature-induced lipocalin is required for basal and acquired thermotolerance in Arabidopsis. *Plant Cell and Environment.* **32:**917–927.

Chisholm, S.T., G. Coaker, B. Day and B.J. Staskawicz. 2006. Host-microbe interactions: shaping the evolution of the plant immune response. *Cell.* **124:**803–814.

Culver, J.N. and M.S. Padmanabhan. 2007. Virus-induced disease: altering host physiology one interaction at a time. *Annual Review of Phytopathology.* **45:**221–243.

Czosnek, H. 2007. *Tomato yellow leaf curl virus disease: management, molecular biology, breeding for resistance.* Dordrecht, The Netherlands: Springer.

Czosnek, H., R. Ber, Y. Antignus, S. Cohen, N. Navot and D. Zamir. 1988. Isolation of tomato yellow leaf curl virus, a geminivirus. *Phytopathology.* **78:**508–512.

Dangl, J.L. and D.G. Jones. 2001. Plant pathogens and integrated defense responses to infection. *Nature.* **411:**826–833.

Davis, R.J. 2000. Signal Transduction by the JNK Group of MAP Kinases. *Cell.* **103:**239–252.

Díaz-Pendón, J.A., M.C. Cañizares, E. Moriones, E.R. Bejarano, H. Czosnek and J. Navas-Castillo. 2010. Tomato yellow leaf curl viruses: *ménage à trois* between the virus complex, the plant and the whitefly vector. *Molecular Plant Pathology.* **11:**441–450.

Dibley, S.J., M.L. Gear, X. Yang, E.G. Rosche, C.E. Offler, D.W. McCurdy and J.W. Patrick. 2005. Temporal and spatial expression of hexose transporters in developing tomato (*Lycopersicon esculentum*) fruit. *Functional Plant Biology.* **32:**777–785.

Dong, X. 2001. Genetic dissection of systemic acquired resistance. Current Opinions in *Plant Biology.* **4:**309–314.

Dong, Y., T.M. Burch-Smith, Y. Liu, P. Mamillapalli and S.P. Dinesh-Kumar. 2007. A ligation-independent cloning Tobacco Rattle Virus vector for high-throughput virus-induced gene silencing identifies roles for *NbMADS4-1* and *-2* in floral development. *Plant Physiology.* **145:**1161–1170.

Dorokhov, Y.L., P.A. Ivanov, T.V. Komarova, M.V. Skulachev and J.G. Atabekov. 2006. An internal ribosome entry site located upstream of the crucifer-infecting tobamovirus coat protein (CP) gene can be used for CP synthesis *in vivo*. *Journal of General Virology.* **87:**2693–2697.

Dunoyer, P., C.H. Lecellier, E.A. Parizotto, C. Himber and O. Voinnet. 2004. Probing the microRNA and small interfering RNA pathways with virus-encoded suppressors of RNA silencing. *Plant Cell.* **16:**1235–1250.

Eybishtz, A., Y. Peretz, D. Sade, F. Akad and H. Czosnek. 2009. Silencing of a single gene in tomato plants resistant to *Tomato yellow leaf curl virus* renders them susceptible to the virus. *Plant Molecular Biology.* **71:**157–171.

Eybishtz, A., Y. Peretz, D. Sade, R. Gorovits and H. Czosnek. 2010. *Tomato yellow leaf curl virus* (TYLCV) infection of a resistant tomato line with a silenced sucrose transporter gene *LeHT1* results in inhibition of growth, enhanced virus spread and necrosis. *Planta.* **231:**537–548.

Faurobert, M., C. Mihr, N. Bertin, T. Pawlowski, L. Negroni, N. Sommerer and M. Causse. 2007. Major proteome variations associated with Cherry tomato pericarp development and ripening. *Plant Physiology.* **143:**1328–1346. Flower, D.R., A.C. North and C.E. Sansom. 2000. The lipocalin protein family: structural and sequence overview. *Biochimica et Biophysica Acta.* **1482:**9–24.

Friedmann, M., M. Lapidot, S. Cohen and M. Pilowsky. 1998. A novel source of resistance to tomato yellow leaf curl virus exhibiting a symptomless reaction to viral infection. *Journal of the American Society for Horticultural Sciences.* **123:**1004–1007.

Fotopoulos, V., M.J. Gilbert, J.K. Pittman, A.C. Marvier, A.J. Buchanan, N. Sauer, J.L. Hall and L.E. Williams. 2003. The Monosaccharide transporter gene, AtSTP4 and the cell-wall invertase, Atβfruct1, are induced in *Arabidopsis* during infection with the fungal. *Plant Physiology.* **132:**821–829.

Gear, M., M. McPhillips, J. Patrick and D. McCurdy. 2000. Hexose transporters of tomato: molecular cloning, expression analysis and functional characterization. *Plant Molecular Biology.* **44:**687–697.

Genoud, T. and J.P. Métraux. 1999. Crosstalk in plant cell signaling: structure and function of the genetic network. *Plant Science.* **4:**503–507.

Gilchrist, E and G. Haughn. 2010. Reverse genetics techniques: engineering loss and gain of gene function in plants. *Briefings in Functional Genomics.* **9:**103–110.

Gilliland, A., A.M. Murphy, C.E. Wong, R.A.J. Carson and J.P. Carr. 2006. Mechanisms involved in induced resistance to plant viruses. In: Tuzun S., Bent E. (eds.) *Multigenic and induced systemic resistance in plants.* Dordrecht, The Netherlands: Springer; pp. 335–359.

Goldbach, R., E. Bucher and M. Prins. 2003. Resistance mechanisms to plant viruses: an overview. *Virus Research.* **92:**207–212.

Greenberg, J.T. and N. Yao. 2004. The role and regulation of programmed cell death in plant-pathogen interactions. *Cellular Microbiology.* **6:**201–211.

Gronenborn, B. 2007. The tomato yellow leaf curl virus: genome and function of its proteins. In: Czosnek H editor *Tomato yellow leaf curl virus disease: management, molecular biology and breeding for resistance.* Dordrecht, The Netherlands: Springer; pp. 67–84.

Gunawardena, A.H.L.A.N., J.S. Greenwood and N.G. Dengler. 2004. Programmed cell death remodels lace plant leaf shape during development. *Plant Cell.* **16:**60–73.

Gutierrez, C. 2000. DNA replication and cell cycle in plants: learning from geminiviruses. *The EMBO Journal.* **19:**792–799.

Hamilton, A.J. and D.C. Baulcombe. 1999. A species of small antisense RNA in posttranscriptional gene silencing in plants. *Science.* **286:**950–952.

Hanley-Bowdoin, L., S.B. Settlage, B.M. Orozco, S. Nagar and D. Robertson. 2000. Geminiviruses: models for plant DNA replication, transcription and cell cycle regulation. *Critical Reviews in Biochemistry and Molecular Biology.* **35:**105–140.

Hanson, P.M., D. Bernacchi, S. Green, S.D. Tanksley, V. Muniyappa, S. Padmaja, H.M. Chen, G. Kuo, D. Fang and J.T. Chen. 2000. Mapping a wild tomato introgression associated with tomato yellow leaf curl virus resistance in a cultivated tomato line. *Journal of the American Society for Horticultural Sciences.* **131:**15–20.

Hofius, D., D.I. Tsitsigiannis, J.D. Jones and J. Mundy. 2007. Inducible cell death in plant immunity. *Seminars in Cancer Biology.* **17:**166–187.

Hulbert, S.H., C.A. Webb, S.M. Smith and Q. Sun. 2001. Resistance gene complexes: evolution and utilization. *Annual Review of Phytopathology.* **39:**285–312.

Ji, Y. and J.W. Scott. 2006. *Ty-3*, a begomovirus resistance locus linked to *Ty-1* on chromosome 6. Report of the Tomato Genetics Co-operative. **56:**22–25.

Ji, Y., J.W. Scott, P. Hanson, E.Graham and D.P. Maxwell. 2007. Sources of resistance, inheritance and location of genetic loci conferring resistance to members of the tomato-infecting begomoviruses. In: Czosnek, H. editor. *Tomato yellow leaf curl virus disease: management, molecular biology and breeding for resistance.* Dordrecht, The Netherlands: Springer; pp. 343–362.

Ji, Y., J.W. Scott, D.P. Maxwell and D.J. Schuster. 2008. Ty-4, a tomato *yellow leaf curl virus* resistance gene on chromosome 3 of tomato. Report of the Tomato Genetics Cooperative. **58:**29–31.

Ji, Y., J.W. Scott and D.P. Maxwell. 2009. Molecular mapping of *Ty-4*, a new Tomato yellow leaf curl virus resistance locus on chromosome 3 of tomato. *Journal of the American Society for Horticultural Sciences.* **134:**281–288.

Kim, M., J.H. Lim, C.S. Ahn, K. Park, G.T. Kim, W.T. Kim and H.S. Pai. 2006. Mitochondria-associated hexokinases play a role in the control of programmed cell death in *Nicotiana benthamiana. Plant Cell.* **18:**2341–2355.

Kelley, L.A. and M.J.E. Sternberg. 2009. Protein structure prediction on the Web: a case study using the Phyre server. *Nature Protocols.* **4:**363–371.

Kempema, L.A., X. Cui, F.M. Holzer and L.L. Walling. 2007. Arabidopsis transcriptome changes in response to phloem-feeding silverleaf whitefly nymphs. Similarities and distinctions in responses to aphids. *Plant Physiology.* **143:**849–865.

Kumagai, M.H., J. Donson, G. della-Cioppa, D. Harvey, K. Hanley and L.K. Grill. 1995. Cytoplasmic inhibition of carotenoid biosynthesis with virus-derived RNA. Proceedings of the National Academy of Science, USA. **92:**1679–1683.

Kunkel, B.N. and D.M. Brooks. 2002. Cross talk between signaling pathways in pathogen defense. *Current Opinion in Plant Biology.* **5:**325–331.

Labate, J.A., L.D. Robertson and A.M. Balso. 2009. Multilocus sequence data reveal extensive departures from equilibrium in domesticated tomato (*Solanum lycopersicum* L.). *Heredity.* **103:**257–267.

Lecourieux, F., D. Lecourieux, C. Vignault and S. Delrot. 2010. A Sugar-Inducible protein kinase, VvSK1, regulates hexose transport and sugar accumulation in grapevine cells. *Plant Physiology.* **152:**1096–1106.

Levesque-Tremblay, G., M. Havaux, and F. Ouellet. 2009.The chloroplastic lipocalin AtCHL prevents lipid peroxidation and protects *Arabidopsis* against oxidative stress. *The Plant Journal.* **60:**691–702.

Levine, A., R. Tenhaken, R. Dixon and C. Lamb. 1994. H_2O_2 from the oxidative burst orchestrates the plant hypersensitive disease resistance response. *Cell.* **79:**583–593.

Lindbo, J.A., L. Silva-Rosales, W.M. Proebsting and W.G. Dougherty. 1993. Induction of a highly specific antiviral state in transgenic plants: implications for regulation of gene expression and virus resistance. *Plant Cell.* **5:**1749–1759.

Liu, Y., M. Schiff and S.P. Dinesh-Kumar. 2002. Virus-induced gene silencing in tomato. *The Plant Journal.* **31:**777–786.

Lu, R., A.M. Martin-Hernandez, J.R. Peart, I. Malcuit and D.C. Baulcombe. 2003. Virus induced gene silencing in plants. *Methods.* **30:**296–3030.

Lucas, W.J. 2006. Plant viral movement proteins: agents for cell-to-cell trafficking of viral genomes. *Virology.* **344:**169–184.

McCarty, D.R. and J. Chory. 2000. Conservation and innovation in plant signaling pathways. *Cell.* **103:**201–209.

McDowell, J.M. and B.J. Woffenden. 2003. Plant disease resistance genes: recent insights and potential applications. *Trends in Biotechnology.* **21:**178–183.

Menda, N., Y. Semel, P. Dror, Y. Eshed and D. Zamir. 2004. *In silico* screening of a saturated mutation library of tomato. *The Plant Journal.* **38**:861–72.

Michelson, I., M. Zeidan, E. Zamski, D. Zamir and H. Czosnek. 1997. Localization of tomato yellow leaf curl virus (TYLCV) in susceptible and tolerant nearly isogenic tomato lines. *Acta Horticulturae.* **447**:407–414.

Murphy, A.M., A. Gilliland, C.E. Wong, J. West, D.P. Singh, J.P. Carr. 2001. Signal transduction in resistance to plant viruses. *European Journal of Plant Pathology.* **107**:121–128.

Navot, N., E. Pichersky, M. Zeidan, D. Zamir and H. Czosnek. 1991. Tomato yellow leaf curl virus: a whitefly-transmitted geminivirus with a single genomic molecule. *Virology* **185**:151–161.

Nørholm, M.H., H.H. Nour-Eldin, P. Brodersen, J. Mundy and B.A. Halkier. 2006. Expression of the *Arabidopsis* high-affinity hexose transporter STP13 correlates with programmed cell death. *FEBS Letters.* **580**:2381–2387.

Page, D., B. Gouble, B. Valot, J.P. Bouchet, C. Callot, A. Kretzschmar, M. Causse, C.M.C.G. Renard and M. Faurobert. 2010. Protective proteins are differentially expressed in tomato genotypes differing for their tolerance to low-temperature storage. *Planta.* **232**:483–500.

Pratt, A.J. and I.J. MacRae. 2009. The RNA-induced silencing complex: a versatile gene-silencing machine. *Journal of Biological Chemistry.* **284**:17897–17901.

Price, J., A. Laxmi, St. Martin and J.C. Jang. 2004. Global transcription profiling reveals multiple sugar signal transduction mechanisms in *Arabidopsis. Plant Cell.* **16**:2128–2150.

Purkayastha, A. and I. Dasgupta. 2009. Virus-induced gene silencing: a versatile tool for discovery of gene functions in plants. *Plant Physiology and Biochemistry.* **47**:967–976.

Ratcliff, F.G., S.A. MacFarlane and D.C. Baulcombe. 1999. Gene silencing without DNA: RNA-mediated cross-protetion between viruses. *Plant cell.* **11**:1207–1216.

Ratcliff, F., A.M. Martin-Hernandez and D.C. Baulcombe. 2001. Tobacco rattle virus as a vector for analysis of gene function by silencing. *The Plant Journal.* **25**:237–245.

Restrepo, S., K. Myers, O. del Pozo, G. Martin, A. Hart, C. Buell, W. Fry and C. Smart. 2005. Gene profiling of a compatible interaction between *Phytophthora infestans* and *Solanum tuberosum* suggests a role for carbonic anhydrase. *Molecular Plant-Microbe Interactions.* **18**:913–922.

Rivas, S., A. Rougon-Cardoso, M. Smoker, L. Schauser, H. Yoshioka and J.D. Jones. 2000. CITRX thioredoxin interacts with the tomato Cf-9 resistance protein and negatively regulates defense. *EMBO Journal.* **23**:2156–2165.

Robertson, D. 2004. VIGS vectors for gene silencing: many targets, many tools. *Annual Review of Plant Biology.* **55**:495–519.

Roth, B.M., G.J. Pruss and V.B. Vance. 2004. Plant viral suppressors of RNA silencing. *Virus Research.* **102**:97–108.

Sasabe, M., K. Takeuchi, S. Kamoun, Y. Ichinose, F. Govers, K. Toyoda, T. Shiraishi and T. Yamada. 2000. Independent pathways leading to apoptotic cell death, oxidative burst and defense gene expression in response to elicitin in tobacco cell suspension culture. *European Journal of Biochemistry.* **267:**5005–13.

Schofield, R.A., Y. Bi, S. Kant and S.J. Rothstein. 2009. Over-expression of STP13, a hexose transporter, improves plant growth and nitrogen use in *Arabidopsis thaliana* seedlings. *Plant Cell and Environment.* **32:**271–285.

Senthil-Kumar, M., H.V. Rame Gowda, R. Hema, K.S. Mysore and M. Udayakumar. 2008. Virus-induced gene silencing and its application in characterizing genes involved in water-deficit-stress tolerance. *Journal of Plant Physiology.* **165:**1404–1421.

Sim, S.C., M.D. Robbins, C. Chilcott, T. Zhui and D.M. Francis 2009. Oligonucleotide array discovery of polymorphisms in cultivated tomato (*Solanum lycopersicum* L.) reveals patterns of SNP variation associated with breeding. *BMC Genomics.* **10:**466.

Sinha, A.K., M.G. Hofmann, U. Romer, W. Kockenberger, L. Elling and T. Roitsch. 2002. Metabolizable and non-metabolizable sugars activate different signal transduction pathways in tomato. *Plant Physiology.* **128:**1480–1489.

Smeekens, S. 1998. Sugar regulation of gene expression in plants. *Current Opinion in Plant Biology.* **1:**230–234.

Smeekens, S. and F. Rook. 1997. Sugar sensing and sugar-mediated signal transduction in plants. *Plant Physiology.* **115:**7–13.

Spitzer, B., M. Moyal Ben Zvi, M. Ovadis, R. Marhevka, O. Barkai, O. Edelbaum, I. Marton, T. Masci, M. Alon, S. Morin, I. Rogachev, A. Aharoni and Vainstein, A. 2007. Reverse genetics of floral scent: application of Tobacco rattle virus-based gene silencing in petunia. *Plant Physiology.* **145:**1241–1250.

Stange, C. 2006. Plant-virus interactions during the infective process. *Ciencia Investigación Agraria.* **33:**1–18.

Takken, F. and M. Rep. 2010. The arms race between tomato and *Fusarium oxysporum*. *Molecular Plant Pathology.* **11:**309–314.

Van Doorn, W.G. and E.J. Woltering. 2005. Many ways to exit? *Cell.* death categories in plants. *Trends in Plant Science.* **10:**117–122.

Vidavski, F. and H. Czosnek. 1998. Tomato breeding lines immune and tolerant to Tomato yellow leaf curl virus (TYLCV) issued from *Lycopersicon hirsutum*. *Phytopathology.* **88:**910–914.

Wahl, R., K. Wippel, S. Gioos, J. Kämper and N. Sauer. 2010. A novel high-affinity sucrose transporter is required for virulence of the plant pathogen Ustilago maydis. *PLoS Biology.* **8:**e10003030.

Walling, L.L. 2000. The myriad plant responses to herbivores. *Journal of Plant Growth Regulation.* **19:**195–216.

Wellmer, F. and J.L. Riechmann. 2005. Gene network analysis in plant development by genomic technologies. *International Journal of Developmental Biology.* **49:**745–759.

Whitham, S.A., C. Yang and M.M. Goodin. 2006. Global impact: elucidating plant responses to viral infection. *Molecular Plant-Microbe Interactions.* **19:**1207–1215.

Xu, C., J. Fan, W. Riekhof, J. Froehlich and C. Benning. 2003. A permease-like protein involved in ER to thylakoid lipid transfer in Arabidopsis. *The EMBO Journal.* **22:**2370–2379.

Zakay, Y., N. Navot, M. Zeidan, N. Kedar, H. Rabinowitch, H. Czosnek and D. Zamir. 1991. Screening of *Lycopersicon* accessions for resistance to tomato yellow leaf curl virus: presence of viral DNA and symptom development. *Plant Disease.* **75:**279–281.

Zamir, D., I. Ekstein-Michelson, Y. Zakay, N. Navot, M. Zeidan, M. Sarffati, Y. Eshed, E. Harel, T. Pleban, H. Vanoss, N. Kedar, H.D. Rabinowitch and H. Czosnek. 1994. Mapping and introgression of a tomato Yellow leaf curl virus tolerance gene, TY-1. *Theoretical and Applied Genetics.* **88:**141–146.

Zeidan, M. and H. Czosnek. 1991. Acquisition of *Tomato yellow leaf curl virus* by the whitefly *Bemisia tabaci*. *Journal of General Virology.* **72:**2607–2614.

Zhang, S. and D.F. Klessig. 2001. MAPK cascades in plant defense signaling. *Trends in Plant Science.* **6:**520–527.

Zrachya, A., E. Glick, Y. Levy, T. Arazi, V. Citovsky and Y. Gafni. 2007. Suppressor of RNA silencing encoded by Tomato yellow leaf curl virus-Israel. *Virology.* **358:**159–165.

CHAPTER 10

RNAi for Crop Improvement

Imran Amin, Muhammad Saeed*, Shahid Mansoor
and Rob W. Briddon

ABSTRACT

RNA interference (RNAi) is a sequence-specific gene silencing technology that is mediated by small 21 to 25 nucleotide single-stranded RNAs, known as small interfering RNAs (siRNAs). These originate from a double-stranded (ds)RNA elicitor and act as guides for RNA cleavage, DNA methylation and translation repression. The eliciting dsRNA can be readily expressed in plants using a new class of RNAi vectors designed to express hairpin RNAs that are processed into siRNA. RNAi technology has been applied to a wide range of species to silence the expression of both specific endogenous genes and genes of invading pathogens. RNAi has been used in the genetic engineering of important plant metabolites including starch, improvement of seed oil, reduction in plant allergens and male sterility. RNAi is also showing promise as a means of protecting plants from pathogens. The demonstration that siRNAs (or their precursor dsRNAs) can be transferred from plants to plant-feeding pests has opened-up the possibility of engineering resistance *in planta* to insects and nematodes. The possible role of RNAi technology in the improvement of crops and future prospects are discussed.

Keywords: RNAi, DCL enzymes, RISC, crop improvement.

INTRODUCTION

RNA silencing

RNA silencing has only emerged as a topic of general interest in the past ten years but, as a matter of fact, the first paper dealing with the effects of RNA silencing was published as long ago as 1928. In that research paper tobacco plants were described in which only the initially infected leaves were necrotic and diseased owing to infection with

---

* *Corresponding author e-mail*: saeed_hafeez@yahoo.com

Tobacco ring spot virus (TRSV; Wingard 1928)). The upper emerging leaves were asymptomatic and resistant to secondary infection, a phenomenon which was described as "recovery". At the time this "recovery" was a mystery, since there was no obvious way to explain this phenomenon.

The details of the TRSV example were not resolved until recently but it is now known that recovery from virus disease involves RNA silencing that is targeted specifically against the viral RNA (Baulcombe 2004, Covey et al. 1997, Ratcliff et al. 1997). There was no information about this mechanism in 1928, nor was it realized that the viral genome of TRSV and many other viruses, is RNA. However, Wingard's paper can be considered an appropriate starting point for the current interest in RNA silencing because it showed a role for RNA silencing in defense against viruses, which may have been one of its original functions in primitive eukaryotes (Baulcombe 2004, Hannon 2002, Voinnet 2009).

The phenomenon of RNA silencing, also referred to as RNA interference (RNAi), was first discovered in the nematode worm *Caenorhabditis elegans* as a response to double-stranded RNA (dsRNA), which results in sequence-specific gene silencing (Fire et al. 1998). RNA silencing pathways are involved in several fundamental biological processes that include cellular defense against pathogens, control of transposon mobility, gene regulation via micro (mi)RNAs, *de novo* histone and DNA methylation and the establishment of heterochromatin (Carrington and Ambros 2003, Lippman and Martienssen 2004, Voinnet 2005).

THE MACHINERY OF RNAi

There are several players in the RNA silencing machinery including the ribonucleases known as Dicers, RNA-dependent RNA polymerases (RDRs) and Argonautes (AGOs). The RNAi machinery in plants is better understood than in the fungal or animal systems. The *Arabidopsis* genome encodes four Dicer-like (DCL) enzymes, six RDRs and 10 AGO proteins (Baulcombe 2004). Various genetic studies have shown that all these factors interact in a specific manner to effect distinct, but partially overlapping, pathways that are commonly triggered by dsRNA (Hannon 2002). Indeed, dsRNA is a more efficient elicitor of RNAi than either sense or antisense RNA alone. RNA silencing works on at least three different levels in plants which will be discussed in later sections. A common feature of these RNA silencing mechanisms is the trigger, dsRNA, which is recognized by DCLs. These then cleave the long dsRNA into small, 21–25 nucleotide (nt) fragments. The small RNAs are subsequently incorporated into a complex known as the 'RNA-induced silencing complex' (RISC) which acts to either degrade the mRNA, repress translation from the mRNA or methylate homologous DNA sequences, depending upon the pathway involved (Figure 1).

RNAi PATHWAYS

The first RNAi pathway identified was post-transcriptional gene silencing (PTGS), which has also been referred to as cytoplasmic RNA silencing. PTGS results in the degradation of target mRNA (or the genome of an RNA virus). A characteristic feature of PTGS is the production of 21–24 nt short interfering RNA (siRNA) species which are generated from the inducing dsRNA (Hamilton and Baulcombe 1999). Apart from some functional redundancy among the DCL proteins, it is believed that siRNAs are possibly generated by DCL-2 and DCL-4 (Gasciolli et al. 2005, Xie et al. 2005). The dsRNA that leads to PTGS may come from exogenous or endogenous sources, such as RNA virus replication intermediates synthesized by viral RDRs, structured single-stranded (ss)RNAs, annealed overlapping transcripts of opposite polarity that can serve as Dicer substrates or products of RDRs acting on transcripts that are aberrant or over-expressed (Gazzani et al. 2004, Molnar et al. 2005, Szittya et al. 2002). Duplex siRNA is subsequently unwound and one strand is incorporated into the RNase-containing effector complex RISC, which contains at least one AGO protein (Figure 1) (Hammond et al. 2000, Hannon 2002).

Cleavage specificity of RISC is the result of complementary base-pairing between the siRNA and the target mRNA. By analogy to the mammalian system, an AGO protein in RISC is most likely the "slicer" that carries out transcript cleavage (Liu et al. 2004). RDRs play multiple roles in the RNAi pathway. Apart from the initial generation or processing of the dsRNA trigger, RDRs also play a major role in the amplification and transitive spreading of siRNAs outside the sequence of the original dsRNA trigger (Himber et al. 2003, Vaistij et al. 2002). One of the remarkable features of RNAi is its ability to spread from cell-to-cell and systemically throughout the plant (Palauqui et al. 1997, Voinnet and Baulcombe 1997). Recently it has been shown that siRNAs act as the mobile silencing signal (Dunoyer et al. 2010b, Molnar et al. 2010).

The second RNAi pathway involves endogenously produced, 21–22 nt RNAs known as miRNAs. These are processed by DCL-1 from larger miRNA precursors (pre-miRNA) and negatively regulate their target mRNAs, either by inhibiting translation (primarily in animal systems) or by degradation (Figure 1) (primarily in plant systems; Bartel 2004). The third RNAi pathway (Figure 1) leads to siRNA-directed transcriptional gene silencing (TGS) and heterochromatic silencing (Lippman and Martienssen 2004). TGS may be triggered by transcription of inverted repeats or tandemly repeated sequences and can also be induced experimentally

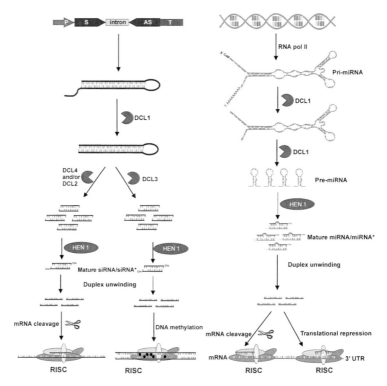

Figure 1. Silencing of gene expression by the RNAi pathway. An RNAi construct is composed of a promoter (P) an inverted repeat (containing sense [S] and anti-sense [AS] sequences) of the gene to be targeted separated by an intron and a terminator (T). A mature hairpin is produced by the action of DCL1. siRNAs of 21nt are produced by the action of DCL2/ DCL4 and 24-nt siRNA are generated by the action of DCL3 which are then methylated by HEN 1 yielding siRNA* that are protected from degradation. A single stranded siRNA is incorporated into the RNA induced silencing complex (RISC) and the AGO1 (or AGO4) component of this complex mediates sequence specific mRNA cleavage or DNA methylation leading to post-transcriptional gene silencing or transcriptional gene silencing, respectively. miRNAs are transcribed in the nucleus by RNA polymerase II or III and are processed by a nuclear type III RNase DCL1 to form primary (pri)-miRNA. The pri-miRNAs are cleaved by DCL1 to form pre-miRNAs. The sequences in the stem of the stem-loop structure of the pri-miRNA that ultimately form the mature miRNA are shown in red. A construct for producing an artificial miRNA would consist of a natural pre-miRNA sequence with the targeting sequences, that form the mature miRNA, replaced with sequences derived from the gene to be silenced. Pre-miRNAs are transported to the cytoplasm with the help of HEN1. After unwinding they are incorporated in the RISC complex where they serve as a guide to target mRNAs and silence their expression, either by translation repression (when the miRNA has imperfect base pairing with the target mRNA 3′ un-translated region) or by mRNA degradation (when the miRNA has perfect base pairing with the target).

(Color image of this figure appears in the color plate section at the end of the book.)

by the ectopic expression of RNA corresponding to promoter regions (Jones et al. 1999, 2001, Mette et al. 2000). TGS differs from PTGS in involving slightly larger siRNAs (24–26 nt) that are generated by DCL-3 which acts in conjunction with AGO4 and RDR2 (Hamilton et al. 2002, Qi and Hannon 2005, Xie et al. 2004, Zilberman et al. 2004). TGS-mediated suppression of gene activity is a result of the methylation of cytosine residues in DNA (RNA-directed DNA methylation) and specific post-translational modifications of histone proteins, including methylation of histone H3 at lysine 9 (Bender, 2004). Various studies have shown that there are multiple links between siRNA and these two epigenetic modifications (Bisaro 2006). It is believed that TGS maintains genome integrity by preventing rearrangement in centromeric and telomeric repeats and by suppressing transposons and other invasive DNAs (Djupedal and Ekwall 2009, Gracheva et al. 2009).

METHODS OF INDUCING RNAi IN PLANTS

Various approaches have been used to produce dsRNA in plants. Initially this was achieved by the independent transformation of plants with constructs for producing sense and anti-sense RNA and subsequently crossing these to yield hybrid lines in which dsRNA is produced. Phytopathogenic viruses have also been used to silence host genes; a process known as virus-induced gene silencing (VIGS). Both RNA and DNA viruses have been used for this purpose (Brigneti et al. 2004, Cai et al. 2007, Fofana et al. 2004, Huang et al. 2009, Lacomme and Chapman 2008). PTGS, when targeted against transcribed sequences, is only transient. However, the silencing of promoter sequences using VIGS is heritable, with methylated promoter sequences remaining methylated in subsequent generations (Jones et al. 2001). However, the VIGS method is somewhat limited by the host range of the virus upon which the VIGS vector is based.

Bacterially expressed siRNAs, applied topically as a spray of crude bacterial extracts, have been shown to elicit silencing (Tenllado et al. 2003). For example, resistance to *Pepper mild mottle virus* (PMMoV) and *Plum pox virus* (PPV) (both viruses with RNA genomes) has been achieved by spraying bacterially expressed, virus-specific siRNA on plants five days before challenge with the viruses. This is an alternative method of inducing RNAi, without the need to produce transgenic plants. Similarly, dsRNA against two coat protein fragments of *Sugarcane mosaic virus* (SCMV) were expressed in *Escherichia coli* and were applied to plants as a spray and induced resistance in maize to SCMV infection (Gan et al. 2010).

So far the most effective and widely used approach to induce the production of dsRNA in plants has been to express constructs with both sense and antisense sequences of the gene to be silenced separated by an intron (Figure 1). Upon transcription, these sequences form a hairpin RNA (hpRNA) molecule that triggers gene silencing (Smith et al. 2000). The silencing efficiency of hpRNA and sense/antisense RNA in a range of plant species has been compared. The hpRNA strategy generally increases gene silencing by 90–100% over the sense/antisense approach (Wesley et al. 2001) and is now the most widely used system for silencing genes in plants (Helliwell et al. 2005, Miki and Shimamoto 2004). Wesley et al., (2001) developed a vector system (known as pHELLSGATE) that uses Gateway™ (Invitrogen) recombinational cloning to allow the rapid production of expression constructs for RNAi in plants. In this system, PCR fragments with sense and antisense orientations are amplified by using primers having the recombinase recognition sites, *att*B1 and *att*B2 sequences. These PCR fragments are cloned into a plasmid containing *att*P1 and *att*P2 sites. The *Cauliflower mosaic virus* (CaMV) 35S promoter is used to drive transcription while a catalase intron is used as the hairpin spacer. A similar high-throughput RNAi vector system (pANDA) was developed (Miki and Shimamoto 2004) using the maize ubiquitin promoter, the intron from the β-*glucuronidase* gene and Gateway™ cloning technology (Miki and Shimamoto 2004). More recently inducible RNAi vectors have been developed (Earley et al. 2006, Hirai and Kodama 2008, Lo et al. 2005).

APPLICATIONS OF RNAi FOR CROP IMPROVEMENT

RNAi-based metabolic engineering

Plants are an important natural resource, providing numerous products, including food, fibre, wood, oils, dyes and pharmaceuticals. Many other valuable secondary metabolites are also produced in small amounts and are thus very difficult to obtain in sufficient quantities. Metabolic engineering using RNAi provides a possible means of overcoming this limitation. Naturally occurring, hpRNA-mediated silencing of a rice gene (*low glutelin content 1* [*Lgc1*]) was identified by Kusaba et al. (2003). Glutelin is a major seed storage protein and the low glutelin content of the resulting rice lines connect beneficial for patients with kidney diseases who must reduce their protein intake. *Lgc1* is a dominant mutation that results in a reduced glutelin content of rice grains. *Lgc1* homozygotes have a deletion of ~3.5 kb between two highly similar glutelin genes, forming a tail-to-tail inverted repeat, which results in a hpRNA that induces gene silencing. This has been confirmed by producing transgenic plants containing the inverted repeat and by detecting siRNAs derived from that region in natural mutants, as well as in transgenic plants (Kusaba et al. 2003). The transgenic trait was stable for 20 generations, demonstrating the stability of the hpRNA producing inverted repeat sequence.

The experimental efficacy of the hpRNAi approach has been demonstrated by silencing two key enzymes in the fatty acid biosynthesis pathway of cotton (*Gossypium hirsutum*): ghSAD-1 and ghFAD2-1. RNAi-mediated down regulation of ghSAD-1 elevated the stearic acid content in cotton seed (44% compared with a normal level of 2%) and silencing ghFAD2-1 increased the oleic acid content (77% compared with a normal level of 15%) (Liu et al. 2002). This approach has since been used to achieve immunity in plants to a diverse range of RNA viruses (Table 1).

RNAi-MEDIATED MODIFICATION OF STARCH

Starch consists of 20 to 25% amylose and 75 to 80% amylopectin. In contrast to amylopectin, amylose is insoluble in water and is more resistant to digestion. Amylose reduces the gel strength of starch and amylose-free (waxy) starch is used as thickener, water binder, emulsion stabilizer and gelling agent in industrial and food processes. Heilersig et al. (2006) produced a number of RNAi constructs containing different regions, lengths and orientations of a cDNA of the potato granule-bound starch synthase (GBSSI) gene. The constructs used the GBSSI promoter and parts of the GBSSI gene (rather than an intron) as a spacer (Heilersig et al. 2006). Analysis of transformed potato lines showed that silencing was independent of the orientation of the construct. High silencing efficiencies were observed with constructs using the 5′ or middle region of the gene, but not the 3′ region. Construct with the stem size of 500–600 bp and a spacer of about 150 bp in length were more efficient than constructs with larger stems and a larger spacer. The results suggested that there was no effect of inverted repeat orientation on silencing efficiency but that the sequence used did affect the efficiency. Although a significant reduction in amylose content was obtained, a complete absence of GBSSI was not achieved (Heilersig et al. 2006). In contrast, a similar study of the RNAi-mediated inhibition of GBSSI in sweet potato yielded plants producing starch with no amylose (Shimada et al. 2006).

High-amylose starch also has industrial applications, including as a binding agent in candy and gums as well as in expanded products used as packaging materials. RNAi was used for gene inhibition of the starch branching enzymes *SBE1* and *SBE2* in potato. Two constructs with 100 bp segments (pHAS2) or 200 bp segments (pHAS3) of the two genes were cloned as inverted repeats controlled by the potato GBSSI promoter. Construct pHAS3 was shown to be very efficient, yielding high-amylose starch in more than 50% of the transgenic lines. The resulting transgenic potato plants showed a high-amylose phenotype with amylose content ranging from 38% to 87%, whereas the amylose contents of parental plants

Table 1. Published reports of metabolic engineering using RNAi.

Target gene	Plant species	Potential Benefit	Reference
Polyphenyl Oxidase gene	*Solanum tuberosum* (Potato)	Extended storage life	(Wesley et al. 2001)
ghSAD-1and ghFAD2-1 genes	*Gossypium hirsutum* (Cotton)	Useful for cooking applications without the need for hydrogenation	(Liu et al. 2002)
CaMxMt 1 gene	*Coffea arabica* (Coffee)	Decaffeinated coffee	(Ogita et al. 2004)
BP1 gene	*Brassica napus* (Oilseed rape)	Improved photosynthesis	(Byzova et al. 2004)
Codeine reductase (COR) gene	*Papaver somniferum* (Opium poppy)	non-narcotic poppy	(Allen et al. 2004)
DET-1 gene	*Solanum lycopersicum* (Tomato)	Consumer health benefits	(Davuluri et al. 2005)
Starch branching enzyme	*Zea mays* (Maize)	Amylose content increased	(Chai et al. 2005)
Lol p1 and Lol p2	*Lolium multiflorum* (Ryegrass)	Hypo-allergic	(Petrovska et al. 2005)
1-aminocyclo propane-1-carboxylate oxidase	*Solanum lycopersicum* (Tomato)	Longer shelf life (slower ripening)	(Xiong et al. 2004)
ACR2 gene	*Arabidopsis thaliana*	Phytoremediation of soils	(Dhankher et al. 2006)
GBSSI	*Solanum tuberosum* (Potato)	Strong reduction in amylose core	(Heilersig et al. 2006)
Sbe1 and Sbe2	*Solanum tuberosum* (Potato)	Increase in amylose contents	(Andersson et al. 2006)
SBEIIa and SBEIIb	*Triticum aestivumat* (Wheat)	Increase in amylose contents	(Regina et al. 2006)
GmFAD3	*Glycine max* (Soybean)	Reduced levels of α-linolenic acids	(Flores et al. 2008)
FAB1	*Arabidopsis thaliana*	Industrial applications of oil	(Pidkowich et al. 2007)
LTPG1, LTPG2	*Solanum lycopersicum* (Tomato)	Hypo-allergic	(Le et al. 2006a)
Ara h 2	*Arachis hypogaea* (Peanut)	Hypo-allergic	(Dodo et al. 2008)
Dau c 1.01, Dau c 1.02	*Daucus carota* (Carrot)	Hypo-allergic	(Peters et al. 2010)

was about 20% (Andersson et al. 2006). Similarly high amylose wheat (*Triticum aestivum*) was developed by silencing the two different isoforms of SBE2 (*SBE2a* and *SBE2b*) in wheat endosperm to raise its amylose content. No effect on amylose content was observed upon suppression of *SBE2b* alone. However, suppression of both *SBE2a* and *SBE2b* expression resulted in starch containing more than 70% amylose. The analysis of total starch content in the endosperm showed a slight reduction in the transgenic wheat plant (43%) when compared with control plants (52%). When this high-amylose wheat grain was fed to rats, the digestion of rats was improved indicating that it serves as a source of "resistant starch", which has the potential to improve human digestive health (Regina et al. 2006).

INCREASING THE OIL CONTENT OF SEED

Vegetable oils produced from oil seed crops such as sunflower, corn, soybean and olives contain oleic, linoleic and α-linolenic acid. It is recommended that those oils be preferentially used in the human diet for healthy nourishment. Mammals lack the ability to introduce double bonds in fatty acids beyond carbons 9 and 10. Hence linoleic acid and α-linolenic acid are essential fatty acids for humans. In the body, essential fatty acids are primarily used to produce hormone-like substances that regulate a wide range of functions, including blood pressure, blood clotting, blood lipid levels, the immune response and the inflammation response to injury and infection (Simopoulos et al. 2000).

Polyunsaturated fatty acids (linoleic and α-linolenic acids) are too unstable to be used in cooking (Napier 2007). RNAi has been used to silence the ω-3 fatty acid desaturase (FAD3) gene family of soybean. The FAD3 enzyme is responsible for the synthesis of α-linolenic acid in the polyunsaturated fatty acid pathway. A conserved nucleotide sequence, 318-nt in length, common to the three members of the gene family in the pathway was used in a hp-construct driven by a seed-specific promoter. Transgenic plants silenced for FAD3 showed a significant reduction in α-linolenic acid content (ranging from 1.0% to 3.1% of that in non-silenced plants). This silencing was heritable in soybean lines (Flores et al. 2008).

Liu et al. (2002) used RNAi to down-regulate the expression of two key fatty acid desaturase genes, ghSAD-1 (encoding stearoyl-acyl-carrier protein Δ9-desaturase) and ghFAD2-1 (encoding oleoyl-phosphatidylcholine ω6-desaturase) in cotton seed. RNAi constructs were transformed into cotton (*G. hirsutum* cv. Coker 315). The resulting down-regulation of the ghSAD-1 gene substantially increased stearic acid content in seed from the normal levels of 2% to 3% up to as high as 40%, whereas silencing of the ghFAD2-1 resulted in greatly elevated oleic acid content, up to 77% compared with about 15% in seeds of untransformed plants. In addition, palmitic acid was significantly lowered in both high-stearic and high-oleic lines (Liu et al. 2002).

Seeds oils, particularly palmitic acid, can provide the starting materials for many industrial applications. While tropical oil crops contain high palmitic acid content, temperate oil seed crops tend to contain higher levels of linoleic acid. The potential for engineering seed oils has been demonstrated by the RNAi-based engineering of *Arabidopsis* to contain extremely high palmitic acid content by silencing the β-ketoacyl-acyl carrier protein synthase II (which elongates the backbone of fatty acids; making the oilseed composition of a temperate species more like that of a tropical oilseed crop; Pidkowich et al. 2007).

REDUCTION IN PLANT ALLERGENS

The major cause of hay fever and seasonal allergic asthma, which affects ~25% of the population in temperate climates, is ryegrass pollen (*Lolium* spp.). RNAi has shown promise in the development of hypoallergenic grasses. The main allergens are the pollen proteins Lol p1 and Lol p2–90% of allergy sufferers are sensitive to these proteins. Levels of Lol p1 and Lol p2 were shown to be down-regulated by expressing antisense cDNA sequences under the control of a maize pollen-specific promoter (Petrovska et al. 2005). Analysis of these transgenic grasses is also providing information about the function of the allergenic proteins in pollen, which is so far unclear.

Allergy to tomato (*Solanum lycopersicum*) is due to profilins (Lyc-1), β-fructofuranosidase (Lyc e 2), a non-specific lipid transfer protein (LTP; Lyc e 3) and other minor allergens. Several transgenic tomato lines were produced with reduced allergens (Le et al. 2006a, 2006b). Two paralogous genes encoding Lyc e 3, LTPG1 and LTPG2, were identified when an RNAi cassette targeting LTPG1 was expressed and both were efficiently silenced in transgenic tomato fruit. Examination of the allergenic potential of the Lyc e 3 reduced fruit revealed that histamine release from human basophils was strongly decreased in comparison to that seen in wild type tomato fruit (Le et al. 2006a). In addition, the same team also reduced the tomato minor allergen Lyc e 1 content by RNAi (Le et al. 2006b).

Many fruits and vegetables contain pathogenesis-related protein-10 (PR10), which is a ubiquitous small plant protein induced by microbial pathogens and abiotic stresses, that is an allergen. Two genes, *Dau c 1.01* and *Dau c 1.02*, have been shown to encode PR10. RNAi technology has been used to silence these two genes, with the aim of generating PR10-reduced, hypoallergenic transgenic carrots. Analysis of transgenic plants showed that PR10 accumulation was strongly decreased compared with untransformed controls. Treatment of carrot plants with the PR protein-inducing chemical salicylic acid resulted in an increase of PR10 isoforms only in wild-type but not in *Dau c 1*-silenced mutants. Skin prick allergy tests on individuals with carrot sensitivity showed a reduced allergic response towards the carrots from transgenic plants (Peters et al. 2010).

Peanut allergy is one of the most severe and potentially life threatening food allergies that is a major challenge for the food industry. In order to eliminate the major allergen (the Ara h 2 protein) transgenic peanuts were produced to express a hpRNA of a 256 bp fragment of the *Ara h 2* gene. Progeny plants were shown to have significantly reduced Ara h 2 content (21–25% of that seen in wild type peanuts) and the sera from patients with peanut allergy showed a marked reduction in IgE binding capacity to transgenic peanuts in comparison to that of controls (Dodo et al. 2008).

ENGINEERING SECONDARY METABOLISM

The first real breakthrough for RNAi-mediated metabolic engineering came in 2004 when RNAi was used to silence enzymes in the codeine reductase (COR) gene family in the opium poppy (*Papaver somniferum*) (Allen et al. 2004). This was achieved using a hpRNA construct containing sequences from multi ple cDNAs of genes in the morphine biosynthetic pathway. The non-narcotic alkaloid (S)-reticuline, which is a precursor and occurs upstream of codeine in the pathway, accumulated at the expense of morphine, codeine, opium and thebaine in transgenic plants. This study was the first to report metabolic engineering of the opium poppy using RNAi and the first to interfere with multiple steps in a complex biochemical pathway.

Transgenic *Coffea* spp. expressing hpRNA constructs containing CaMXMT1 sequences, a cDNA encoding one of the genes involved in the caffeine biosynthetic pathway, showed theobromine and caffeine accumulation between 30–50% of that normally found in these species (Ogita et al. 2004). This finding showed that theobromine is involved in the major biosynthetic pathway of coffee and that the technology can be applied to engineer decaffeinated coffee plants. For medical and health reasons some people avoid caffeine, which is a psychoactive alkaloid. The use of coffee beans from transgenic, low caffeine plants could greatly reduce the cost of producing caffeine free coffee.

RNAi has also been used to improve an edible crop. In tomato, a hpRNA construct was used to suppress an endogenous photo-morphogenesis regulatory gene, *DE-ETIOLATED1* (*DET1*), using a fruit-specific promoter (Davuluri et al. 2005). Mutations of *DET1* lead to elevated levels of flavonoids and carotenoids. Transgenic plants showed specific degradation of DET1 and an increase in carotenoid and flavonoid content, compared with that shown in wild-type tomato, while all other parameters for fruit quality remained unchanged. This was the first demon stration of metabolic engineering being used to increase the content of two compounds with health benefits, without detrimental effects on either quality or yield.

Silencing the ACR2 gene encoding arsenic reductase in *Arabidopsis* using hpRNA constructs resulted in plants that accumulate 10- to 16-fold more arsenic in shoots (350–500 ppm) but retain less arsenic in roots compared with wild-type plants. This technology could potentially be used for bioremediation of heavy metal-contaminated soils (Dhankher et al. 2006).

RNAi has been used to disrupt gossypol biosynthesis in cotton seed tissue by interfering with the expression of the δ-cadinene synthase gene under the control of a highly seed specific promoter. Results from enzyme activity and molecular analyses on developing transgenic embryos were consistent with the observed phenotype in the mature seeds. It was also found that the levels of gossypol and related terpenoids in the foliage and floral parts were not affected and thus their potential function in plant defense against insects and diseases remained unaffected (Sunilkumar et al. 2006). This gossypol free cotton seeds can potentially provide the protein requirements of a large number of people which were previously underutilized because of the presence of toxic gossypol.

RNAi-MEDIATED ENGINEERING OF PLANT DEVELOPMENT

Plant reproduction, particularly male gametogenesis, is not only an area of interest in plant developmental biology but also of significant commercial interest in controlling crop fertility (Goldberg et al. 1993, Gorman and McCormick 1997). An important point is the pollen development for F1 hybrid seed production. Numerous genes have been identified in a diverse range of plant species that show anther-specific expression. The silencing of some of these genes have been used to interfere with male fertility. The commercial production of hybrid plant lines relies on an efficient and effective mechanism for inducing male sterility in one of the parental lines so as to ensure purity of the resultant hybrid seed. To prevent self-pollination in plants such as tomato, maize and cabbage, laborious and expensive manual emasculation of the female line is required (Peacock 1990).

TAZ1 is an anther-specific protein which is involved in tapetum development. An RNAi construct was produced to target the TAZ1 gene in petunia. Transgenic plants showed a general degeneration of the tapetum and extensive microspore abortion, which initiated soon after their release from pollen tetrads. The few pollen grains that were retained show reduced flavonol accumulation, defects in pollen wall formation and poor germination (Kapoor et al. 2002).

In another study RNAi was used to silence a male-specific gene (TA 29) in the model host tobacco. About 10 out of 13 tobacco lines transformed with an RNAi construct containing TA29 sequences were male sterile. Transgenic plants were phenotypically indistinguishable from non-transgenic plants. At the anthesis stage, pollen grains from transgenic, male-sterile plants were aborted and lysed in comparison to the round and fully developed pollen in non-transgenic plants. Microscopic analysis of anthers showed selective degradation of the tapetum in transgenic plants with no microspore development. One week after self-pollination, the ovules of non-transgenic plants were double the size of those in transgenic plants, due to successful self-fertilization. It was also observed that male sterile transgenic plants set seed normally, when cross-pollinated with pollen from non-transgenic plants, confirming no adverse effect on the female parts of the flower (Nawaz-ul-Rehman et al. 2007).

Another way of using RNAi to create male sterility is to silence anther-specific promoters using hairpin structures directed against promoter sequences. The maize gene MS45 is expressed exclusively in the tapetal layer of anthers during microspore development and mutation results in a male-sterile phenotype. A high frequency of male-sterile plants was obtained by constitutively expressing a hp construct containing fragments of the MS45 promoter (Cigan et al. 2005).

One drawback of the RNAi mediated production of male sterile plants mentioned above is that the trait is constitutive and there is thus no simple way of multiplying the "female line". This could be overcome by the use of an inducible promoter such that, for multiplication of the female line, the male-sterile trait is not expressed. For production of the hybrid lines, the male sterility would be switched on allowing only progeny resulting from cross-pollination to be produced.

A somewhat different approach was adopted to improve the productivity of oilseed rape (*Brassica napus*). This species produces a bright-yellow canopy of flowers that can absorb nearly 60% of photosynthetically active radiation, result ing in reduced yield. A plant with reduced (or no) petals was achieved using a hpRNA construct targeted at the BPI gene family (MADS-box floral organ identity genes) under the control of a chimeric, petal-specific promoter derived from *Arabidopsis* (Byzova et al. 2004). The resultant plants produced male fertile flowers in which petals were converted into sepals (*Arabidopsis*) or into sepaloid petals (*B. napus*).

DISEASE RESISTANCE BY RNAi

Gene silencing can be considered to be the first line of defense of plants against invading pathogens (Baulcombe 2003). The effects of gene silencing in plants were first used in efforts to develop resistance to diseases, particularly those caused by viruses, although the mechanism was not clear at the time. This so-called "pathogen-derived resistance" (PDR) was achieved expressing pathogen genes (sequences) in plants, with the aim of blocking a specific step in the life/infection cycle of the pathogen. Many of the strategies used for PDR were subsequently shown to be mediated by RNA, rather than protein and led directly to the identification of RNAi (Goldbach et al. 2003 , Voinnet 2001). A very important finding, which was first recognize in plants, was that once triggered, the silencing spreads throughout the organism by virtue of a silencing signal, thus providing systemic rather than just localized resistance (Voinnet et al. 1998).

The effectiveness of RNAi technology for generating virus resistance in plants was first demonstrated in 1998. Immunity to *Potato virus Y* (PYY) in potato was achieved by the simultaneous expression of both sense and antisense transcripts of the viral helper-component proteinase (HC-Pro) gene (Waterhouse et al. 1998). This approach has since been used to achieve immunity in plants to a diverse range of RNA viruses (Table 2).

A hpRNA construct was designed to target the nuclear inclusion protein 'a' (NIa) gene of *Wheat streak mosaic virus* (WSMV). Wheat was stably transformed with the hpRNA and the T_1 progeny were assessed for susceptibility to WSMV. Ten of sixteen lines showed complete resistance to the virus with no disease symptoms evident, no virus detected by ELISA, no viral sequences detected by RT-PCR from leaf extracts and leaf extracts failed to give infections when inoculated to susceptible plants. The accumulation of small RNAs derived from the hpRNA transgene sequence was shown to positively correlate with the level of resistance (Fahim et al. 2010).

A more recent development has been the use of miRNAs to develop resistance against plant viruses. Niu et al., (2006) modified an *Arabidopsis* miR159 precursor to express artificial miRNAs (amiRNAs) containing viral RNA sequences derived from the genes of two gene silencing suppressors; P69 of *Turnip yellow mosaic virus* (TYMV) and HC-Pro of *Turnip mosaic virus* (TuMV). Transgenic *Arabidopsis* plants expressing amiR-P69 (159) and amiR-HC-Pro(159) were specifically resistant to TYMV and TuMV, respectively. Expression of amiR-TuCP(159), targeting the TuMV coat protein gene, similarly conferred specific TuMV resistance. Transgenic plants that express both amiR-P69(159) and amiR-HC-Pro(159) from a dimeric pre-amiR-P69(159)/amiR-HC-Pro(159) transgene were resistant to both viruses (Niu et al. 2006). A similar study using amiRNA, targeting sequences encoding the silencing suppressor 2b of *Cucumber mosaic virus* (CMV), efficiently inhibited 2b gene expression and suppressor function in transient expression assays and conferred resistance to CMV infection in transgenic tobacco. Moreover, the resistance level conferred by the transgenic miRNA correlated with the expression level of the amiRNA (Qu et al. 2007).

Several studies have shown the control of ssDNA viruses, specifically the geminiviruses, by RNAi. The non-coding intergenic region of the geminivirus *Mungbean yellow mosaic India virus* (MYMIV) was expressed as a hp-construct under the control of the 35S promoter and used to biolistically inoculate MYMIV-infected black gram (*Vigna mungo*) plants. Plants treated with the construct showed a complete recovery from infection that lasted until senescence. This work showed that phytopathogenic DNA viruses can potentially be controlled by RNAi and that promoter sequences (which are not usually transcribed but can possibly be covered by fortuitous read-through) make a suitable target for silencing (Pooggin et al. 2003). In several reports the coding sequences of geminiviruses, particularly the replication-associated protein (Rep; a rolling-circle initiator protein essential for viral DNA replication), have been used as a target for PDR (Bendahmane and Gronenborn 1997, Ilyas et al. 2010, Sangaré et al. 1999). The potential for silencing geminiviruses by RNAi using a transient protoplast assay has also been shown (Vanitharani et al. 2003). Protoplasts were co-transfected with a siRNA designed to the Rep coding sequence of *African cassava mosaic virus* (ACMV) and the genomic DNA of ACMV resulting in a 91% reduction in Rep transcript and 66% reduction in viral DNA. This siRNA was able to silence a closely related strain of ACMV but not a more distantly related virus. Subsequently the complete Rep gene of ACMV was transformed, in sense orientation, into cassava (*Manihot esculenta*). Resistance was shown following challenge with infectious clones of ACMV and viral DNA levels were reduced by 98%. Despite the presence of virus, these plants remained symptomless. In addition, the ACMV Rep transgene provided good protection against several distantly related geminiviruses, showing the potential of RNAi for developing broad-spectrum resistance, something that has not been possible for geminiviruses using other means.

Recently RNAi based resistance against a geminivirus, *Bean golden mosaic virus* (BGMV; family *Geminiviridae*), has been reported that has progressed to field testing in Brazil. An hp-construct was produced targeting Rep sequences. A total of 22 transgenic common bean lines were obtained and evaluated under greenhouse conditions. Two of thee lines showed high resistance (~93% of the plants were symptom free) under high inoculation pressure (>300 viruliferous vector whiteflies per

Table 2. Published studies reporting the use of RNAi to achieve virus resistance in plants.

Name of virus	Family	Genome	Region targeted	Results	Host Plant	Reference
Potato Virus Y	Potyviridae	RNA	Helper component protease gene	Immunity	Solanum tuberosum (Potato)	(Waterhouse et al. 1998)
Pepper mild mottle virus	Tobamoviridae	RNA	Replicase gene	Resistance	Nicotiana tabacum (Tobacco)	(Tenllado et al. 2003)
Plum pox virus	Potyviridae	RNA	Helper component and coat protein genes	Resistance	Nicotiana benthamiana	(Tenllado et al. 2003)
Alfalfa mosaic virus	Bromoviridae	RNA	Arbitrary sequence	Tolerance	Nicotiana tabacum (Tobacco)	(Tenllado and Díaz-Ruiz 2001)
Beet necrotic yellow vein virus	Benyviridae	RNA	Coat protein gene	Tolerance	Nicotiana tabacum (Tobacco)	(Andika et al. 2005)
Tobacco mosaic virus	Tobamoviridae	RNA	Replication-associated protein gene	Resistance	Nicotiana tabacum (Tobacco)	(Zhao et al. 2006)
Turnip yellow mosaic virus	Tymoviridae	RNA	P69	Resistance	Arabidopsis thaliana	(Niu et al. 2006)
Turnip mosaic virus	Potyviridae	RNA	Helper component protease gene	Resistance	Arabidopsis thaliana	(Niu et al. 2006)
Mungbean yellow mosaic India virus	Geminiviridae	DNA	Bidirectional promoter	Tolerance	Vigna mungo (Blackgram)	(Pooggin et al. 2003)
African cassava mosaic virus	Geminiviridae	DNA	Replication-associated protein gene	Tolerance	Nicotiana tabacum (Tobacco protoplasts)	(Vanitharani et al. 2003)
Bean golden mosaic virus	Geminiviridae	DNA	Replication-associated protein gene	Resistance	Phaseolus vulgaris (Common bean)	(Aragao and Faria 2009)

plant). In the field, 2–10 whiteflies per plant led to 100% virus infection of wild-type plants. In further studies homozygous plants were crossed with non-transgenic plants to generate a hemizygous population. Both homozygous and hemizygous plants were inoculated using viruliferous whiteflies. Two weeks after inoculation, 100% of homozygous plants remained symptomless, whereas 28.7% of hemizygous plants showed mild symptoms and all non-transgenic plants showed severe symptoms. In field trials in areas of both high and low inoculum pressure, transgenic plants remained symptomsless, whereas non-transgenic plants showed 18% and 83% infection, respectively (Aragão and Faria 2009). This was the first report of RNAi based resistance to a geminivirus being tested in the field. A comparison of the transgenic and non-transgenic common bean lines, following challenge with BGMV, is shown in Figure 2.

RNAi is hypothesized to be a plant defense against invading nucleic acids such as viruses. Evidence in support of this is the fact that viruses have evolved a counter defense against RNAi by encoding proteins that can overcome this resistance that are referred to as RNA silencing suppressors (RSS) (Silhavy and Burgyan 2004). Many RSSs have been identified but these have no obvious sequence homology, appear to have evolved independently and act at differing steps in the RNAi pathway. Some viral RSSs bind siRNAs and sequester siRNA from host RNA silencing machinery (Goto et al. 2007, Lozsa et al. 2008, Vargason et al. 2003, Ye et al. 2003). RSSs are often involved in viral pathogenicity and mediate synergism among plant viruses, a phenomenon whereby two viruses, each providing an essential factor for the synergism, induce a more severe disease than either on their own (Pruss et al. 1997, Vanitharani et al. 2004).

Attempts to obtain RNAi-mediated resistance against the geminivirus *Tomato yellow leaf curl Sardinia virus*, also by targeting Rep sequences, resulted in either no or limited resistance (Noris et al. 2004). This suggests that RNAi-mediated resistance might not work against all geminiviruses. The reason for this might be the differential expression of RNAi against viruses. In plants with RNAi-mediated resistance to the coat protein of *Beet necrotic yellow vein virus* the expression of resistance differed between tissues, levels of resistance in leaves being higher than in roots (Andika et al. 2005). Recent studies have shown that geminiviruses use different strategies to overcome RNA silencing. A better understanding of suppression of RNA silencing by these viruses might thus be needed for the effective use of RNA silencing-mediated resistance (Bisaro 2006, Lucioli et al. 2008, Raja et al. 2010).

RNAi has similarly been used to provide protection against phytopathogenic bacteria. *Agrobacterium tumefaciens* causes crown gall disease that affects many perennial fruit, nut and ornamental

Figure 2. Comparison of non-transgenic (left) and transgenic (right) common bean (*Phaseolus vulgaris*) plants following exposure to viruliferous whiteflies (*Bemisia tabaci*) harbouring *Bean golden mosaic virus* (BGMV) in the glasshouse. The transgenic line was transformed with a hairpin RNAi construct containing BGMV replication-associated protein (Rep) gene sequence. The non-transgenic line shows severe disease symptoms (yellow mosaic, and reduced leaf size) whereas the transgenic plant is phenotypically normal. Note that the transgenic line has produced normal bean pods whereas the non-transgenic line has not. The photo was kindly provided by Francisco J.L. Aragão (Embrapa-Recursos Genéticos e Biotecnologia.Brasília-DF, Brazil) and Josias C. Faria (Embrapa Arroz and Feijao, Santo Antonio De Goias, Brazil).

(Color image of this figure appears in the color plate section at the end of the book.)

crops. There are two genes that play an important role in this disease: iaaM, encoding a tryptophan monooxygenase that converts tryptophan to the auxin precursor indoleacetamide (Depicker et al. 1978) and ipt, encoding a product catalysing the condensation of AMP and isopentenyl pyrophosphate to form the cytokinin zeatin (Lichtenstein et al. 1984). Expression of both of these oncogenes is required for wild-type tumour formation (Ooms et al. 1981). Transgenic *A. thaliana* and *S. lycopersicum* transformed with RNAi constructs targeting iaaM and ipt showed resistance to crown gall disease. RNA silencing plays an important role in establishing crown gall disease and plants deficient in silencing are hyper-susceptible to *A. tumefaciens* (Dunoyer et al. 2006). Successful infection relies on a potent anti-silencing state established in tumours whereby siRNA synthesis is specifically inhibited.

Plant parasitic nematodes are important crop pest throughout the world that cause heavy economic losses (Lilley et al. 2007). RNAi has been used effectively to target nematode genes important for pathogenicity on plants. Expression of silencing constructs against these genes in transgenic plants might provide protection against phytopathogenic nematodes (Bakhetia et al. 2005). A nematode parasitism gene 16D10 encodes a conserved root-knot nematode (RKN) secretory peptide that stimulates root growth and functions as a ligand for a putative plant transcription factor. *In vitro* and *in vivo* RNAi approaches were used to silence this gene in RKN. Ingestion of 16D10 dsRNA *in vitro* silenced the gene in RKN and resulted in reduced nematode infectivity. *In vivo* expression of 16D10 dsRNA in *Arabidopsis* resulted in resistance effective against the four major RKN species (Huang et al. 2006). Yadev et al. (2006) introduced an RNAi vector harbouring target sequence for nematode splicing factor and integrase genes into tobacco plants. The mRNA levels for the splicing factor and integrase genes markedly decrease in nematodes, but mRNA levels of other genes did not significantly change (Yadav et al. 2006). These results indicate that inactivation of target genes in nematodes probably results from an RNAi like mechanism.

RNAi AND INSECT PESTS THAT AFFECT CROPS

A somewhat surprising finding has been that RNAi can be exploited to control insect pests by *in planta* expression of dsRNA specific to target insect gene sequences. In 2007 it was shown that RNAi can be induced by micro-injection of dsRNA into the pea aphid, *Acyrthosiphon pisum*. Two marker genes were selected, the ubiquitously expressed Ap-crt gene encoding a calreticulin and the gut specific Ap-cath-L gene encoding a cathepsin-L and dsRNA was produced for each by co-*in vitro* translation of sense and anti-sense sequences. Analysis of injected aphids showed that gene expression decrease by 40% (Jaubert-Possamai et al. 2007). This study proved that siRNA can be used directly to silence target genes in insect pests. A similar study in the whitefly *Bemisia tabaci*, targeting genes uniquely expressed in the midgut and salivary glands, reduced expression levels by 70% in comparison to whiteflies injected with buffer or with a green fluorescent protein-specific dsRNA (Ghanim et al. 2007). A number of other insect species have since been shown to be susceptible to gene silencing by orally administered dsRNA (Bautista et al. 2009, Zhao et al. 2008, Zhou et al. 2008).

Whyard et al. (2009) using orally-delivered dsRNAs, showed that even closely related species can be killed selectively by feeding on dsRNAs that target the more variable regions of genes, such as the 3' un-translated regions. The study used the E-subunit of the H^+ transporting lysosomal vacuolar proton pump (vATPase) gene from four distinct insect species; fruit flies (*Drosophila melanogaster*), flour beetles (*Tribolium castaneum*), pea aphids (*Acyrthosiphon pisum*) and tobacco hornworms (*Manduca sexta*). The same study achieved even greater selectivity in specifically killing distinct species in the genus *Drosophila* by silencing the γ-tubulin gene, highlighting the usefulness of the approach they used.

The real break-through came with the realization that insect control can be achieved by the expression of dsRNA *in planta*, leading to siRNAs that silence genes in insects feeding on plants. Baum et al. (2007) reported a reduction in the damage cause to maize by the western corn rootworm (*Diabrotica virgifera*) in transgenic plants expressing dsRNA to the vATPase gene. The model plants *Nicotiana tabacum* and *A. thaliana* were modified with a construct for the expression of a hpRNA containing the cytochrome P450 gene sequences of the cotton bollworm, *Helicoverpa armigera* (Mao et al. 2007). Cotton bollworm larvae fed leaves from transformed plants showed reduced levels of cytochrome P450 mRNA in mudgut cells and retarded growth. This effect was significantly enhanced when the leaves were supplemented with gossypol, a toxin produced by cotton.

OFF-TARGET EFFECTS

The specificity of RNAi is determined by the sequence similarity between the siRNA generated by silencing constructs and incorporated into RISC of the target gene with. However, one of the limitations of the technology could be off-target effects of siRNA that might result in silencing of non-target genes (Jackson et al. 2003, Malik et al. 2006, Mansoor et al. 2006). A system that was developed to identify possible off-target effects in plants found no off-target effects when used to investigate the silencing of the salicylic acid-binding protein 2 gene (Kumar et al. 2006). Another study to investigate off-target effects of RNAi in plants reported that siRNAs with at least 22 nt of continuous identity are required for silencing of non-target genes. A microarray-based comparison of the transcriptome of silenced and non-silenced *A. thaliana* plants identified no differences amongst known transcripts (Aelbrecht et al. 2006).

Most reports of off-target effects of siRNAs are of translational repression of non-target genes resulting from imperfect complementarity to the 3′ un-translated regions of genes in animals (Birmingham et al. 2006, Lin et al. 2005). A possible reason for the absence of such effects in plants is that miRNA in plants cause silencing because of their complementarity to coding sequences that result in cleavage of mRNA (Bartel 2004). Some modification may be made in siRNA stability and delivery *in vivo*, by chemical modification of 2′-O-methyl ribosyl at position 2 in the guide strand, which reduces the silencing of most off-target transcripts (Jackson et al. 2006). Such modified siRNAs are compatible with the cellular siRNA machinery and are useful for reducing undesirable, sequence-related off-target effects but are limited to mammalian systems (Elmen et al. 2005). The use of synthetic siRNAs in plants is still limited to experimental applications and is not applicable to the transgenic hpRNA strategy—the most widely used strategy for generating siRNAs in plants. Nevertheless, the possibility of off-target effects in plants cannot be ruled out and therefore requires further investigation. Caution is warranted in interpreting gene function and phenotype information resulting from RNAi-mediated gene knock-down experiments. RNAi data should always be validated using the theoretical and practical tools available to predict and identify potential off-target effects of siRNAs in plants.

CONCLUSION

Since the discovery of RNAi, it has become the technology of choice for plant scientists investigating gene function and manipulating plants to generate novel traits. RNAi-based methods have proven to be important tools for addressing fundamental questions in the biology of living organisms. So far no data is available about the stability of synthetic, engineered RNAi based traits in transgenic plants, although a naturally occurring mutant that mimics the hpRNA effect has been shown to be stable for 20 generations, suggesting that engineered traits should be similarly stable (Kusaba et al. 2003). The use of tissue-specific and inducible promoters should improve the ability to silence gene expression in only the target tissues and when required, thus minimizing possible 'off-target' effects. Another possibility, that so far has not been explored, is the silencing of undesirable or pathogen genes in only the root-stock of grafted crops, such as fruit trees and relying on the systemic silencing signal to deliver siRNAs to the scion. This could positively affect public opinion regarding the use of established transgenic technologies, since resulting fruit would not be transgenic.

A major benefit of RNAi technology is that a relatively small sequence derived from the target gene is required for silencing, meaning that multiple genes (or multiple pathogens) can be targeted in a single construct. Currently efforts are underway to silence multiple genes using such a single construct approach. This would reduce the amount of manipulation and time required to achieve the desired trait. Because a 21nt sequence can trigger RNAi, so this ability should enable the targeting of shorter and conserved sequences to develop broad-spectrum resistance against pathogens with high levels of variability, such as viruses. Conversely, such a small sequence allows RNAi to be targeted very selectively, as in the study of Whyard et al. (2009) that discriminated between species in a single genus. However, it is important to note that we do not yet fully understand the complexities of RNAi and a wealth of information about this important process has yet to be discovered. It is only recently, for example, that the Nature of the mobile silencing signal has been elucidated (Dunoyer et al. 2010a, Molnar et al. 2010, Searle et al. 2010). It is for this reason that we cannot yet fully predict the likelihood, or effects, of off-target silencing. Nevertheless, it is also clear that full potential of RNAi for crop improvement has yet to be explored in full.

REFERENCES

Aelbrecht, T., M. Vuylsteke, M. Bauwens, H. Van Houdt and A. Depicker. 2006. Introduction of silencing-inducing transgenes does not affect expression of known transcripts. *FEBS Lett*. **580**:4154–4159.

Allen, R.S., A.G. Millgate, J.A. Chitty, J. Thisleton, J.A. Miller, A.J. Fist, W.L. Gerlach and P.J. Larkin. 2004. RNAi-mediated replacement of morphine with the nonnarcotic alkaloid reticuline in opium poppy. *Nat. Biotech*. **22**:1559–1566.

Andersson, M., M. Melander, P. Pojmark, H. Larsson, L. Bulow and P. Hofvander. 2006. Targeted gene suppression by RNA interference: an efficient method for production of high-amylose potato lines. *J. Biotech*. **123**:137–148.

Andika, I.B., H. Kondo and T. Tamada. 2005. Evidence that RNA silencing-mediated resistance to *Beet necrotic yellow vein virus* is less effective in roots than in leaves. *Mol. Plant-Microbe Interact*. **18**:194–204.

Aragão, F.J.L. and J.C. Faria. 2009. First transgenic geminivirus-resistant plant in the field. *Nat. Biotech*. **27**:1086–1088.

Bakhetia, M., W. Charlton, H.J. Atkinson and M.J. McPherson. 2005. RNA intereference of dual oxidase in the plant nematode *Meloidogyne incognita*. *Mol. Plant-Microbe Interact*. **18:**1099–1106.

Bartel, D.P. 2004. MicroRNAs: genomics, biogenesis, mechanism and function. *Cell*. **16:**281–297.

Baulcombe, D. 2003. Overview of RNA interference and related processes. *Curr. Protoc. Mol. Biol*. Chap.**26:** Unit 26.1.

Baulcombe, D. 2004. RNA silencing in plants. *Nature*. **431:**356–363.

Baum, J.A., T. Bogaert, W. Clinton, G.R. Heck, P. Feldmann, O. Ilagan, S. Johnson, G. Plaetinck, T. Munyikwa, M. Pleau, T. Vaughn and J. Roberts. 2007. Control of coleopteran insect pests through RNA interference. *Nat. Biotechnol*. **25:**1322–1326.

Bautista, M.A., T. Miyata, K. Miura and T. Tanaka. 2009. RNA interference-mediated knockdown of a cytochrome P450, CYP6BG1, from the diamondback moth, *Plutella xylostella*, reduces larval resistance to permethrin. *Insect Biochem. Mol. Biol*. **39:**38–46.

Bendahmane, M. and B. Gronenborn. 1997. Engineering resistance against tomato yellow leaf curl virus (TYLCV) using antisense RNA. *Plant Mol. Biol*. **33:**351–357.

Birmingham, A., A.M. Anderson, A. Reynold, D. Ilsley-Tyree, D. Leake, Y. Fedorov, S. Baskerville, E. Maksimova, K. Robinson, J. Karpilow, W.S. Marshall and A. Khvorova. 2006. 3′ UTR seed matches, but not overall identity are associated with RNAi off-targets. *Nat. Methods*. **3:**188–204.

Bisaro, D.M. 2006. Silencing suppression by geminivirus proteins. *Virology*. **344:**158–168.

Brigneti, G.and A.M. Martin-Hernandez, H. Jin, J. Chen, D.C. Baulcombe, B. Baker and J.D.G. Jones. 2004. Virus-induced gene silencing in *Solanum* species. *Plant J*. **39:**264–272.

Byzova, M., C. Verduyn, D. De Brouwer and M. Block. 2004. Transforming petals into sepaloid organs in *Arabidopsis* and oilseed rape: implementation of the hairpin RNA-mediated gene silencing technology in an organ-specific manner. *Planta*. **218:**379–387.

Cai, X., C. Wang, Y. Xu, Q. Xu, Z. Zheng and X. Zhou. 2007. Efficient gene silencing induction in tomato by a viral satellite DNA vector. *Virus Res*. **125:**169–75.

Carrington, J.C. and V. Ambros. 2003. Role of microRNAs in plant and animal development. *Science*. **301:**336–338.

Chai, X.J., P.W. Wang, S.Y. Guan and Y.W. Xu. 2005. Reducing the maize amylopectin content through RNA interference manipulation. *Zhi Wu Sheng Li Yu Fen Zi Sheng Wu Xue Xue Bao*. **31:**625–630.

Cigan, A.M., E. Unger-Wallace and K. Haug-Collet. 2005. Transcriptional gene silencing as a tool for uncovering gene function in maize. *Plant J*. **43:**929–940.

Covey, S.N., N.S. Al-Kaff, A. Langara and D.S.P. Turner. 1997. Plants combat infection by gene silencing. *Nature*. **385:**781–782.

Davuluri, G.R., A.V. Tuinen, P.D. Fraser, A. Manfredonia, R. Newman, D. Burgess, D.A. Brummell, S.R. King, J. Palys, J. Uhlig, P.M. Bramley, H.M.J.

Pennings and C. Bowler. 2005. Fruit-specific RNAi-mediated suppression of DET1enhances carotenoid and flavonoid content in tomatoes. *Nat. Biotechnol.* **23**:890–895.

Depicker, A., M. Van Montagu and J. Schell. 1978. Homologous DNA sequences in different Ti-plasmids are essential for oncogenicity. *Nature.* **275**:150–153.

Dhankher, O.P., B.P. Rosen, E.C. McKinney and R.B. Meagher. 2006. Hyperaccumulation of arsenic in the shoots of *Arabidopsis* silenced for arsenate reductase (ACR2). *Proc. Natl. Acad. Sci. USA.* **103**:5413–5418.

Djupedal, I. and K. Ekwall. 2009. Epigenetics: heterochromatin meets RNAi. *Cell. Res.* **19**:282–295.

Dodo, H.W., K.N. Konan, F.C. Chen, M. Egnin and O.M. Viquez. 2008. Alleviating peanut allergy using genetic engineering: the silencing of the immunodominant allergen Ara h 2 leads to its significant reduction and a decrease in peanut allergenicity. *Plant Biotechnol. J.* **6**:135–45.

Dunoyer, P., C.A. Brosnan, G. Schott, Y. Wang, F. Jay, A. Alioua, C. Himber and O. Voinnet. 2010a. An endogenous, systemic RNAi pathway in plants. *EMBO J.* **29**:1699–1712.

Dunoyer, P., C. Himber and O. Voinnet. 2006. Induction, suppression and requirement of RNA silencing pathways in virulent *Agrobacterium tumefaciens* infections. *Nat. Genet.* **38**:258–263.

Dunoyer, P., G. Schott, C. Himber, D. Meyer, A. Takeda, J.C. Carrington and O. Voinnet. 2010b. Small RNA duplexes function as mobile silencing signals between plant cells. *Science.* **328**:912–916.

Earley, K.W., J.R. Haag, O. Pontes, K. Opper, T. Juehne, K. Song and C.S. Pikaard. 2006. Gateway-compatible vectors for plant functional genomics and proteomics. *Plant J.* **45**:616–629.

Elmen, J., H. Thonberg, K. Ljungberg, M. Frieden, M. Westergaard, Y. Xu, B. Wahren, Z. Liang, H. Orum, T. Koch and C. Wahlestedt. 2005. Locked nucleic acid (LNA) mediated improvements in siRNA stability and functionality. *Nucleic Acids Res.* **33**:439–47.

Fahim, M., L. Ayala-Navarrete, A.A. Millar and P.J. Larkin. 2010. Hairpin RNA derived from viral NIa gene confers immunity to wheat streak mosaic virus infection in transgenic wheat plants. *Plant Biotechnol. J.* **8**:821–834.

Fire, A., S. Xu, M.K. Montgomery, S.A. Kostas, S.E. Driver and C.C. Mellow. 1998. Potent and specific genetic interference by double-stranded RNA in *Caenorhabditis elegans. Nature.* **391**:806–811.

Flores, T., O. Karpova, X. Su, P. Zeng, K. Bilyeu, D.A. Sleper, H.T. Nguyen and Z.J. Zhang. 2008. Silencing of GmFAD3 gene by siRNA leads to low alpha-linolenic acids (18:3) of fad3-mutant phenotype in soybean *Glycine max* Merr. *Transgenic Res.* **17**:839–850.

Fofana, I.B., A. Sangare, R. Collier, C. Taylor and C.M. Fauquet. 2004. A geminivirus-induced gene silencing system for gene function validation in cassava. *Plant Mol. Biol.* **56**:613–624.

Gan, D., J. Zhang, H. Jiang, T. Jiang, S. Zhu and B. Cheng. 2010. Bacterially expressed dsRNA protects maize against SCMV infection. *Plant Cell. Rep.* (in press).

Gasciolli, V., A.C. Mallory, D.P. Bartel and H. Vaucheret. 2005. Partially redundant functions of *Arabidopsis* DICER-like enzymes and a role for DCL4 in producing trans-acting siRNAs. *Curr. Biol.* **15**:1494–1500.

Gazzani, S., T. Lawrenson, C. Woodward, D. Headon and R. Sablowski. 2004. A link between mRNA turnover and RNA interference in *Arabidopsis*. *Science.* **306**:1046–8.

Ghanim, M., S. Kontsedalova and A.H. Czosnek. 2007. Tissue-specific gene silencing by RNA interference in the whitefly *Bemisia tabaci (Gennadius). Insect Biochem. Mol. Biol.* **37**:732–738.

Goldbach, R., E. Bucher and M. Prins. 2003. Resistance mechanisms to plant viruses: an overview. *Virus Res.* **92**:207–12.

Goldberg, R., T. Beals and P. Sanders. 1993. Anther development: basic principles and practical applications. *Plant Cell.* **5**:1217–1229.

Gorman, S.W. and S. McCormick. 1997. Male sterility in tomato. *Crit. Rev. Plant Sci.* **16**:31–53.

Goto, K., T. Kobori, Y. Kosaka, T. Natsuaki and C. Masuta. 2007. Characterization of silencing suppressor 2b of *Cucumber mosaic virus* based on examination of its small RNA-binding abilities. *Plant Cell. Physiol.* **48**:1050–1060.

Gracheva, E., M. Dus and S.C. Elgin. 2009. *Drosophila* RISC component VIG and its homolog Vig2 impact heterochromatin formation. *PLoS One.* **4**:e6182.

Hamilton, A.J. and D.C. Baulcombe. 1999. A species of small antisense RNA in post-transcriptional gene silencing. *Science.* **286**:950–952.

Hamilton, A.J., O. Voinnet, L. Chappell and D. Baulcombe. 2002. Two classes of short interfering RNA in RNA silencing. *EMBO J.* **21**:4671–4679.

Hammond, S.M., E. Bernstein, D. Beach and G.J. Hannon. 2000. An RNA-directed nuclease mediates post-transcriptional gene silencing in *Drosophila* cells. *Nature.* **404**:293–296.

Hannon, G.J. 2002. RNA interference. *Nature.* **418**:244–251.

Heilersig, H.J., A. Loonen, M. Bergervoet, A.M. Wolters and R.G. Visser. 2006. Post-transcriptional gene silencing of GBSSI in potato: effects of size and sequence of the inverted repeats. *Plant Mol. Biol.* **60**:647–662.

Helliwell, C.A., P.M. Waterhouse, D.R. Engelke and J.J. Rossi. 2005. Constructs and methods for hairpin RNA-mediated gene silencing in plants. *Methods Enzymol.* **392**:24–35.

Himber, C., P. Dunoyer, G. Moissiard, C. Ritzenthaler and O. Voinnet. 2003. Transitivity-dependent and -independent cell-to-cell movement of RNA silencing. *EMBO J.* **22**:4523 4533.

Hirai, S. and H. Kodama. 2008. RNAi vectors for manipulation of gene expression in higher plants. *Open Plant Science J.* **2**:31–40.

Huang, C., Y. Xie and X. Zhou. 2009. Efficient virus-induced gene silencing in plants using a modified geminivirus DNA1 component. *Plant Biotechnol. J.* **7**:254–65.

Huang, G., R. Allen, E.L. Davis, T.J. Baum and R.S. Hussey. 2006. Engineering broad root-knot resistance in transgenic plants by RNAi silencing of a conserved and essential root-knot nematode parasitism gene. *Proc. Natl. Acad. Sci. USA.* **103**:14302–14306.

Ilyas, M., I. Amin, S. Mansoor, R.W. Briddon and M. Saeed. 2010. Challenges for transgenic resistance to control geminiviral diseases. *In* "Emerging geminivirial diseases and their management" by Nova Science Publishers, Inc., Suite 1600, Hauooauge, NY 1788 USA (in press).

Jackson, A.L., S.R. Bartz, J. Schelter, S.V. Kobayashi, J. Burchard, M. Mao, B. Li, G. Cavet and P.S. Linsle. 2003. Expression profiling reveals off-target gene regulation by RNAi. *Nat. Biotchnol.* **21**:635–637.

Jackson, A.L., J. Burchard, D. Leake, A. Reynolds, J. Schelter, J. Guo, J.M. Johnson, L. Lim, J. Karpilow, K. Nichols, W. Marshall, A. Khvorova and P.S. Linsley. 2006. Position-specific chemical modification of siRNAs reduces "off-target" transcript silencing. *RNA.* **12**:1197–1205.

Jaubert-Possamai, S., G. Le Trionnaire, J. Bonhomme, G.K. Christophides, C. Rispe and D. Tagu. 2007. Gene knockdown by RNAi in the pea aphid *Acyrthosiphon pisum. BMC Biotechnol.* **7**:63.

Jones, L., A.J. Hamilton, O. Voinnet, C.L. Thomas, A.J. Maule and D.C. Baulcombe. 1999. RNA–DNA interactions and DNA methylation in post-transcriptional gene silencing. *Plant Cell.* **11**:2291–2301.

Jones, L., F. Ratcliff and D.C. Baulcombe. 2001. RNA-directed transcriptional gene silencing in plants can be inherited independently of the RNA trigger and requires Met1 for maintenance. *Current Biol.* **11**:747–757.

Kapoor, S., A. Kobayashi and H. Takatsuji. 2002. Silencing of the tapetum-specific zinc finger gene TAZ1 causes premature degeneration of tapetum and pollen abortion in petunia. *Plant Cell.* **14**:2353–2367.

Kumar, D., C. Gustafsson and D.F. Klessig. 2006. Validation of RNAi silencing specificity using synthetic genes: salicylic acid-binding protein 2 is required for innate immunity in plants. *Plant J.* **45**:863–868.

Kusaba, M., K. Miyahara, S. Iida, H. Fukuoka, T. Takano, H. Sassa, M. Nishimura and T. Nishio. 2003. Low glutelin content1: a dominant mutation that suppresses the glutelin multigene family via RNA silencing in rice. *Plant Cell.* **15**:1455–1467.

Lacomme, C. and S. Chapman. 2008. Use of *Potato virus X* (PVX)-based vector for gene expression and virus-induced gene silencing (VIGS). *Curr. Protoc. Microbiol.* Chap. **16**:Unit 16I.1.

Le, L.Q., Y. Lorenz, S. Scheurer, K. Fotisch, E. Enrique, J. Bartra, S. Biemelt, S. Vieths and U. Sonnewald. 2006a. Design of tomato fruits with reduced allergenicity by dsRNAi-mediated inhibition of ns-LTP (Lyc e 3) expression. *Plant Biotechnol. J.* **4**:231–242.

Le, L.Q., V. Mahler, Y. Lorenz, S. Scheurer, S. Biemelt, S. Vieths and U. Sonnewald. 2006b. Reduced allergenicity of tomato fruits harvested from Lyc e 1-silenced transgenic tomato plants. *J. Allergy Clin. Immunol.* **118**:1176–1183.

Lichtenstein, C., H. Klee, A. Montoya, D. Garfinkel, S. Fuller, C. Flores, E. Nester and M.J. Gordon. 1984. Nucleotide sequence and transcript mapping of the tmr gene of the pTiA6NC octopine Ti-plasmid: a bacterial gene involved in plant tumorigenesis. *Mol. Appl. Genet.* **2**:354–362.

Lilley, C.J., M. Bakhetia, W.L. Charlton and P.E. Urwin. 2007. Recent progress in the development of RNA interference for plant parasitic nematodes. *Mol. Plant Pathol.* **8**:701–11.

Lin, X., X. Ruan, M.G. Anderson, J.A. McDowell, P.E. Kroeger, S.W. Fesik and Y. Shen. 2005. siRNA-mediated off-target gene silencing triggered by a 7 nt complimentation. *Nucliec Acids Res.* 33:4527–4535. Lippman, Z. and R. Martienssen. 2004. The role of RNA interference in heterochromatic silencing. *Nature.* **431**:364–370.

Liu, J., M.A. Carmell, F.V. Rivas, C.G. Marsden, J.M. Thomson, J. Song, S.M. Hammond, L. Joshua-Tor and G.J. Hannon. 2004. Argonaute2 is the catalytic engine of mammalian RNAi. *Science Express.* **305**:1437–1441.

Liu, Q., S.P. Singh and A.G. Green. 2002. High-stearic and high-oleic cottonseed oils produced by hairpin RNA-mediated post-transcriptional gene silencing. *Plant Physiol.* **129**:1732–1743.

Lo, C., N. Wang and E. Lam. 2005. Inducible double-stranded RNA expression activates reversible transcript turnover and stable transla-tional suppression of a target gene in transgenic tobacco. *FEBS Lett.* **579**:1498–1502.

Lozsa, R., T. Csorba, L. Lakatos and J. Burgyan. 2008. Inhibition of 3′ modification of small RNAs in virus-infected plants require spatial and temporal co-expression of small RNAs and viral silencing-suppressor proteins. *Nucleic Acids Res.* **36**:4099–4107.

Lucioli, A., D.E. Sallustio, D. Barboni, A. Beradni, V. Papacchioli, R. Tavazza and M. Tavazza. 2008. A cautionary note on pathogen derived sequences. *Nat. Biotechnol.* **26**:617–619.

Malik, I., M. Garrido, M. Bahr, S. Kugler and U. Michel. 2006. Comparison of test systems for RNA interference. *Biochem. Biophys. Res. Commun.* **341**:245–253.

Mansoor, S., I. Amin, M. Hussain, Y. Zafar and R.W. Briddon. 2006. Engineering novel traits in plants through RNA interference. *Trends Plant Sci.* **11**:559–565.

Mao, Y.B., W.J. Cai, J.W. Wang, G.J. Hong, X.Y. Tao, L.J. Wang, Y.P. Huang and X.Y. Chen. 2007. Silencing a cotton bollworm P450 monooxygenase gene by plant-mediated RNAi impairs larval tolerance of gossypol. *Nat. Biotechnol.* **25**:1307–1313.

Mette, M.F., W. Aufsatz, J. van der Winden, M.A. Matzke and A.J. Matzke. 2000. Transcriptional silencing and promoter methylation triggered by double-stranded RNA. *EMBO J.* **19**:5194–5201.

Miki, D. and K. Shimamoto. 2004. Simple RNAi vectors for stable and transient suppression of gene function in rice. *Plant Cell. Physiol.* **45**:490–495.

Molnar, A., T. Csorba, L. Lakatos, E. Varallyay, C. Lacomme and J. Burgyan. 2005. Plant virus-derived small interfering RNAs originate predominantly from highly structured single-stranded viral RNAs. *J. Virol.* **79**:7812–7818.

Molnar, A., C.W. Melnyk, A. Bassett, T.J. Hardcastle, R. Dunn and D.C. Baulcombe. 2010. Small silencing RNAs in plants are mobile and direct epigenetic modification in recipient cells. *Science.* **328**:872–875.

Moritoh, S., D. Miki, M. Akiyama, M. Kawahara , T. Izawa, H. Maki and S.K. 2005. RNAi-mediated silencing of OsGEN-L (OsGEN-like), a new member of the

RAD2/XPG nuclease family, causes male sterility by defect of microspore development in rice. *Plant Cell. Physiol.* **46**:699–715.

Napier, J.A. 2007. The production of unusual fatty acids in transgenic plants. *Annu. Rev. Plant. Biol.* **58**:295–319.

Nawaz-ul-Rehman, M.S., S. Mansoor, A.A. Khan, Y. Zafar and R.W. Briddon. 2007. RNAi-mediated male sterility of tobacco by silencing TA29. *Mol. Biotechnol.* **36**:159–165.

Niu, Q.W., S.S. Lin, J.L. Reyes, K.C. Chen, H.W. Wu, S.D. Yeh and N.H. Chua. 2006. Expression of artificial microRNAs in transgenic *Arabidopsis thaliana* confers virus resistance. *Nat. Biotechnol.* **24**:1420–8.

Noris, E., A. Lucioli, R. Tavazza, P. Caciagli, G.P. Accotto and M. Tavazza. 2004. *Tomato yellow leaf curl Sardinia virus* can overcome transgene-mediated RNA silencing of two essential viral genes. *J. Gen. Virol.* **85**:1745–1749.

Ogita, S., H. Uefuji, M. Morimoto and H. Sano. 2004. Application of RNAi to confirm theobormine as the major intermediate for efficient biosynthesis in coffee plants with potential for construction of decaffeinated varieties. *Plant Mol. Biol.* **54**:931–941

Ooms, G., P.J.J. Hooykaas, G. Moolenaar and R.A. Schilperoort. 1981. Crown gall plant tumors of abnormal morphology, induced by *Agrobacterium tumefaciens* carrying mutated octopine Ti plasmids; analysis of T-DNA functions. *Gene.* **14**:33–50.

Palauqui, J.-C., T. Elmayan, J.-M. Pollien and H. Vaucheret. 1997. Systemic acquired silencing: transgene-specific post-transcriptional silencing is transmitted by grafting from silenced stocks to non-silenced scions. *EMBO J.* **16**:4738–4745.

Peacock, J. 1990. Ways to pollen sterility. *Nature.* **347**:714–715.

Peters, S., J. Imani, V. Mahler, K. Foetisch, S. Kaul, K.E. Paulus, S. Scheurer, S. Vieths and K.H. Kogel. 2010. Dau c 1.01 and Dau c 1.02-silenced transgenic carrot plants show reduced allergenicity to patients with carrot allergy. Transgenic Res. (in press).

Petrovska, N., X. Wu, R. Donato, Z. Wang, E. Ong, E. Jones, J. Forster, M. Emmerling, A. Sidoli, R.O. Hehir and G. Spangenberg. 2005. Transgenic ryegrasses (*Lolium* spp.) with down-regulation of main pollen allergens. *Mol. Breeding.* **14**:489–501.

Pidkowich, M.S., H.T. Nguyen, I. Heilmann, T. Ischebeck and J. Shanklin. 2007. Modulating seed beta-ketoacyl-acyl carrier protein synthase II level converts the composition of a temperate seed oil to that of a palm-like tropical oil. *Proc. Natl. Acad. Sci. USA.* **104**:4742–4747.

Pooggin, M., P.V. Shivaprasad, K. Veluthambi and T. Hohn. 2003. RNAi targeting of DNA virus in plants. *Nat. Biotechnol.* **21**:131–132.

Pruss, P., X. Ge, X.M. Shi, J.C. Carrington and V.B. Vance. 1997. Plant viral synergism: the potyviral genome encodes a broad-range pathogenicity enhancer that transactivates replication of heterologous viruses. *Plant Cell.* **9**:859–868.

Qi, Y. and G.J. Hannon. 2005. Uncovering RNAi mechanisms in plants: biochemistry enters the foray. *FEBS Lett.* **579**:5899.

Qu, J., J. Ye and R. Fang. 2007. Artificial microRNA-mediated virus resistance in plants. *J. Virol.* **81:**6690–6699.

Raja, P., J.N. Wolf and D.M. Bisaro. 2010. RNA silencing directed against geminiviruses: post-transcriptional and epigenetic components. *Biochim. Biophys. Acta.* **1799:**337–351.

Ratcliff, F.and B.D. Harrison and D.C. Baulcombe. 1997. A similarity between viral defense and gene silencing in plants. *Science.* **276:**1558–1560.

Regina, A., A. Bird, D. Topping, S. Bowden, J. Freeman, T. Barsby, B. Kosar-Hashemi, Z. Li, S. Rahman and M. Morell. 2006. High-amylose wheat generated by RNA interference improves indices of large-bowel health in rats. *Proc. Natl. Acad. Sci. USA.* **103:**3546–3551.

Sangaré, A., D. Deng, C.M. Fauquet and R.N. Beachy. 1999. Resistance to *African cassava mosaic virus* conferred by a mutant of the putative NTP-binding domain of the Rep Gene (AC1) in *Nicotiana benthamiana. Mol. Breeding.* **5:**95-102.

Searle, I.R., O. Pontes, C.W. Melnyk, L.M. Smith and D.C. Baulcombe. 2010. JMJ14, a JmjC domain protein, is required for RNA silencing and cell-to-cell movement of an RNA silencing signal in *Arabidopsis. Genes Dev.* **24:**986-991

Shimada, T., M. Otani, T. Hamada, K.S. Kim, Y. Takahata, K. Katayama, K. Kitahara and T. Suganuma. 2006. Trangenic sweet potato with amylose-free starch. *Acta Horticulturae.* **703:**141–144.

Silhavy, D. and J. Burgyan. 2004. Effects and side-effects of viral RNA silencing suppressors on short RNAs. *Trends Plant Sci.* **9:**76–83.

Simopoulos, A.P., A. Leaf and N. Salem, Jr. 2000. Workshop statement on the essentiality of and recommended dietary intakes for Omega-6 and Omega-3 fatty acids. Prostaglandins Leukot. *Essent. Fatty Acids.* **63:**119–121.

Smith, N.A., S. S.P., M.-B. Wang, P.A. Stoutjesdijk, A.G. Green and P.M. Waterhouse. 2000. Total silencing by intron-spliced hairpin RNAs. *Nature.* **407:**319–320.

Sunilkumar, G., L.M. Campbell, L. Puckhaber, R.D. Stipanovic and K.S. Rathore. 2006. Engineering cottonseed for use in human nutrition by tissue-specific reduction of toxic gossypol. *Proc. Natl. Acad. Sci. USA.* **103:**18054–18059.

Szittya, G., A. Molnár, D. Silhavy, C. Hornyik and J. Burgyán. 2002. Short defective interfering RNAs of tombusviruses are not targeted but trigger post-transcriptional gene silencing against their helper virus. *Plant Cell.* **14:**359–372.

Tenllado, F. and J.R. Díaz-Ruíz. 2001. Double-stranded RNA-mediated interference with plant virus infection. *J. Virol.* **75:**12288–12297.

Tenllado, F., B. Martinez-Garcia, M. Vargas and J. Diaz-Ruiz. 2003. Crude extracts of bacterially expressed dsRNA can be used to protect plants against virus infections. *BMC Biotechnol.* **3:**3.

Vaistij, F.E., L. Jones and D.C. Baulcombe. 2002. Spreading of RNA targeting and DNA methylation in RNA silencing requires transcription of the target gene and a putative RNA-dependent RNA polymerase. *Plant Cell.* **14:**857–867.

Vanitharani, R., P. Chellappan and C.M. Fauquet. 2003. Short interfering RNA-mediated interference of gene expression and viral DNA accumulation in cultured plant cells. *Proc. Natl. Acad. Sci. USA.* **100:**9632–9636.

Vanitharani, R., P. Chellappan, J.S. Pita and C.M. Fauquet. 2004. Differential roles of AC2 and AC4 of cassava geminiviruses in mediating synergism and suppression of posttranscriptional gene silencing. *J. Virol.* **78**:9487–9498.

Vargason, J.M., G. Szittya, J. Burgyan and T.M. Hall. 2003. Size selective recognition of siRNA by an RNA silencing suppressor. *Cell.* **115**:799–811.

Voinnet, O. 2001. RNA silencing as a plant immune system against viruses. *Trends Genet.* **17**:449–459.

Voinnet, O. 2005. Induction and suppression of RNA silencing: insights from viral infections. *Nat. Genet.* **6**:206–221.

Voinnet, O. 2009. Origin, biogenesis and activity of plant microRNAs. *Cell.* **136**:669–687.

Voinnet, O. and D.C. Baulcombe. 1997. Systemic signalling in gene silencing. *Nature.* **389**:553.

Voinnet, O., P. Vain, S. Angell and D.C. Baulcombe. 1998. Systemic spread of sequence-specific transgene RNA degradation in plants is initiated by localized introduction of ectopic promoterless DNA. *Cell.* **95**:177–187.

Waterhouse, P., M, M.W. Graham and M.-B. Wang. 1998. Virus resistance and gene silencing in plants can be induced by simultaneous expression of sense and antisense RNA. *Proc. Natl. Acad. Sci. USA.* **95**:13959–13964.

Wesley, S.V., C. Helliwell, N.A. Smith, M.-B. Wang, D. Rouse, Q. Liu, P.S. Gooding, S.P. Singh, D. Abbott, P.A. Stoutjesdijk, S.P. Robinson, A.P. Gleave, A.G. Green and P.M. Waterhouse. 2001. Constructs for efficient, effective and high throughput gene silencing in plants. *Plant J.* **27**:581–590.

Whyard, S., A.D. Singh and S. Wong. 2009. Ingested double-stranded RNAs can act as species-specific insecticides. *Insect Biochem. Mol. Biol.* **39**:824–832.

Wingard, S.A. 1928. Hosts and symptoms of ring spot, a virus disease of plants. *J. Agri. Res.* **37**:127–153. Xie, Z., E. Allen, A. Wilken and J.C. Carrington. 2005. DICER-LIKE 4 functions in trans-acting small interfering RNA biogenesis and vegetative phase change in *Arabidopsis thaliana. Proc. Natl. Acad. Sci. USA.* **102**:12984–12989.

Xie, Z., L.K. Johansen, A.M. Gustafson, K.D. Kasschau, A.D. Lellis, D. Zilberman, S.E. Jacobsen and J.C. Carrington. 2004. Genetic and functional diversification of small RNA pathways in plants. *PLoS Biol.* **2**:e104.

Xiong, S., Q. Yao, R. Peng, X. Li, P. Han and H. Fan. 2004. Different effects on ACC oxidase gene silencing triggered by RNA interference in transgenic tomato. *Plant Cell. Rep.* **23**:639–646.

Yadav, B.C., K. Veluthambi and K. Subramaniam. 2006. Host-generated double stranded RNA induces RNAi in plant-parasitic nematodes and protects the host from infection. *Mol. Biochem. Parasitol.* **148**:219–222.

Ye, K., L. Malinina and D.J. Patel. 2003. Recognition of small interfering RNA by a viral suppressor of RNA silencing. *Nature.* **426**:874–878.

Zhao, M.M., D.R. An, J. Zhao, G.H. Huang, Z.H. He and J.Y. Chen. 2006. Transiently expressed short hairpin RNA targeting 126 kDa protein of *Tobacco mosaic virus* interferes with virus infection. *Acta. Biochim. Biophys. Sin.* (Shanghai). **38**:22–28.

Zhao, Y.Y., G. Yang, G. Wang-Pruski and M.S. You. 2008. *Phyllotreta striolata* (Coleoptera:Chrysomelidae): arginine kinase cloning and RNAi-based pest control. *Eur. J. Entomol.* **105:**815–822.

Zhou, X., M.M. Wheeler, F.M. Oi and M.E. Scharf. 2008. RNA interference in the termite *Reticulitermes flavipes* through ingestion of double-stranded RNA. *Insect Biochem. Mol. Biol.* **38:**805–815.

Zilberman, D., X. Cao, L.K. Johansen, Z. Xie, J.C. Carrington and S.E. Jacobsen. 2004. Role of *Arabidopsis* ARGONAUTE4 in RNA-directed DNA methylation triggered by inverted repeats. *Curr. Biol.* **14:1214–1220.**

CHAPTER 11

RNA Silencing and Viral Encoded Silencing Suppressors

Masato Ikegami,* Tatsuya Kon and Pradeep Sharma

ABSTRACT

RNA silencing is a sequence-specific degradation of RNA and a means of regulation of gene expression via small RNA molecules. In plants this process is commonly referred to as post-transcriptional gene silencing (PTGS). PTGS is characterized by accumulation of small-interfering RNAs (siRNAs) and microRNAs (miRNAs), degradation of target mRNAs and methylation of homologous gene sequences. In addition to gene regulation, PTGS also plays an important role as an immunity system in plants, which acts against diverse molecular parasites or foreign genetic elements including transposons, transgenes, viruses and viroids. Most plant viruses have evolved the capacity to counteract this defence response by encoding suppressors of PTGS. Viral suppressors are highly diverse in sequence and act at distinct steps in PTGS. Furthemore, the relative ability of viral suppressors to interfere with PTGS in plants can be associated with pathogenicity. In this review, we focus on recent advances in understanding RNA silencing and its suppression by plant viruses.

Keywords: RNA silencing, PTGS, virus-induced gene silencing (VIGS), plant viruses, suppressor.

INTRODUCTION

RNA silencing is a mechanism that results in sequence-specific degradation of RNA in eukaryotes. This phenomenon has been termed RNA interference (RNAi) in animals, quelling in fungi and post-transcriptional gene silencing (PTGS) in plants (Baulcombe 2004, Fulci and Macino 2007, Umbach and Cullen 2009). A common feature of RNA silencing is that it is

---

* Corresponding author e-mail: m3ikegam@nodai.ac.jp

most strongly triggered by double-stranded RNA (dsRNA); however, it is also triggered by aberrant RNAs associated with transposons, transgenes and viruses. RNAs with hairpin structures are particularly effective inducers of PTGS in plants. All of these RNAs are cleaved by Dicer and dicer-like (DCL) enzymes, which are members of the RNase III family of dsRNA-specific endonucleases, to generate RNAs of 21–25 nucleotides (nt). These are generally referred to as small interfering RNAs (siRNAs) (Chapman and Carrington 2007). siRNAs become incorporated into an RNA-induced silencing complex (RISC), which includes Argonaute (AGO)-type family proteins that contain small RNA-binding PAZ and RNaseH-like PIWI domains. The RISC/siRNA complex binds and degrades RNA that is complementary in sequence to the bound siRNA (Matranga et al. 2005, Vaucheret 2008).

RNA silencing appears to be an ancient immunity system in plant that acts as a natural defense mechanism against pathogens, such as viruses and viroids and mobile genetic elements, such as transposons (Aliyari and Ding 2009). Plant viruses counteract this defense response by encoding suppressors of PTGS. Since the discovery of the first silencing suppressors encoded by plant viruses (Anandalakshmi et al. 1998, Brigneti et al. 1998, Kasschau and Carrington 1998), a remarkable diversity of such suppressors has been identified. This has revealed that most plant viruses have evolved one or more suppressors of RNA silencing and that suppressors are critical for viruses to evade host defenses and establish systemic infection. The identification of a wide diversity of suppressors in most plant viruses is consistent with RNA silencing being an ancient defense mechanism (Li and Ding 2006, Voinnet 2005). This review focuses on recent information regarding RNA silencing pathways in plants and suppression of these pathways by plant viruses. We also discuss the roles and interactions of viral suppressors and small RNAs in inhibition of the small RNA pathways and induction of symptoms of viral infection in plants.

MOLECULAR MECHANISMS OF PTGS IN PLANTS

PTGS was first discovered in the course of two studies of the expression of the *chalcone synthase* (*CHS*) gene in transgenic petunia (*Petunia hybrida*). These plants had been modified to overexpress *CHS*, with the aim of intensifying the purple coloration of flowers. Unexpectedly, the flowers of these transgenic plants were wholly or partially white and this was shown to be due to decreased abundance of both transgenic and endogenous *CHS* mRNA. This phenomenon, in which overexpression of an endogenous gene, via a transgene, triggers reduced gene expression, was subsequently observed in other types of transgenic plants and was termed co-suppression (Napoli et al. 1990, Van der Krol et al. 1990). Genetic analysis of *Arabidopsis* mutant lines deficient in PTGS has revealed several components of the PTGS pathway, including CARPEL FACTORY (CAF); SILENCING DEFECTIVE 1 (SDE1); SDE3; SUPPRESSOR OF GENE SILENCING 2 (SGS2) and SGS3; and a number of AGO proteins (Dalmay et al. 2000, Dalmay et al. 2001, Fagard et al. 2000, Jacobsen et al. 1999, Mourrain et al. 2000). CAF is a DCL enzyme, as it is a member of the family of plant-specific RNase III endonucleases. The *Arabidopsis* genome encodes four DCL and ten AGO proteins. Other factors involved in PTGS include RNA-dependent RNA polymerases (RDR) and dsRNA-binding (DRB) proteins, which work in concert with DCL and AGO proteins (Vaucheret 2008). Table 1 summarizes plant proteins involved in small RNA-mediated PTGS. Genetic studies have indicated that these key players in RNA silencing functionally interact with each other and are also involved in overlapping pathways, all of which are triggered by double-stranded (ds) and/or aberrant RNAs (Broderson and Voinnet 2006).

PTGS processes share three biochemical features: (1) generation of dsRNA, (2) processing of the dsRNA into various types of small RNAs and (3) inhibition of gene expression. The PTGS pathways can also be divided into initiation and maintenance steps (Figure 1). The initiation step is characterized by its dependence on a dsRNA trigger. A DCL generates siRNAs from this dsRNA. The AGO component(s) of the silencing complexes binds to small RNAs, which initiates the maintenance step (Baumberger and Baulcombe 2005, Henderson et al. 2006, Qi et al. 2005, Qi et al. 2006). The maintenance step is independent of the dsRNA trigger and is responsible for persistent silencing, which continues well after the trigger dsRNA has been degraded. RNA silencing is maintained through secondary synthesis of dsRNA by SDE1, using the siRNA-complementary target RNA as a template (Dalmay et al. 2000). In *Arabidopsis*, the RDR is also essential for sustained PTGS of transgenes, presumably because it is involved in synthesis of dsRNA (Vaistij et al 2002). SGS2 is required for RNA silencing of transgenes and some viruses in plants, but is not required for silencing induced by inverted repeat dsRNA (Mourrain et al. 2000). Thus, diverse small RNA pathways in plants have a common requirement for RDR.

Several lines of evidence indicate that siRNAs are involved in cell-to-cell and systemic movement of the silencing signal (Hamilton et al. 2002, Himber et al. 2003). In the absence of signal amplification, silencing can spread over 10–15 cells via movement of the primary 21-nt siRNA; however, extensive cell-to-cell movement requires generation of 21-nt siRNA via *de novo* synthesis of dsRNA by the action of RDR and the helper component, SGS3 (Dunoyer et al. 2005). Small RNAs can be detected in the phloem of silenced plants, consistent with these molecules serving as signals for initiation of silencing (Yoo et al. 2004).

Table 1. Plant proteins involved in RNA silencing.

Protein	Function	Reference
ARGONAUTE 1 (AGO1)	RNA slicer	(Baumberger and Baulcombe 2005, Brodersen et al. 2008, Qi et al. 2005)
AGO2	siRNA-binding	(Takeda et al. 2008)
AGO4	RNA slicer, *de novo* DNA methylation and ra-siRNA binding	(Herr et al.2005, Qi et al. 2006, Zilberman et al. 2003)
AGO5	siRNA-binding	(Takeda et al. 2008)
AGO6	ra-siRNA-directed heterochromatin formation	(Zheng et al. 2007)
AGO7	RNA slicer	(Adenot et al. 2006, Montgomery et al. 2008, Montgomery et al. 2008)
DEFECTIVE IN RNA-DIRECTED DNA MRTHYLATION 1 (DRD1)	RNA-directed *de novo* DNA methylation	(Huettel et al. 2006, Kanno et al. 2005)
DICER-LIKE 1 (DCL1)	miRNA biogenesis	(Kurihara and Watanabe 2004)
DCL2	nat-siRNA biogenesis	(Borsani et al. 2005, Henderson et al. 2006)
DCL3	ra-siRNA biogenesis	(Henderson et al. 2006, Herr et al. 2005, Qi et al. 2005)
DCL4	ta-siRNA biogenesis	(Dunoyer et al. 2005, Henderson et al. 2006, Xie et al. 2005, Yoshikawa et al. 2005)
DOMAINS REARRANGED METHYLTRASFERASE 2 (DRM2)	*De novo* methylation	(Chan et al. 2004, Li et al. 2006)
DOUBLE STRANDED RNA-BINDING PROTEIN 4 (DRB4)	miRNA and ta-siRNA biogenesis	(Hiraguri et al. 2005)
HUN ENHANCER 1(HEN1)	Small RNA-specific methylation	(Yang et al. 2006, Yu et al. 2006)
HYPONASTIC LEAVES 1 (HYL1)	miRNA and ta-siRNA biogenesis	(Hiraguri et al. 2005, Kurihara et al. 2006, Vazquez et al. 2004)
NUCLEAR RNA	ra-siRNA biogenesis	(Herr et al. 2005, Onodera et al. 2005, Zhang et al. 2007)

POLYMERASE D 1A (NRDP1a)		
NRPD2a	ra-siRNA biogenesis	(Herr et al. 2005, Zhang et al. 2007)
NRPD1b	*De novo* DNA methylation	(Huettel et al. 2006, Kanno et al. 2005, Zhang et al. 2007)
RNA-DEPENDENT RNA POLYMERASE 2 (RDR2)	Synthesis of dsRNA in ra-siRNA pathway	(Herr et al. 2005, Xie et al. 2004, Zhang et al. 2007)
RDR6	ta-siRNA and nat-siRNA biogenesis and spreading RNA silencing	(Borsani et al 2005, Himber et al. 2003, Peragine et al. 2004, Vaistij et al. 2002, Vazquez et al. 2004)
SERRATE (SE)	miRNA biogenesis	(Dong et al. 2008)
SILENCING DEFECTIVE 3 (SDE3)	RNA helicase	(Dalmay et al. 2001)
SUPPRESSOR OF GENE SILENCING 3 (SGS3)	nat-siRNA and ta-siRNA biogenesis	(Borsani et al. 2005, Peragine et al. 2004, Yoshikawa et al. 2005)

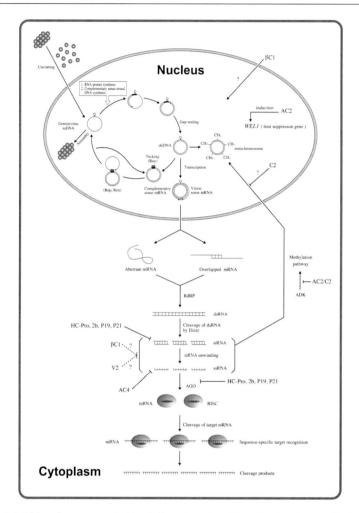

Figure 1. PTGS pathways encoded by different viral proteins. Geminivirus replication cycle as shown in the nucleus, where dsDNA (RF) serves as a potential target of methyltransferases, which modify the DNA and histone proteins. Cytoplasmic RNA-silencing (PTGS) ultimately degrades the target mRNA, and siRNA-directed methylation leads to TGS. During *trans*methylation AC2 and C2 (*Curtovirus*) proteins interfere with the methyl cycle by inhibiting ADK. Dicer cleaves dsRNA into siRNA, and RISC then distinguishes different strands of siRNA forms. The sense strand is degraded (not shown), while the anti-sense strand is used to target the genes for silencing. AC4 protein binds single-stranded siRNA forms. βC1 of TYLCCNV-[Y10] suppresses silencing by acting in the nucleus, while AC2 protein of MYMV-(V$_{ig}$) inactivates transcription of host genes (*WEL1*), In contrast, begomoviruses like βC1 protein of ToLCJV DNAβ and V2 protein of ToLCJV and TYLCV-[IL] suppress silencing and are localized in the cytoplasm. Still precise mechanisms of these suppressors are not known. HC-Pro of tobacco etch virus, 2b of cucumber mosaic virus, P19 of carnation Italian ringspot virus and P21 of beet yellows virus bind ds-RNA and inhibit RISC formation.

DIVERSITY OF RNA SILENCING PATHWAYS IN PLANTS

Several lines of evidence have established that small RNAs play an important role in plant development by triggering RNA degradation, translational inhibition or chromatin modification (Vaucheret 2006). First, small RNAs were detected in transgenic plants undergoing cosuppression. Second, plants undergoing silencing of the transgenes β-*glucuronidase* or *green fluorescent protein* (*GFP*), which have no homology to endogenous genes, accumulated small RNAs with homology to these foreign gene sequences. Finally, the finding that plants infected with potato virus X (PVX, genus *Potexvirus*) accumulated siRNAs homologous to PVX demonstrated that plant viruses can be targets of PTGS (Hamilton and Baulcombe 1999).

Characterization of small RNAs revealed two major types, which are referred to as siRNA and miRNA, based on their mode of biogenesis and function. Furthermore, numerous types of siRNAs have been recognized, including trans-acting siRNA (ta-siRNA), natural-anti-sense siRNA (nat-siRNA) and repeat-associated siRNA (ra-siRNA) (Ramachandran and Chen 2008).

VIRUS INFECTION TRIGGERS RNA SILENCING

In plants, RNA silencing plays an important role as a cellular defense mechanism against molecular parasites (Aliyari and Ding 2009). Experiments with transgenic plants engineered for resistance to virus infection first revealed that plants could target specific virus or transgene mRNA sequences for degradation by a cosuppression-like mechanism (Lindbo et al. 1993). Here, tobacco plants transformed with a non-translatable form of the *capsid protein* (*CP*) gene of the tobacco etch virus (TEV, genus *Potyvirus*) were initially susceptible to TEV infection. However, a few weeks after infection, these plants subsequently recovered from infection, such that newly emerging leaves were symptomless. Recovered leaves could not be re-infected with TEV, but were susceptible to infection by another potyvirus (potato virus Y, PVY). Northern blot analysis showed that TEV CP transgene mRNA and TEV genomic RNA levels were substantially reduced in recovered tissues. Furthermore, this RNA degradation activity was virus-specific because PVY RNA accumulated and systemic PVY symptoms developed in leaves that had recovered from TEV infection. Because this recovery phenomenon was triggered by a non-translatable gene, it was hypothesized to have been activated by RNA rather than protein.

Subsequent examination of virus infection in non-transgenic plants also revealed a recovery phenotype associated with PTGS. For example, plants with infected with cauliflower mosaic virus (CaMV, genus *Caulimovirus*) or tomato black ring virus (TBRV, genus *Nepovirus*) strain W22 showed a recovery phenotype (Covey et al. 1997, Ratcliff et al. 1997). Here, virus-infected plants initially showed severe symptoms and high levels of virus accumulation, with subsequently leaves emerging showing few or no symptoms and low levels of virus accumulation. Furthermore, recovered leaves were resistant to subsequent infection with the same virus and this resistance was also virus-specific. Thus, this natural recovery phenotype had properties similar to transgene-induced gene silencing.

The recovery phenomenon indicated that plant viruses can trigger RNA silencing in plants. The majorities of plant viruses have RNA genomes and encode an RDR, which generates a dsRNA replication form in the initial steps of viral replication. It has been suggested that this dsRNA molecule triggers virus-induced RNA silencing (Dalmay et al. 2000). Thus, the first step of the virus-induced PTGS pathway is using virus-derived dsRNA as a template to generate virus-specific siRNAs (vsiRNA), via the activity of the DCL proteins (Herr et al. 2005). Furthermore, the identification of three distinct size classes of vsiRNAs (21-nt, 22-nt and 24-nt) in plants infected with DNA or RNA viruses indicated that more than one DCL is involved in the biogenesis of vsiRNAs (Blevins et al. 2006, Diaz-Pendon et al. 2007). For example, in *Arabidopsis* plants infected with RNA or DNA viruses, DCL4 predominantly produces 21-nt vsiRNA. However, in *Arabidopsis dcl4* or *dcl2/dcl4* mutants infected with cucumber mosaic virus (CMV, genus *Cucumovirus*), 21-nt vsiRNA were not detected; instead, 22-nt vsiRNA were detected in the *dcl4* mutant plants, whereas 24-nt vsiRNA were detected in the *dcl2/dcl4* mutant plants (Bouché et al. 2006, Fusaro et al. 2006). This suggested that, in the absence of DCL4, DCL2 and DCL3 produce 22-nt and 24-nt vsiRNA, respectively, thereby providing a system of redundancy for antiviral defense. Further evidence of this came from the finding that the 22-nt vsiRNA is the dominant form detected in *Arabidopsis* plants infected with turnip crinkle virus (TCV, genus *Carmovirus*). This is due to the fact that the TCV-encoded CP suppresses DCL4 activity (Deleris et al. 2006).

DNA viruses also can trigger RNA silencing. Geminiviruses, which are single-stranded DNA (ssDNA) viruses, do not utilize a dsRNA in replication. However, because the double-stranded DNA (dsDNA) replicative forms of these viruses are bidirectionally transcribed, the viral- and complementary-sense transcripts can overlap and have the potential to form dsRNA (Chellappan et al. 2004). CaMV is a dsDNA plant pararetrovirus that expresses its genome through the polycistronic 35S RNA. This large viral RNA has an unusually extensive secondary structure, which can induce RNA silencing in plants (Moissiard and Voinnet 2006). Three distinct size classes (21-, 22- and 24-nt) of vsiRNAs have been identified in plants infected with DNA viruses. Genetic evidence has shown that 24-nt vsiRNAs, which are generated by DCL3, were more abundant than the 21-nt and 22-nt vsiRNA produced by DCL4 and DCL2, respectively, in plants infected with cabbage leaf curl virus (CaLCuV, genus *Begomovirus*) or CaMV (Blevins et al. 2006). Moreover, DCL1, which produces miRNA, generates 21-nt vsiRNA from the CaMV 35S RNA leader region in *dcl2/dcl3/dcl4* mutants (Blevins et al. 2006, Moissiard and Voinnet 2006). Thus, all four DCLs can be involved in vsiRNA production in VIGS targeting DNA viruses.

In *Arabidopsis*, AGO1 plays an important role in the miRNA pathway. However, an AGO1 mutant (*ago1*) was hypersusceptible to infection by CMV, indicating that AGO1 is also involved in virus resistance (Morel et al 2002). Recent studies have shown that AGO1 recruits vsiRNA in plants infected with CMV or turnip yellow mosaic virus (TYMV, genus *Tymovirus*) (Zhang et al. 2006). Similarly, AGO7 also plays an important role in defense against TCV infection based upon the observation that higher amounts of TCV RNAs accumulated in *ago7* mutant plants (Qu et al. 2008). In addition, recent findings have also revealed that AGO2 and AGO5 interact with vsiRNA of CMV. Thus, it is clear that multiple AGOs play a role in antiviral defense in plants (Takeda et al. 2008).

At the initiation stage of virus infection, the primary dsRNA is generated by the viral RDR. However, SGS2 (also known as RDR6) and SDE3 are essential for effective antiviral defense (Dalmay et al. 2001, Mourrain et al. 2000). For example, CMV RNAs accumulate at much higher levels in *Arabidopsis sgs2* mutants plants (Mourrain et al. 2000); and *Nicotiana benthamiana* plants silenced for RDR6 are hypersusceptible to infection by TCV, PVX and tobacco mosaic virus (TMV, genus *Tobamovirus*) (Qu et al. 2005, Schwach et al. 2005). When infected by TMV and PVX, silenced NtRDRP1 tobacco lines accumulated higher levels of viral RNA and developed more severe disease symptoms than wild-type plants (Xie et al. 2001). Furthermore, the tobacco ortholog of *Arabidopsis* RDR1, NtRDRP1, was induced by virus infection. This suggests that the plant RDR plays an important role in antiviral defense. Similarly, vsiRNA biogenesis and antiviral defense in plants infected by tobacco rattle virus (TRV, genus *Tobravirus*) are dependent on the combined activity of RDR1, RDR2 and RDR6 (Donaire et al. 2008). Here, the dsRNA replicative form of TRV, generated by the viral RDR, may be targeted by three DCLs (DCL2, DCL3 and DCL4) to produce the primary vsiRNA at the initiation step of PTGS. However, RDR1, RDR2 and RDR6 must be required for secondary vsiRNA production, because TRV vsiRNAs are substantially reduced in an *Arabidpsis rdr1/rdr2/rdr6* triple mutant compared with wild-type plants (Donaire et al. 2008). In *Arabidopsis* plants infected with the crucifer-infecting TMV-Cg strain, 21-nt and 24-nt vsiRNAs can be detected, suggesting a role for DCL4 and DCL2, respectively (Qi et al. 2009). Consistent with this, *rdr1* and *rdr6* mutants infected with TMV-Cg showed reduced levels of vsiRNAs. SGS2 activity is also required for the CaLCuV-mediated defense pathway in *Arabidopsis*, because viral DNA accumulation is slightly higher and viral symptoms are more severe in *sgs2* mutant plants (Muangsan et al. 2004). Interestingly, the *Arabidopsis* RDR6 can use ssDNA as a template, suggesting that the viral ssDNA may be used by RDR6 to generate ssRNA, which can then be used to generate dsRNA (Curaba and Chen 2008). Together, these findings strongly suggest that plant RDRs play an important role in antiviral defense; however, the precise mechanisms of secondary vsiRNA production are not yet fully understood. The *Arabidopsis* genome encodes six putative RDRs; studies with RDR3, RDR4 and RDR5 mutants should provide further insight into the role of these proteins in vsiRNA biogenesis.

A DIVERSITY OF PLANT VIRAL SUPPRESSORS AND MECHANISMS OF SUPPRESSION OF RNA SILENCING IN PLANTS

Further evidence that RNA silencing plays an important role as a defense mechanism against virus infection in plants comes from the finding that most, if not all, RNA viruses have evolved strategies to counteract this defense mechanism. The potyvirus helper component protease (HC-Pro) and the CMV 2b were the first plant viral proteins shown to suppress RNA silencing in plants (Anandalakshmi et al. 1998, Brigneti et al. 1998, Kasschau and Carrington 1998). Mutant viruses that do not have express these suppressors induce mild symptoms. Thus, in these early studies, these proteins were called pathogenicity determinants. Now, a wide diversity of silencing suppressors have been identified in viruses from most families/ groups (Table 2) and, as they often share no obvious homology at the amino acid level, it is likely these have evolved independently.

Plant Viral Suppressors Inhibit siRNA Biogenesis and Bind siRNA

The potyviral HC-Pro is a multifunctional protein that is involved in aphid transmission, replication, symptom expression and movement (Maia et al 1996). The mechanism by which HC-Pro suppresses silencing is not known, but it seems to involve multiple steps of the pathway. In transient silencing assays, HC-Pro interferes with the accumulation of siRNAs (Johansen and Carrington 2001). Expression of HC-Pro in transgenic plants is associated with accumulation of unprocessed long dsRNA, indicating that it inhibits dsRNA processing by DCL proteins (Dunoyer et al. 2004,

Mallory et al. 2001). This is also consistent with a recent report indicating that HC-Pro is a dsRNA-binding protein and physically interacts with siRNA duplexes (Lakatos et al. 2006). HC-Pro may also inhibit 3′ modification (i.e., methylation) of small RNAs in virus-infected plants (Lózsa et al. 2008). As methylation appears protect small RNAs from degradation (Li et al. 2005), interference with this process by HC-Pro may lead to reduced levels of vsiRNAs. Together, HC-Pro appears to suppress VIGS by inhibiting vsiRNA biogenesis and assembly of the RISC.

Because dsRNA plays a key role in RNA silencing, binding of dsRNA or small RNAs is a common strategy used by silencing suppressors encoded by plant viruses. Many studies have shown that these suppressors bind small RNAs (Chapman et al. 2004, Chen et al. 2008, Fukunaga and Doudna 2009, Goto et al. 2007, Kurihara et al. 2007, Lakatos et al. 2006, Martínez-Turiño and Hernández 2009, Mérai et al. 2005, Mérai et al. 2006, Valli et al. 2008, Xiong et al. 2008, Ye and Patel 2005, Zhou et al. 2006). The tombusvirus P19 is a pathogenicity determinant and suppresses RNA silencing. It is now well-established that P19 binds small RNAs (Scholthof 2006). The P19 effectively binds ds-21-nt siRNA, but not ss-siRNA, or larger-sized ssRNA or dsRNA (Silhavy et al. 2002). Furthermore, P19 binds ds-siRNA as a homodimer (Ye et al. 2003) and binding is enhanced by siRNAs containing 5′ phosphate groups (Vargason et al. 2003). Thus, P19 is a size-selective dsRNA binding protein. A number of other viral suppressors, including the closterovirus P21, hordeivirus γb, pecluvirus P15 and potyvirus HC-Pro bind ds-siRNAs on the basis of size. In contrast, the aureusvirus P14 and carmoviral CP are size-independent dsRNA-binding proteins (Mérai et al 2005, Mérai et al 2006).

Inhibition of Movement of RNA Silencing Signals by Viral Suppressors

In plants, siRNAs are thought to move cell-to-cell through plasmodesmata and then systemically through the phloem (Yoo et al. 2004). In silencing suppression assays, many plant virus silencing suppressors suppress local and/or systemic RNA silencing (Bucher et al. 2003, Cañizares et al. 2008, Cao et al 2005, Chen et al. 2004, Cui et al. 2005, Ding et al. 2004, Ghazala et al. 2008, Gopal et al. 2007, Kreuze et al. 2005, Kubota et al. 2003, Liu et al. 2004, Mangwende et al. 2009, Martín-Hernández and Baulcombe 2008, Martínez-Priego et al. 2008, Meng et al. 2006, Powers et al. 2008, Qu et al. 2003, Reed et al. 2003, Sarmiento et al. 2007, Senshu et al. 2009, Silhavy et al. 2002, Takeda et al. 2002, Te at al. 2005, Valli et al. 2006, Voinnet et al. 1999, Yaegashi et al. 2007, Yelina et al. 2002). For example, the CMV 2b protein cannot inhibit the initiation step of RNA silencing, but it effectively prevents systemic silencing (Brigneti et al. 1998). This may

Table 2. RNA silencing suppressors of plant viruses.

Virus genus	Virus	Suppressor	Mechanism of suppression	Reference
Single-stranded DNA viruses				
Begomovirus	African cassava mosaic virus-[Cameroon]	AC4	Ss-siRNA binding	(Chellappan et al. 2005, Vanitharani et al. 2004)
	African cassava mosaic virus-[Kenya]	AC2	Suppression of systemic silencing	(Voinnet et al. 1999)
	East African cassava mosaic Cameroon virus	AC2		(Vanitharani et al. 2004)
	Indian cassava mosaic virus	AC2		(Vanitharani et al. 2004)
	Mungbean yellow mosaic virus	AC2		(Trinks et al. 2005)
	Sri Lankan cassava mosaic virus	AC4		(Vanitharani et al. 2004)
	Tomato golden mosaic virus	AC2	Inhibition of ADK activity and methylation	(Buchmann et al. 2009, Wang et al. 2005)
	Tomato leaf curl virus-[Australia]	C2		(Selth et al. 2004)
	Tomato leaf curl Java virus	C2	Suppression of local silencing	(Kon et al. 2009)
	Tomato yellow leaf curl China virus	C2		(van Wezel et al. 2002)
	Ageratum yellow vein virus-[ID]	C2	Suppression of systemic silencing	(Sharma et al. 2010)
	Tomato yellow leaf curl virus	V2	Ds-siRNA binding and inhibition of SGS3 activity	(Fukunaga et al. 2009, Glick et al. 2008, Zrachya et al. 2007)
	Tomato leaf curl Java virus	V2		(Sharma et al. 2010)
Betasatellite	Bhendi yellow vein mosaic betasatellite	βC1		(Gopal et al. 2007)
	Tomato leaf curl Java betasatellite	βC1	Suppression of systemic silencing	(Kon et al. 2007)
	Tomato yellow leaf curl China betasatellite	βC1		(Cui et al. 2005)
	Ageratum yellow vein betasatellite	βC1	Suppression of systemic silencing	(Sharma et al. 2010)
Curtovirus	Beet curly top virus	C2	Inhibition of ADK activity and methylation	(Buchmann et al. 2009, Wang et al. 2005)

Double-stranded DNA virus

Caulimovirus	Cauliflower mosaic virus	P6	Inhibition of DRB4 activity	(Haas et al. 2008, Love et al. 2007)
Positive single-stranded RNA viruses				
Aureusvirus	Pothos latent virus	P14	Long dsRNA and ds-siRNA binding	(Mérai et al. 2005, Mérai et al. 2006)
Carmovirus	Hibiscus chlorotic ringspot virus	CP	Ds-siRNA binding	(Meng et al. 2006)
	Pelargonium flower break virus	CP	Ds-siRNA binding	(Martínez-Turiño and Hernández 2009)
	Turnip crinkle virus	CP	Ds-siRNA binding and inhibition of DCL4 activity	(Deleris et al. 2006, Mérai et al. 2005, Qu et al. 2003)
Cheravirus	Apple latent spherical virus	Vp20	Suppression of systemic silencing	(Yaegashi et al. 2007)
Closterovirus	Beet yellows virus	P21	Ds-siRNA binding and inhibition of RISC formation	(Chapman et al. 2004, Lakatos et al. 2006, Mérai et al. 2005, Reed et al. 2003, Ye et al. 2005)
	Citrus tristeza virus	P20	Suppression of local and systemic silencing	(Lu et al. 2004)
		P23	Suppression of local silencing	(Lu et al. 2004)
		CP	Suppression of systemic silencing	(Lu et al. 2004)
Crinivirus	Sweet potato chlorotic stunt virus	P22	Suppression of local and systemic silencing	(Kreuze et al. 2005)
		RNase3	Enhancement of p22-mediated suppression	(Kreuze et al. 2005)
	Tomato chlorosis virus	P22		(Cañizares et al. 2008)
		CP		(Cañizares et al. 2008)
		CPm		(Cañizares et al. 2008)
Comovirus	Cowpea mosaic virus	S-CP		(Liu et al. 2004)
Cucumovirus	Cucumber mosaic virus	2b	Long dsRNA and ds-siRNA binding, inhibition of RISC	(Brigneti et al. 1998, Díaz-Pendon et al. 2007,

contd. ... Table

Table ... contd.

Virus genus	Virus	Suppressor	Mechanism of suppression	Reference
Dianthovirus	Tomato aspermy virus	2b	formation and suppression of systemic silencing	Goto et al. 2007, Guo et al. 2002, Zhang et al. 2006)
	Red clover necrotic mosaic virus	P27+ P88+ viral RNA	Ds-siRNA binding	(Chen et al. 2008)
		MP	Inhibition of host RNA silencing factor?	(Takeda et al. 2005)
Furovirus	*Soilborne wheat mosaic virus*	19K		(Powers et al. 2008)
Hordeivirus	*Barley stripe mosaic virus*	γb	Ds-siRNA binding	(Te et al. 2005) (Mérai et al. 2005, Yelina et al. 2002)
Ipomovirus	Cucumber vein yellowing virus	P1b	Ds-siRNA binding	(Valli et al. 2008, Valli et al. 2006)
Pecluvirus	Peanut clump virus	P15	Ds-siRNA binding	(Mérai et al. 2005)
Polerovirus	Beet western yellows virus	P0	Inhibition of AGO activity	(Baumberger et al. 2007, Bortolamiol et al. 2007, Pazhouhandeh et al. 2006, Pfeffer et al. 2002)
	Cucurbit aphid-borne yellows virus	P0		(Pfeffer et al. 2002)
	Sugarcane yellow leaf virus	P0	Suppression of local and systemic silencing	(Mangwende et al. 2009)
Potexvirus	Asparagus virus 3	TGBp1		(Senshu et al. 2009)
	Plantago asiatica mosaic virus	TGBp1		(Senshu et al. 2009)
	Potato virus X	P25	Suppression of systemic silencing	(Senshu et al. 2009, Voinnet et al. 1999)
	Tulip virus X	TGBp1		(Senshu et al. 2009)
	White clover mosaic virus	TGBp1		(Senshu et al. 2009)
Potyvirus	Potato virus Y	HC-Pro		(Brigneti et al. 1998)
	Tobacco etch virus	HC-Pro	Ds-siRNA binding and inhibition of RISC formation	(Anandalakshmi et al. 1998, Kasschau et al. 1998, Lakatos et al. 2006, Mérai et al. 2005)

Sobemovirus	Rice yellow mottle virus	P1	Suppression of local and systemic silencing	(Sarmiento et al. 2007, Voinnet et al. 2005)
	Cocksfoot mottle virus	P1	Suppression of local and systemic silencing	(Sarmiento et al. 2007)
Tobamovirus	Tobacco mosaic virus	122K/126K	Ds-siRNA binding	(Csorba et al. 2007, Fukunaga and Doudna 2009, Martín-Hernández and Baulcombe 2008)
	Tomato mosaic virus	130K		(Kubota et al. 2003)
Tobravirus	Tobacco rattle virus	16K	Suppression of local and systemic silencing	(Ghazala et al. 2008, Martín-Hernández and Baulcombe2008,Martínez-Priego et al. 2008)
Tombusvirus	Carnation Italian ringspot virus	P19	Ds-siRNA binding and inhibition of RISC formation	(Lakatos et al. 2006, Vargason et al. 2003)
	Cymbidium ringspot virus	P19	Ds-siRNA binding and suppression of systemic silencing	(Lakatos et al. 2006, Mérai et al. 2005, Silhavy et al. 2002)
	Tomato busby stunt virus	P19	Ds-siRNA binding and suppression of systemic silencing	(Voinnet et al. 1999, Ye et al. 2003)
Trichovirus	Apple chlorotic leaf spot virus	P50	Suppression of systemic silencing	(Yaegashi et al. 2007, Yaegashi et al. 2008)
Tymovirus	Turnip yellow mosaic virus	P69		(Chen et al. 2004)
Vitivirus	Grapevine virus A	P10	Ss- and ds-siRNA binding and suppression of systemic silencing	(Zhou et al. 2006)
Negative single-stranded RNA viruses				
Tenuivirus	Rice hoja blanca virus	NS3	Ds-siRNA binding and inhibition of RISC formation	(Bucher et al. 2003, Hemmes et al. 2007)
	Rice stripe virus	NS3	Long ssRNA, ss- and ds-	(Xiong et al. 2008)

contd. ... Table

Table ... contd.

Virus genus	Virus	Suppressor	Mechanism of suppression	Reference
Tospovirus	Tomato spotted wilt virus	NSs	siRNA binding and suppression of systemic silencing Suppression of local and systemic silencing	(Bucher et al. 2003, Takeda et al. 2002)
Double-stranded RNA virus				
Phytoreovirus	Rice dwarf virus	Pns10	Suppression of local and systemic silencing	(Cao et al. 2005)

occur via an inactivation of the systemic silencing signal, as suggested by results of grafting experiments with 2b-expressing transgenic plants (Guo and Ding 2002). Because 21-nt siRNA is involved in the cell-to-cell spread of silencing in plants, it is possible that the 2b protein inhibits biogenesis of RDR1-dependent secondary siRNAs. The genomic RNA of the apple chlorotic leaf spot virus (ACLSV, genus *Trichovirus*) encodes three proteins (P216, P50 and CP). None of these proteins can suppress local silencing (Yaegashi et al. 2007). However, grafting experiments established that the P50 protein can inhibit systemic silencing (Yaegashi et al. 2008). Thus, the ACLSV P50 protein is an unusual suppressor in that it inhibits systemic silencing without interfering with local silencing. Citrus tristeza virus (CTV, genus *Closterovirus*) has a large genome (~20 kb) that encodes at least three RNA silencing suppressors (P20, P23 and CP) (Lu et al. 2004). In transient expression assays, the P20 and P23 proteins, but not the CP, showed RNA silencing suppression activity. Grafting experiments revealed that the CP suppresses systemic silencing. It has been further established that P23 suppresses cell-to-cell silencing, whereas P20 suppresses cell-to-cell and systemic silencing. These findings suggest that CTV employs a strategy that targets the antiviral silencing pathway at multiple steps. This type of strategy may be necessary to protect viruses having large RNA genomes.

The AC2/C2 proteins encoded by begomoviruses are nuclear proteins that activate viral transcription and also have suppressor activity (Selth et al. 2004, Shivaprasad et al. 2005, Trinks et al. 2005, van Wezel at al. 2002, Vanitharani et al 2004, Voinnet et al. 1999). For example, the tomato leaf curl Java virus (ToLCJV) C2 protein suppresses local but not systemic silencing (Kon et al. 2009, Kon et al. 2007). In addition, a satellite DNA (tomato leaf curl Java betasatellite (ToLCJB)) associated with ToLCJV encodes the βC1 protein, which suppresses systemic, but not local silencing (Kon et al. 2009, Kon et al. 2007). Whereas AC2/C2 are nuclear proteins, the βC1 appears to accumulate at the cell periphery (Kon et al. 2009, Saeed et al. 2007). Thus, the ToLCJV/ToLCJB complex has at least two RNA silencing suppressors that, together, suppress both local and systemic silencing and likely target the antiviral silencing pathway at multiple steps. Small RNAs, including the 21-nt vsiRNAs are detectable in pumpkin (*Cucurbita maxima*) phloem sap (Hagen et al. 2008, Yoo et al. 2004). This finding also led to the identification of the *C. maxima* PHLOEM SMALL RNA BINDING PROTEIN1 (CmPSRP1) (Yoo et al. 2004). CmPSRP1 is expressed in vascular tissues and binds siRNAs. Moreover, CmPSRP1 mediates plasmodesmata trafficking of siRNAs, indicating that it plays an important role in the transport of small RNAs in the plant. Interestingly, no PSRP1 orthologue has been found in *Arabidopsis* or *Nicotiana* species.

Interaction between Plant Viral Suppressors and Host Factors Involved in RNA Silencing

Several plant viral RNA silencing suppressors have been shown to interact with host proteins during virus infection. For example, the strong suppressor HC-Pro inhibits multiple steps in the RNA silencing pathways and interacts with endogenous negative cellular regulators of RNA silencing, such as rgs-CaM, a calmodulin-related protein (Anandalakshmi et al. 2000). HC-Pro induces rgs-CaM expression, leading to suppression of RNA silencing; this suggests that HC-Pro interacts with an endogenous silencing pathway. The CMV 2b protein directly interacts with AGO1, binding to the portion of the protein with the PIWI-box motif, which forms a structure similar to that of RNase H (Vaucheret 2008, Zhang et al. 2006). *Arabidopsis* AGO1 has RNA slicer activity and recruits small RNAs (Baumberger and Baulcombe 2005); it also plays an important role in anti-CMV defense (Morel et al. 2002). Therefore, by inhibiting AGO1 activity, the CMV 2b disables a key component of the anti-viral RNA silencing pathway. In contrast, the polerovirus-encoded P0 protein carries an F-box, which is required for suppression of RNA silencing (Pazhouhandeh et al. 2006, Pfeffer et al. 2002). The P0 protein interacts with AGO1 and the capacity of the protein to suppress silencing and degrade AGO1 is dependent on the F-box motif. Furthermore, P0 mediates destabilization of multiple *Arabidopsis* AGOs including AGO1, AGO2, AGO4, AGO5, AGO6 and AGO9 (Baumberger et al. 2007, Bortolamiol et al. 2007). F-box proteins are components of E3 ubiquitin ligase complexes (Pazhouhandeh et al. 2006), suggesting that AGO degradation mediated by P0 may occurs through ubiquitylation.

The V2 (pre-coat protein) of the tomato yellow leaf curl virus (TYLCV, genus *Begomovirus*), has RNA silencing suppression activity (Sharma and Ikegami 2010, Sharma et al. 2010, Zrachya et al. 2007) and interacts with *Solanum lycopersicon* SGS3 (SlSGS3), an orthologue of the *Arabidopsis* SGS3 that is a partner of RDR6 in the RNA silencing pathway (Glick et al. 2008). Mutational analysis revealed that interaction with SlSGS3 was required for RNA silencing suppression activity and that the zinc-finger motif of the V2 protein was involved. Thus, the V2/SlSGS3 interaction plays an important role in allowing TYLCV to counter antiviral defense. Recent findings have suggested that V2 inhibits RNA silencing by competitive binding of SGS3 substrates, thereby blocking the RDR6/SGS3-mediated silencing pathway (Fukunaga and Doudna 2009). The AC2 of tomato golden mosaic virus (TGMV, genus *Begomovirus*) and the C2 of the beet curly top virus (BCTV, genus *Curtovirus*) not only suppress RNA silencing, but also inactivate the SNF1-related kinase and adenosine kinase (ADK), which catalyze the synthesis of 5′-AMP from adenosine (Hao et al. 2003, Wang et al. 2003). ADK is a cellular enzyme that plays an important role in adenosine salvage and maintenance of the methylation cycle. Inhibition of ADK also suppresses RNA silencing (Wang et al. 2005). DNA viruses apparently target the RNA-directed methylation pathway, because the DNA virus-derived 24-nt siRNA is involved in the TGS pathway. Indeed, a recent study has shown that AC2/C2 causes a global reduction in cytosine methylation, resulting in increased susceptibility to virus infection. This suggests that DNA methylation of begomovirus genomic DNA is an important antiviral defense strategy (Buchmann et al. 2009, Hagen et al. 2008). CaMV P6 protein is a symptom determinant that suppresses RNA silencing (Love et al. 2007). CaMV-derived 21-nt siRNA is produced by DCL4 from the highly structured 35S RNA leader (Moissiard and Voinnet 2006). The P6 protein inactivates DRB4, a dsRNA binding protein that interacts with DCL4 (Haas et al. 2008). This suggests that the inactivation of a host RNA silencing component by a viral suppressor plays an important role in overcoming the host antiviral system.

VIRAL SUPPRESSORS REGULATE SMALL RNA PATHWAYS AND INDUCE SYMPTOM DEVELOPMENT

In plants, endogenous small RNAs (e.g., miRNA) play important roles in gene expression and plant development (Jones-Rhoades et al. 2006, Xie et al. 2005, Yoshikawa et al. 2006). For example, genetic analysis of *Arabidopsis* mutant lines such as *ago* and *dcl*, which are deficient in the small RNA pathway, display developmental abnormalities, some of which mimic symptoms of virus infection (Fagard et al. 2000, Golden et al. 2002, Morel et al. 2002). As previously mentioned, many plant virus suppressors were initially identified as pathogenicity determinants. Some of these have now been associated with perturbation of normal small RNA activity (Diaz-Pendon and Ding 2008). In *Aradidopsis*, TuMV infection induces a number of defects in embryonic, vegetative and reproductive development (Kasschau et al. 2003). Moreover, overexpression in the TuMV silencing suppressor P1/HC-Pro in transgenic plants results in developmental defects that resemble those of TuMV-infected plants (Kasschau et al. 2003). Interestingly, similar defects were observed in miRNA-deficient DCL1 mutants in the absence of TuMV infection or P1/HC-Pro expression. DCL1 is essential for miRNA biogenesis and miRNAs control plant development through the processing of small RNA hairpins that, in turn, inhibit the translation of target mRNAs (Kurihara and Watanabe 2004, Qi et al. 2005). A connection between viral suppressors and miRNA was established by showing that expression of TuMV P1/HC-Pro triggers inhibition of miRNA-guided cleavage of mRNA targets, which are involved in several developmental processes. Furthermore, miRNA accumulation was higher in P1/HC-Pro transgenic plants than in control non-transgenic plants, suggesting that inhibition of mRNA cleavage is not due to a reduction of levels of miRNA.

Developmental defects that resemble virus-like symptoms have been associated with expression of other plant viral suppressors, including the begomovirus AC4, closterovirus P21 and tymovirus P69; all of these suppressors have been shown to inhibit the miRNA pathway (Chapman et al. 2004, Chellappan et al. 2005, Chen et al. 2004, Dunoyer et al. 2004). The βC1 protein encoded by the betasatellite associated with monopartite begomoviruses is a strong pathgenicity determinant and βC1-expressing plants develop virus-like disease symptoms (Cui et al. 2004, Kon et al. 2007, Saunders et al. 2004). For example, expression of the tomato yellow leaf curl China betasatellite (TYLCCNB) βC1 in transgenic *Arabidopsis* resulted in a phenotype with upward leaf curling, abnormal leaf development and outgrowth of tissues from the abaxial leaf surfaces (Yang et al. 2008). These developmental defects were associated with a reduction in levels of miRNA165/166, which are regulated by the leaf polarity gene, *HD-ZIP III*. TYLCCNB βC1 interacts with ASYMMETRIC LEAVES 1 (AS1), which is encoded by a leaf development-associated gene, to cause alterations in leaf development. Consistent with this result, *as1* mutant plants show altered leaf morphology. Moreover, in a mutational analysis of ASYMMETRIC LEAVES 2 (AS2), which is a partner of AS1 in regulation of leaf development, TYLCCNB βC1 partially complemented the *as2* mutation, suggesting that βC1 may be a molecular mimic of AS2. Therefore, part of the molecular basis of pathogenicity associated with plant viral suppressors can be explained by perturbation of endogenous small RNA pathways. What remains to be established is whether this is an unintended consequence of viral infection or do the resulting abnormalities provide a benefit for the virus.

CONCLUSIONS

Virus infection triggers RNA silencing in plants, which is part of an antiviral defense system. However, viruses have responded by evolving silencing suppressors that can overcome this part of the antiviral response. Suppression of silencing can occur at multiple steps in the process and can involve interfering with movement of the silencing signal, binding small RNAs, inhibiting small RNA biogenesis and interacting with host factors involved in RNA silencing. Many studies have now shown that plant virus suppressors also impact the host endogenous small RNA pathway, although not all of these perturbations are directly involved in silencing suppression. Recent findings have shown that a diversity of host genes can be targeted by vsiRNAs for post-transcriptional silencing, which can lead to symptom development (Qi et al. 2009). Similarly, viroid-derived small RNAs also effect host gene regulation and can result in symptom induction (Gómez et al. 2008, Wang et al. 2004). Thus, these data have established that plant viral suppressors and vsiRNAs are a major contributor to viral pathogenicity. However, several key questions remain unanswered including the nature of small RNA-mediated systemic silencing signals and the precise small RNA pathways and host genes targeted by different viruses. Further studies of the interactions among viral suppressors, small RNAs and host genes should help answer these questions.

ACKNOWLEDGEMENTS

Our research reported in this review was supported by a Grant-in-Aid from the Academic Frontier Promotion Program and Scientific Research of the Ministry of Education, Culture, Sports, Science and Technology of Japan.

REFERENCES

Adenot, X., T. Elmayan, D. Lauressergues, S. Boutet, N. Bouché and V. Gascolliet. 2006. DRB4-dependent *TAS3 trans*-acting siRNAs control leaf morphology through AGO7. *Current Biology.* **16:**927–32.

Aliyari, R. and S.W. Ding. 2009. RNA-based viral immunity initiated by the Dicer family of host immune receptors. *Immunological Reviews.* **227:**176–188.

Anandalakshmi, R., R. Marathe, X. Ge, J.M. Jr. Herr, C. Mau and A. Mallory. 2000. A calmodulin-related protein that suppresses post-transcriptional gene silencing in plants. *Science.* **290:**142–4.

Anandalakshmi, R., G.J. Pruss, X. Ge, R. Marathe, A.C. Mallory and T.H. Smith. 1998. A viral suppressor of gene silencing in plants. Proceedings of the National Academy of Sciences of the United States of America. **95:**13079–84.

Baulcombe, D.C. 2004. RNA silencing in plants. *Nature.* **431:**356–63.

Baumberger, N. and D.C. Baulcombe. 2005. *Arabidopsis* ARGONAUTE1 is an RNA Slicer that selectively recruits microRNAs and short interfering RNAs. Proceedings of the National Academy of Sciences of the United States of America. **102:**11928–33.

Baumberger, N., C.H. Tsai, M. Lie, E. Havecker and D.C. Baulcombe. 2007. The polerovirus silencing suppressor P0 targets ARGONAUTE proteins for degradation. *Current Biology.* **17:**1609–14.

Blevins, T., R. Rajeswaran, P.V. Shivaprasad, D. Beknazariants, A . Si-Ammour and H.S. Park. 2006. Four plant Dicers mediate viral small RNA biogenesis and DNA virus induced silencing. *Nucleic Acids Research.* **34:**6233–46.

Borsani, O., J. Zhu, P.E. Verslues, R. Sunkar and J.K. Zhu. 2005. Endogenous siRNAs derived from a pair of natural *cis*-antisense transcripts regulate salt tolerance in *Arabidopsis*. *Cell.* **123:**1279–91.

Bortolamiol, D., M. Pazhouhandeh, K. Marrocco, P. Genschik and V. Ziegler-Graff. 2007. The polerovirus F box protein P0 targets ARGONAUTE1 to suppress RNA silencing. *Current Biology.* **17:**1615–21.

Bouché, N., D. Lauressergues, V. Gasciolli and H. Vaucheret. 2006. An antagonistic function for *Arabidopsis* DCL2 in development and a new function for DCL4 in generating viral siRNAs. *The EMBO Journal.* **25:**3347–56.

Brigneti, G., O. Voinnet, W.X. Li, L.H. Ji, S.W. Ding and D.C. Baulcobme. 1998. Viral pathogenicity determinants are suppressors of transgene silencing in *Nicotiana benthamiana. The EMBO Journal.* **17:**6739–46.

Brodersen, P., L. Sakvarelidze-Achard, M. Bruun-Rasmussen, P. Dunoyer, Y.Y. Yamamoto and L. Sieburth. 2008. Widespread translational inhibition by plant miRNAs and siRNAs. *Science.* **320:**1185–90.

Broderson, P. and O. Voinnet. 2006. The diversity of RNA silencing pathways in plants. *Trends in Genetics.* **22:**268–80.

Bucher, E., T. Sijen, P. de Haan, R. Goldbach and M. Prins. 2007. Negative-strand tospoviruses and tenuiviruses carry a gene for a suppressor of gene silencing at analogous genomic positions. *Journal of Virology.* **77:**1329–36.

Buchmann, R.C., S. Asad, J.N. Wolf, G. Mohannath and D.M. Bisaro. 2009. Geminivirus AL2 and L2 proteins suppress transcriptional gene silencing and cause genome-wide reductions in cytosine methylation. *Journal of Virology.* **83:**5005–13.

Cañizares, M.C., J. Navas-Castillo and E. Moriones. 2008. Multiple suppressors of RNA silencing encoded by both genomic RNAs of the crinivirus, *Tomato chlorosis virus. Virology.* **379:**168–74.

Cao, X., P. Zhou, X. Zhang, S. Zhu, X. Zhong and Q. Xiao. 2005. Identification of an RNA silencing suppressor from a plant double-stranded RNA virus. *Journal of Virology.* **79:**13018–27.

Chan, S.W., D. Zilberman, Z. Xie, L.K. Johansen, J.C. Carrington and S.E. Jacobsen. 2004. RNA silencing genes control de novo DNA methylation. *Science.* **303:**1336.

Chapman, E.J. and J.C. Carrington. 2007. Specialization and evolution of endogenous small RNA pathways. *Nature Reviews Genetics.* **8:**884–96.

Chapman, E.J., A.I. Prokhnevsky, K. Gopinath, V.V. Dolja and J.C. Carrington. 2004. Viral RNA silencing suppressors inhibit the microRNA pathway at an intermediate step. *Genes & Development.* **18:**1179–86.

Chellappan, P., R. Vanitharani, J. Pita and C.M. Fauquet. 2004. Short interfering RNA accumulation correlates with host recovery in DNA virus-infected hosts and gene silencing targets specific viral sequences. *Journal of Virology.* **78:**7465–77.

Chellappan, P., R. Vanitharani, J. Pita and C.M. Fauquet. 2005. MicroRNA-binding viral protein interferes with *Arabidopsis* development. Proceedings of the National Academy of Sciences of the United States of America. **102:**10381–86.

Chen, J., W.X. Li, D. Xie, J.R. Peng and S.W. Ding. 2004. Viral virulence protein suppresses RNA silencing-mediated defense but upregulates the role of microRNA in host gene expression. *The Plant Cell.* **16:**1302–13.

Chen, H.Y., J. Yang, C. Lin and Y.A. Yuan. 2008. Structural basis for RNA-silencing suppression by *Tomato aspermy virus* protein 2b. *EMBO Reports.* **9**:754–60.

Covey, S.N., N.S. Al-Kaff, A. Lángara and D.S. Turner. 1997. Plants combat infection by gene silencing. *Nature.* **385**:781–2.

Csorba, T., A. Bovi, T. Dalmay and J. Burgyán. 2007. The p122 subunit of *Tobacco mosaic virus* replicase is a potent silencing suppressor and compromises both small interfering RNA- and microRNA-mediated pathways. *Journal of Virology.* **81**:11768–80.

Cui, X., G. Li, D. Wang, D. Hu and X. Zhou. 2005. A begomovirus DNAβ-encoded protein binds DNA, functions as a suppressor of RNA silencing and targets the cell nucleus. *Journal of Virology.* **79**:10764–75.

Cui, X., X. Tao, Y. Xie, C.M. Fauquet and X. Zhou. 2004. A DNAβ associated with *Tomato yellow leaf curl China virus* is required for symptom induction. *Journal of Virology.* **78**:13966–74.

Curaba, J. and X. Chen. 2008. Biochemical activities of *Arabidopsis* RNA-dependent RNA polymerase 6. *The Journal of Biological Chemistry.* **283**:3059–66.

Dalmay, T., A. Hamilton, S. Rudd, S. Angell and D.C. Baulcombe. 2000. An RNA-dependent RNA polymerase gene in *Arabidopsis* is required for post-transcriptional gene silencing mediated by a transgene but not by a virus. *Cell.* **101**:543–53.

Dalmay, T., R. Horsefield, T.H. Braunstein and D.C. Baulcombe. 2001. *SDE3* encodes an RNA helicase required for post-transcriptional gene silencing in *Arabidopsis.* *The EMBO Journal.* **20**:2069–77.

Deleris, A., J. Gallego-Bartolome, J. Bao, K.D. Kasschau, J.C. Carrington and O. Voinnet. 2006. Hierarchical action and inhibition of plant Dicer-like proteins in antiviral defense. *Science.* **313**:68–71.

Diaz-Pendon, J.A. and S.W. Ding. 2008. Direct and indirect roles of viral suppressors of RNA silencing in pathogenesis. *Annual Review of Phytopathology.* **46**:303–26.

Diaz-Pendon, J.A., F. Li, W.X. Li and S.W. Ding. 2007. Suppression of antiviral silencing by *Cucumber mosaic virus* 2b protein in *Arabidopsis* is associated with drastically reduced accumulation of three classes of viral small interfering RNAs. *The Plant Cell.* **19**:2053–63.

Ding, X.S., J. Liu, N.H. Cheng, A. Folimonov, Y.M. Hou. and Y. Bao. 2004. The *Tobacco mosaic virus* 126-kDa protein associated with virus replication and movement suppresses RNA silencing. *Molecular Plant-Microbe Interactions* **17**:583–92.

Donaire, L., D. Barajas, B. Matínez-García, L. Matínez-Priego, I. Pagán and C. Llave. 2008. Structural and genetic requirements for the biogenesis of *Tobacco rattle virus*-derived small interfering RNAs. *Journal of Virology.* **82**:5167–77.

Dong. Z., M. Han and N. Fedoroff. 2008. The RNA-binding proteins HYL1 and SE promote accurate *in vitro* processing of pri-miRNA by DCL1. Proceedings of the National Academy of Sciences of the United States of America. **105**:9970–5.

Dunoyer, P., C. Himber and O. Voinnet. 2005. DICER-LIKE 4 is required for RNA interference and produces the 21-nucleotide small interfering RNA component of the plant cell-to-cell silencing signal. *Nature Genetics.* **37**:1356–60.

Dunoyer, P., C.H. Lecellier, E.A. Parizotto, C. Himber and O. Voinnet. 2004. Probing the microRNA and small interfering RNA pathways with virus-encoded suppressors of RNA silencing. *The Plant Cell.* **16**:1235–50.

Fagard, M., S. Boutet, J.B. Morel, C. Bellini and H. Vaucheret. 2000. AGO1, QDE-2 and RDE-1 are related proteins required for posttranscriptional gene silencing in plants, quelling in fungi and RNA interference in animals. Proceedings of the National Academy of Sciences of the United States of America. **97**:11650–4.

Fukunaga, R. and J.A. Doudna. 2009. dsRNA with 5′ overhangs contributes to endogenous and antiviral RNA silencing pathways in plants. *The EMBO Journal.* **28**:545–55.

Fulci, V. and G. Macino. 2007. Quelling:post-transcriptional gene silencing guided by small RNAs in *Neurospora crassa. Current Opinion in Microbiology.* **10**:199–203.

Fusaro, A.F., L. Matthew, N.A. Smith, S.J. Curtin, J. Dedic-Hagan and G.A. Ellacott. 2006. RNA interference inducing hairpin RNAs in plants act through the viral defence pathway. *EMBO Reports.* **7**:1168–75.

Ghazala, W., A. Waltermann, R. Pilot, S. Winter and M. Varrelmann. 2008. Functional characterization and subcellular localization of the 16K cysteine-rich suppressor of gene silencing protein of tobacco rattle virus. *Journal of General Virology.* **89**:1748–58.

Glick, E., A. Zrachya, Y. Levy, A. Mett, D. Gidoni and E. Belausov. 2008. Interaction with host SGS3 is required for suppression of RNA silencing by tomato yellow leaf curl virus V2 protein. Proceedings of the National Academy of Sciences of the United States of America. **105**:157–61.

Golden, T.A., S.E. Schauer, J.D. Lang, S. Pien, A.R. Mushegian and U. Grossniklaus. 2002. *Short integuments1/suspensor1/carpel factory*, a Dicer homolog, is a maternal effect gene required for embryo development in Arabidopsis. *Plant Physiology.* **130**:808–22.

Gómez, G., G. Martínez and V. Pallás. 2008. Viroid-induced symptoms in *Nicotiana benthamiana* plants are dependent on RDR6 activity. *Plant Physiology.* **148**:414–23.

Gopal, P., P.P. Kumar, B. Sinilal, J. Jose, A.K. Yadunandam and R. Usha. 2007. Differential roles of C4 and βC1 in mediating suppression of posttranscriptional gene silencing:evidence for transactivation by the C2 of *Bhendi yellow vein mosaic virus*, a monopartite begomovirus. *Virus Research.* **123**:9–18.

Goto, K., T. Kobori, Y. Kosaka, T. Natsuaki and C. Masuta. 2007. Characterization of silencing suppressor 2b of cucumber mosaic virus based on examination of its small RNA-binding abilities. *Plant & Cell Physiology.* **48**:1050–60.

Guo, H.S. and S.W. Ding. 2002. A viral protein inhibits the long range signaling activity of the gene silencing signal. *The EMBO Journal.* **21**:398–407.

Haas, G., J. Azevedo, G. Moissiard, A. Geldreich, C. Himber and M. Bureau. 2008. Nuclear import of CaMV P6 is required for infection and suppression of the RNA silencing factor DRB4. *The EMBO Journal.* **27**:2102–12.

Hagen, C., M.R. Rojas, T. Kon and R.L. Gilbertson. 2008. Recovery from *Cucurbit leaf crumple virus* (family *Geminiviridae*, genus *Begomovirus*) infection is an adaptive antiviral response associated with changes in viral small RNAs. *Phytopathology.* **98:**1029–37.

Hamilton, A.J. and D.C. Baulcombe. 1999. A species of small antisense RNA in posttranscriptional gene silencing in plants. *Science.* **286:**950–2.

Hamilton, A., O. Voinnet, L. Chappell and D. Baulcombe. 2002. Two classes of short interfering RNA in RNA silencing. *The EMBO Journal.* **21:**4671–9.

Hao, L., H. Wang, G. Sunter and D.M. Bisaro. 2003. Geminivirus AL2 and L2 proteins interact with and inactivate SNF1 kinase. *The Plant Cell.* **15:**1034–48.

Hemmes, H., L. Lakatos, R.Goldbach, J. Burgyán and M. Prins. 2007. The NS3 protein of *Rice hoja blanca tenuivirus* suppresses RNA silencing in plant and insect hosts by efficiently binding both siRNAs and miRNAs. *RNA.* **13:**1079–89.

Henderson, I.R., X. Zhang, C. Lu, L. Johnson, B.C. Meyers and P.J. Green. 2006. Dissecting *Arabidopsis thaliana* DICER function in small RNA processing, gene silencing and DNA methylation patterning. *Nature Genetics.* **38:**721–5.

Herr, A.J., M.B. Jensen, T. Dalmay and D.C. Baulcombe. 2005. RNA polymerase IV directs silencing of endogenous DNA. *Science.* **308:**118–20.

Himber, C., P. Dunoyer, G. Moissiard, C. Ritzenthaler and O. Voinnet. 2003. Transitivity-dependent and -independent cell-to-cell movement of RNA silencing. *The EMBO Journal.* **22:**4523–33.

Hiraguri, A., R. Itoh, N. Kondo, Y. Nomura, D. Aizawa and Y. Murai. 2005. Specific interactions between Dicer-like proteins and HYL1/DRB-family dsRNA-binding proteins in *Arabidopsis thaliana*. *Plant Molecular Biology.* **57:**173–88.

Huettel, B., T. Kanno, L. Daxinger, W. Aufsatz, A.J.M. Matzke and M. Matzke. 2006. Endogenous targets of RNA-directed DNA methylation and Pol IV in *Arabidopsis.* *The EMBO Journal.* **25:**2828–36.

Jacobsen, S.E., M.P. Running and E.M. Meyerowitz. 1999. Disruption of an RNA helicase/RNAse III gene in *Arabidopsis* causes unregulated cell division in floral meristems. *Development.* **126:**5231–43.

Johansen, L.K. and J.C. Carrington. 2001. Silencing on the spot. Induction and suppression of RNA silencing in the *Agrobacterium*-mediated transient expression system. *Plant Physiology.* **126:**930–8.

Jones-Rhoades, M.W., D.P. Bartel and B. Bartel. 2006. MicroRNAs and their regulatory roles in plants. *Annual Review of Plant Biology.* **57:**19–53.

Kanno, T., B.Huettel, M.F. Mette, W. Aufsatz, E. Jaligot and L. Daxinger. 2005. Atypical RNA polymerase subunits required for RNA-directed DNA methylation. *Nature Genetics.* **37:**761–5.

Kasschau, K.D. and J.C. Carrington. 1998. A counterdefensive strategy of plant viruses: suppression of posttranscriptional gene silencing. *Cell.* **95:**461–70.

Kasschau, K.D., Z. Xie, E. Allen, C. Llave, E.J. Chapman and K.A. Krizan. 2003. P1/HC-Pro, a viral suppressor of RNA silencing, interferes with *Arabidopsis* development and miRNA function. *Developmental Cell.* **4:**205–17.

Kon, T. and M. Ikegami. 2009. RNA silencing in plants and its suppression by plant viruses. CAB Review: Perspectives in Agriculture, Veterinary Science, Nutrition and Natural Resources. **064:**1–16 (http://www.cabi.org/ cabreviews).

Kon, T., P. Sharma and M. Ikegami. 2007. Suppressor of RNA silencing encoded by the monopartite tomato leaf curl Java begomovirus. *Archives of Virology.* **152**:1273–82.

Kreuze, J.F., E.I. Savenkov, W. Cuellar, X. Li and J.P.T. Valkonen. 2005. Viral class 1 RNase III involved in suppression of RNA silencing. *Journal of Virology.* **79**:7227–38.

Kubota, K., S. Tsuda, A. Tamai and T. Meshi. 2003. Tomato mosaic virus replication protein suppresses virus-targeted posttranscriptional gene silencing. *Journal of Virology.* **77**:11016–26.

Kurihara, Y., N. Inaba, N. Kutsuna, A. Takeda, Y. Tagami and Y. Watanabe. 2007. Binding of tobamovirus replication protein with small RNA duplexes. *Journal of General Virology.* **88**:2347–52.

Kurihara, Y., Y. Takashi and Y. Watanabe. 2006. The interaction between DCL1 and HYL1 is important for efficient and precise processing of pri-miRNA in plant microRNA biogenesis. *RNA.* **12**:206–12.

Kurihara, Y. and Y. Watanabe. 2004. *Arabidopsis* micro-RNA biogenesis through Dicer-like 1 protein functions. Proceedings of the National Academy of Sciences of the United States of America. **101**:12753–8.

Lakatos, L., T. Csorba, V. Pantaleo, E.J. Chapman, J.C. Carrington and Y.P. Liu. 2006. Small RNA binding is a common strategy to suppress RNA silencing by several viral suppressors. *The EMBO Journal.* **25**:2768–80.

Li, F. and S.W. Ding. 2006. Virus counterdefense: diverse strategies for evading the RNA-silencing immunity. *Annual Review of Microbiology.* **60**:503–31.

Li, C.F., O. Pontes, M. El-Shami, I.R. Henderson, Y.V. Bernatavichute and S.W.L. Chan. 2006. An ARGONAUTE4-containing nuclear processing center colocalized with Cajal bodies in *Arabidopsis thaliana. Cell.* **126**:93–106.

Li, J., Z. Yang, B. Yu, J. Liu and X. Chen. 2005. Methylation protects miRNAs and siRNAs from a 3'-end uridylation activity in *Arabidopsis. Current Biology.* **15**:1501–7.

Lindbo, J.A., L. Silva-Rosales, W.M. Proebsting and W.G. Dougherty. 1993. Induction of a highly specific antiviral state in transgenic plants: implications for regulation of gene expression and virus resistance. *The Plant Cell.* **5**:1749–59.

Liu, L., J. Grainger, M.C. Cañizares, S.M. Angell and G.P. Lomonossoff. 2004. *Cowpea mosaic virus* RNA-1 acts as an amplicon whose effects can be counteracted by a RNA-2-encoded suppressor of silencing. *Virology.* **323**:37–48.

Love, A.J., J. Laird, J. Holt, A.J. Hamilton, A. Sadanandom and J.J. Milner. 2007. Cauliflower mosaic virus protein P6 is a suppressor of RNA silencing. *Journal of General Virology.* **88**:3439–44.

Lózsa, R., T. Csorba, L. Lakatos and J. Burgyán. 2008. Inhibition of 3' modification of small RNAs in virus-infected plants require spatial and temporal co-expression of small RNAs and viral silencing-suppressor proteins. *Nucleic Acids Research.* **36**:4099–107.

Lu, R., A. Folimonov, M. Shintaku, W.X. Li, B.W. Falk and W.O. Dawson. 2004. Three distinct suppressors of RNA silencing encoded by a 20-kb viral RNA genome. Proceedings of the National Academy of Sciences of the United States of America. **101**:15742–7.

Maia I.G., A.L. Haenni and F. Bernardi. 1996. Potyviral HC-Pro: a multifunctional protein. *Journal of General Virology.* **77:**1335–41.

Mallory, A.C., L. Ely, T.H. Smith, R. Marathe, R. Anandalakshmi and M. Fagardet. 2001. HC-Pro suppression of transgene silencing eliminates the small RNAs but not transgene methylation or the mobile signal. *The Plant Cell.* **13:**571–83.

Mangwende, T., M.L. Wang, W. Borth, J. Hu, P.H. Moore and T.E. Mirkovd. 2009. The P0 gene of *Sugarcane yellow leaf virus* encodes an RNA silencing suppressor with unique activities. *Virology.* **384:**38–50.

Martín-Hernández, A.M. and D.C. Baulcombe. 2008. Tobacco rattle virus 16-kilodalton protein encodes a suppressor of RNA silencing that allows transient viral entry in meristems. *Journal of Virology.* **82:**4064–71.

Martínez-Priego, L., L. Donaire, D. Barajas and C. Llave. 2008. Silencing suppressor activity of the *Tobacco rattle virus*-encoded 16-kDa protein and interference with endogenous small RNA-guided regulatory pathways. *Virology.* **376:**346–56.

Martínez-Turiño, S. and C. Hernández. 2009. Inhibition of RNA silencing by the coat protein of *Pelargonium* flower break virus: distinctions from closely related suppressors. *Journal of General Virology.* **90:**519–25.

Matranga, C., Y. Tomari, C. Shin, D.P. Bartel and P.D. Zamore. 2005. Passenger-strand cleavage facilitates assembly of siRNA into Ago2-containing RNAi enzyme complexes. *Cell.* **123:**607–20.

Meng, C., J. Chen, J. Peng and S.M. Wong. 2006. Host-induced avirulence of hibiscus chlorotic ringspot virus mutants correlates with reduced gene-silencing suppression activity. *Journal of General Virology.* **87:**451–9.

Mérai, Z., Z. Kerényi, A. Molnár, E. Barta, A. Válóczi and G. Bisztray. 2005. Aureusvirus P14 is an efficient RNA silencing suppressor that binds double-stranded RNAs without size specificity. *Journal of Virology.* **79:**7217–26.

Mérai, Z., Z. Kerényi, S. Kertész, M. Magna, L. Lakatos and D. Silhavy. 2006. Double-stranded RNA binding may be a general plant RNA viral strategy to suppress RNA silencing. *Journal of Virology.* **80:**5747–56.

Moissiard, G. and O. Voinnet. 2006. RNA silencing of host transcripts by cauliflower mosaic virus requires coordinated action of the four *Arabidopsis* Dicer-like proteins. Proceedings of the National Academy of Sciences of the United States of America. **103:**19593–8.

Montgomery, T.A., M.D. Howell, J.T. Cuperus, D. Li, J.E. Hansen and A.L. Alexander. 2008. Specificity of ARGONAUTE7-miR390 interaction and dual functionality in *TAS3 trans*-acting siRNA formation. *Cell.* **133:**128–41.

Montgomery T.A., S.J. Yoo, N. Fahlgren, S.D. Gilbert, M.D. Howell and C.M. Sullivan. 2008. AGO1-miR173 complex initiates phased siRNA formation in plants. Proceedings of the National Academy of Sciences of the United States of America. **105:**20055–62.

Morel, J.B., C. Godon, P. Mourrain, C. Béclin, S. Boutet and F. Feuerbach. 2002. Fertile hypomorphic *ARGONAUTE* (*ago1*) mutants impaired in post-transcriptional gene silencing and virus resistance. *The Plant Cell.* **14:**629–39.

Mourrain, P., C. Béclin, T. Elmayan, F. Feuerbach, C. Godon and J.B. Morel. 2000. *Arabidopsis SGS2* and *SGS3* genes are required for posttranscriptional gene silencing and natural virus resistance. *Cell.* **101:**533–42.

Muangsan, N., C. Beclin, H. Vaucheret and D. Robertson. 2004. Geminivirus VIGS of endogenous genes requires SGS2/SDE1 and SGS3 and defines a new branch in the genetic pathway for silencing in plants. *The Plant Journal.* **38:**1004–14.

Napoli, C., C. Lemieux and R. Jorgensen. 1990. Introduction of a chimeric chalcone synthase gene into petunia results in reversible co-suppression of homologous genes *in trans. The Plant Cell.* **2:**279–89.

Onodera, Y., J.R. Haag, T. Ream, P.C. Nunes, O. Pontes and C.S. Pikaard. 2005. Plant nuclear RNA polymerase IV mediates siRNA and DNA methylation-dependent heterochromatin formation. *Cell.* **120:**613–22.

Pazhouhandeh, M., M. Dieterle, K. Marrocco, E. Lechner, B. Berry and V. Brault. 2006. F-box-like domain in the polerovirus protein P0 is required for silencing suppressor function. Proceedings of the National Academy of Sciences of the United States of America. **103:**1994–9.

Peragine, A., M. Yoshikawa, G. Wu, H.L. Albrecht and R.S. Poethig. 2004. *SGS3* and *SGS2/SDE1/RDR6* are required for juvenile development and the production of *trans*-acting siRNAs in *Arabidopsis. Genes & Development.* **18:**2368–79.

Pfeffer, S., P. Dunoyer, F. Heim, K.E. Richards, G. Jonard and V. Ziegler-Graff. 2002. P0 of beet western yellows virus is a suppressor of posttranscriptional gene silencing. *Journal of Virology.* **76:**6815–24.

Powers, JG., T.L. Sit, C. Heinsohn, C.G. George, K.H. Kim and S.A. Lommel. 2008. The *Red clover necrotic mosaic virus* RNA-2 encoded movement protein is a second suppressor of RNA silencing. *Virology.* **381:**277–86.

Qi, X., F.S. Bao and Z. Xie. 2009. Small RNA deep sequencing reveals role for *Arabidopsis thaliana* RNA-dependent RNA polymerases in viral siRNA biogenesis. *PLoS One.* **4:**e4971.

Qi, Y., A.M. Denli and G.J. Hannon. 2005. Biochemical specialization within *Arabidopsis* RNA silencing pathways. *Molecular Cell.* **19:**421–8.

Qi, Y., X. He, X.J. Wang, O. Kohany, J. Jurka and G.J. Hannon. 2006. Distinct catalytic and non-catalytic roles of ARGONAUTE4 in RNA-directed DNA methylation. *Nature.* **443:**1008–12.

Qu, F., T. Ren and T.J. Morris. 2003. The coat protein of turnip crinkle virus suppresses posttranscriptional gene silencing at an early initiation step. *Journal of Virology.* **77:**511–22.

Qu, F., X. Ye, G. Hou, S. Sato, T.E. Clemente and T.J. Morris. 2005. RDR6 has a broad-spectrum but temperature-dependent antiviral defense role in *Nicotiana benthamiana. Journal of Virology.* **79:**15209–17.

Qu, F., X. Ye and T.J. Morris. 2008. *Arabidopsis* DRB4, AGO1, AGO7 and RDR6 participate in a DCL4-initiated antiviral RNA silencing pathway negatively regulated by DCL1. Proceedings of the National Academy of Sciences of the United States of America. **105:**14732–7.

Ramachandran, V. and X. Chen. 2008. Small RNA metabolism in *Arabidopsis. Trends in Plant Science.* **13:**368–74.

Ratcliff, F, B.D. Harrison and D.C. Baulcombe. 1997. A similarity between viral defense and gene silencing in plants. *Science.* **276:**1558–60.

Reed, J.C., K.D. Kasschau, A.I. Prokhnevsky, K. Gopinath, G.P. Pogue and J.C. Carrington. 2003. Suppressor of RNA silencing encoded by *Beet yellows virus. Virology.* **306:**203–9.

Saeed, M., Y. Zafar, J.W. Randles and M.A. Rezaian. 2007. A monopartite begomovirus-associated DNA β satellite substitutes for the DNA B of a bipartite begomovirus to permit systemic infection. *Journal of General Virology.* **88:**2881–9.

Sarmiento, C., E. Gomez, M. Meier, T.A. Kavanagh and E. Truve. 2007. *Cocksfoot mottle virus* P1 suppresses RNA silencing in *Nicotiana benthamiana* and *Nicotiana tabacum. Virus Research.* **123:**95–9.

Saunders, K., A. Norman, S. Gucciardo and J. Stanley. 2004. The DNA β satellite component associated with ageratum yellow vein disease encodes an essential pathogenicity protein (βC1). *Virology.* **324:**37–47.

Scholthof, H.B. 2006. The *Tombusvirus*-encoded P19: from irrelevance to elegance. *Nature Reviews Microbiology.* **4:**405–11.

Schwach, F., F.E. Vaistij, L. Jones and Baulcombe. 2005. An RNA-dependent RNA polymerase prevents meristem invasion by potato virus X and is required for the activity but not the production of a systemic silencing signal. *Plant Physiology.* **138:**1842–52.

Selth, L.A., J.W. Randles and M.A. Rezaian. 2004. Host responses to transient expression of individual genes encoded by *Tomato leaf curl virus. Molecular Plant-Microbe Interactions.* **17:**27–33.

Senshu, H., J. Ozeki, K. Komatsu, M. Hashimoto, K. Hatada and M. Aoyama. 2009. Variability in the level of RNA silencing suppression caused by triple gene block protein 1 (TGBp1) from various potexviruses during infection. *Journal of General Virology.* **90:**1014–24.

Sharma, P. and M. Ikegami. 2010. *Tomato leaf curl Java virus* V2 protein is a determinant of virulence, hypersensitive response and suppression of post transcriptional gene silencing. *Virology.* **396:**85–93.

Sharma, P., M. Ikegami and T. Kon. 2010. Identification of the virulence factors and suppressors of posttranscriptional gene silencing encoded by *Ageratum yellow vein virus*, a monopartite begomovirus. *Virus Research.* **149:**19–27.

Shivaprasad, P.V., R. Akbergenov, D. Trinks, R. Rajeswaran, K. Veluthambi and T. Hohn. 2005. Promoters, transcripts and regulatory proteins of mungbean yellow mosaic geminivirus. *Journal of Virology.* **79:**8149–63.

Silhavy, D., A. Molnár, A. Lucioli, G. Szittya, C. Hornyik and M. Tavazza. 2002. A viral protein suppresses RNA silencing and binds silencing-generated, 21- to 25-nucleotide double-stranded RNAs. *The EMBO Journal.* **21:**3070–80.

Takeda, A., S. Iwasaki, T. Watanabe, M. Utsumi and Y. Watanabe. 2008. The mechanism selecting the guide strand from small RNA duplexes is different among *Argonaute* proteins. *Plant & Cell Physiology.* **49:**493–500.

Takeda, A., K. Sugiyama, H. Nagano, M. Mori, M. Kaido and K. Mise. 2002. Identification of a novel RNA silencing suppressor, NSs protein of *Tomato spotted wilt virus. FEBS Letters.* **532:**75–9.

Takeda, A., M. Tsukuda, H. Mizumoto, K. Okamoto, M. Kaido and K. Mise. 2005. A plant RNA virus suppresses RNA silencing through viral RNA replication. *The EMBO Journal.* **24:**3147–57.

Te, J., U. Melcher, A. Howard and J. Verchot-Lubicz. 2005. Soilborne wheat mosaic virus (SBWMV) 19K protein belongs to a class of cysteine rich proteins that suppress RNA silencing. *Virology Journal.* **2:**18.

Trinks, D., R. Rajeswaran, P.V. Shivaprasad, R. Akbergenov, E.J. Oakeley and K. Veluthambi. 2005. Suppression of RNA silencing by a geminivirus nuclear protein, AC2, correlates with transactivation of host genes. *Journal of Virology.* **79:**2517–27.

Umbach, J.L. and B.R. Cullen. 2009. The role of RNAi and microRNAs in animal virus replication and antiviral immunity. *Genes & Development.* **23:**1151–64.

Vaistij, F.E., L. Jones and D.C. Baulcombe. 2002. Spreading of RNA targeting and DNA methylation in RNA silencing requires transcription of the target gene and a putative RNA-dependent RNA polymerase. *The Plant Cell.* **14:**857–67.

Valli, A., G. Dujovny and J.A. García. 2008. Protease activity, self interaction and small interfering RNA binding of the silencing suppressor P1b from *Cucumber vein yellowing Ipomovirus. Journal of Virology.* **82:**974–86.

Valli, A., A.M. Martín-Hernández, J.J. López-Moya and J.A. García. 2006. RNA silencing suppression by a second copy of the P1 serine protease of *Cucumber vein yellowing Ipomovirus* (CVYV), a member of the family *Potyviridae* that lacks the cysteine protease HCPro. *Journal of Virology.* **80:**10055–63.

Van der Krol, AR., L.A. Mur, M. Beld, J.N.M. Mol and A.R. Stuitje. 1990. Flavonoid genes in petunia:addition of a limited number of gene copies may lead to a suppression of gene expression. *Plant Cell.* **2:**291–99.

van Wezel, R., X. Dong, H. Liu, P. Tien, J. Stanley and Y. Hong. 2002. Mutation of three cysteine residues in *Tomato yellow leaf curl virus-China* C2 protein causes dysfunction in pathogenesis and posttranscriptional gene-silencing suppression. *Molecular Plant-Microbe Interactions.* **15:**203–8.

Vanitharani, R., P. Chellappan, J.S. Pita and C.M. Fauquet. 2004. Differential roles of AC2 and AC4 of cassava geminiviruses in mediating synergism and suppression of posttranscriptional gene silencing. *Journal of Virology.* **78:**9487–98.

Vargason, J.M., G. Szittya, J. Burgyán and T.M. Tanaka Hall. 2003. Size selective recognition of siRNA by an RNA silencing suppressor. *Cell.* **115:**799–811.

Vaucheret, H. 2006. Post-transcriptional small RNA pathways in plants: mechanisms and regulations. *Genes & Development.* **20:**759–71.

Vaucheret, H. 2008. Plant ARGONAUTES. *Trends in Plant Science.* **13:**350–8.

Vazquez, F., H. Vaucheret, R. Rajagopalan, C. Lepers, V. Gasciolli and A.C. Mallory. 2004. Endogenous *trans*-acting siRNAs regulate the accumulation of *Arabidopsis* mRNAs. *Molecular Cell.* **16:**69–79.

Voinnet, O. 2005. Induction and suppression of RNA silencing: insights from viral infections. *Nature Reviews Genetics.* **6:**206–20.

Voinnet, O., Y.M. Pinto and D.C. Baulcombe. 1999. Suppression of gene silencing: a general strategy used by diverse DNA and RNA viruses of plants. Proceedings of the National Academy of Sciences of the United States of America. **96:**14147–52.

Wang, M.B., X.Y. Bian, L.M. Wu, L.X. Liu, N.A. Smith and D. Isenegger. 2004. On the role of RNA silencing in the pathogenicity and evolution of viroids and viral satellites. Proceedings of the National Academy of Sciences of the United States of America. **101**:3275–80.

Wang, H., K.J. Buckley, X. Yang, R.C. Buchmann and D.M. Bisaro. 2005. Adenosine kinase inhibition and suppression of RNA silencing by geminivirus AL2 and L2 proteins. *Journal of Virology.* **79**:7410–8.

Wang, H., L . Hao, C.Y. Shung, G. Sunter and D.M. Bisaro. 2003. Adenosine kinase is inactivated by geminivirus AL2 and L2 proteins. *The Plant Cell.* **15**:3020–32.

Xie, Z., E. Allen, A. Wilken and J.C. Carrington. 2005. DICER-LIKE 4 functions in trans-acting small interfering RNA biogenesis and vegetative phase change in *Arabidopsis thaliana.* Proceedings of the National Academy of Sciences of the United States of America. **102**:12984–9.

Xie, Z., B. Fan, C. Chen and Z. Chen. 2001. An important role of an inducible RNA-dependent RNA polymerase in plant antiviral defense. Proceedings of the National Academy of Sciences of the United States of America. **98**:6516–21.0

Xie, Z., L.K. Johansen, A.M. Gustafson, K.D. Kasschau, A.D. Lellis and D. Zilberman. 2004. Genetic and functional diversification of small RNA pathways in plants. *PLoS Biology.* **2**:e104.

Xiong, R., J. Wu, Y. Zhou and X. Zhou. 2008. Identification of a movement protein of the *Tenuivirus* rice stripe virus. *Journal of Virology.* **82**:12304–11.

Yaegashi, H., T. Takahashi, M. Isogai, T. Kobori, S. Ohki and N. Yoshikawa. 2007. Apple chlorotic leaf spot virus 50 kDa movement protein acts as a suppressor of systemic silencing without interfering with local silencing in *Nicotiana benthamiana. Journal of General Virology.* **88**:316–24.

Yaegashi, H., A. Tamura, M. Isogai and N. Yoshikawa. 2008. Inhibition of long-distance movement of RNA silencing signals in *Nicotiana benthamiana* by *Apple chlorotic leaf spot virus* 50 kDa movement protein. *Virology.* **382**:199–206.

Yaegashi, H., T. Yamatsuta, T. Takahashi, C. Li, M. Isogai and T. Kobori. 2007. Characterization of virus-induced gene silencing in tobacco plants infected with apple latent spherical virus. *Archives of Virology.* **152**:1839–49.

Yang, J.Y., M. Iwasaki, C. Machida, Y. Machida, X. Zhou and N.H. Chua. 2008. βC1, the pathogenicity factor of TYLCCNV, interacts with AS1 to alter leaf development and suppress selective jasmonic acid responses. *Genes & Development.* **22**:2564–77.

Yang, Z., Y.W. Ebright, B. Yu and X. Chen. 2006. HEN1 recognizes 21–24 nt small RNA duplexes and deposits a methyl group onto the 2′ OH of the 3′ terminal nucleotide. *Nucleic Acids Research.* **34**:667–75.

Ye, K., L. Malinina and D.J. Patel. 2003. Recognition of small interfering RNA by a viral suppressor of RNA silencing. *Nature.* **426**:874–8.

Ye, K. and D.J. Patel. 2005. RNA silencing suppressor p21 of beet yellows virus forms an RNA binding octameric ring structure. *Structure.* **13**:1375–84.

Yelina, NE., E.I. Savenkov, A.G. Solovyev, S.Y. Morozov and J.P.T. Valkonen. 2002. Long-distance movement, virulence and RNA silencing suppression controlled by a single protein in hordei- and potyviruses: complementary functions between virus families. *Journal of Virology.* **76**:12981–91.

Yoo, B.C., F. Kragler, E. Varkonyi-Gasic, V. Haywood, S. Archer-Evans and Y.M. Lee. 2004. A systemic small RNA signaling system in plants. *The Plant Cell.* **16**:1979–2000.

Yoshikawa, M., A . Peragine, M.Y. Park and R.S. Poethig. 2005. A pathway for the biogenesis of *trans*-acting siRNAs in *Arabidopsis*. *Genes & Development.* **19**:2164–75.

Yu, B., Z. Yang, J. Li, S. Minakhina, M. Yang and R.W. Padgett. 2005. Methylation as a crucial step in plant microRNA biogenesis. *Science.* **307**:932–5.

Zhang, X., I.R. Henderson, C. Lu, P.J. Green and S.E. Jacobsen. 2007. Role of RNA polymerase IV in plant small RNA metabolism. Proceedings of the National Academy of Sciences of the United States of America. **104**:4536–41.

Zhang X., Y.R. Yuan, Y. Pei, S.S. Lin, T. Tuschl, D.J. Patel et al. 2006. *Cucumber mosaic virus*-encoded 2b suppressor inhibits *Arabidopsis* Argonaute1 cleavage activity to counter plant defense. *Genes & Development.* **20**:3255–68.

Zheng, X., J. Zhu, A. Kapoor and J.K. Zhu. 2007. Role of *Arabidopsis* AGO6 in siRNA accumulation, DNA methylation and transcriptional gene silencing. *The EMBO Journal.* **26**:1691–701.

Zhou, Z.S., M. Dell'Orco, P. Saldarelli, C. Turturo, A. Minafra and G.P. Martelli. 2006. Identification of an RNA-silencing suppressor in the genome of *Grapevine virus A*. *Journal of General Virology.* **87**:2387–95.

Zilberman, D., X. Cao and S.E. Jacobsen. 2003. *ARGONAUTE4* control of locus-specific siRNA accumulation and DNA and histone methylation. *Science.* **299**:716–9.

Zrachya, A., E. Glick, Y. Levy, T. Arazi, V. Citovsky and Y. Gafni. 2007. Suppressor of RNA silencing encoded by *Tomato yellow curl virus*-Israel. *Virology.* **358**:159–65.

CHAPTER 12

Gene Silencing and its Applications in Plants

S. Abdolhamid Angaj* and Somayeh Darvishani

ABSTRACT

Gene silencing can occur either through repression of transcription, termed transcriptional gene silencing (TGS), or through mRNA degradation, termed post-transcriptional gene silencing (PTGS). TGS results in reduction of transcription whereas PTGS results in sequence specific mRNA degradation in cytoplasm without dramatic changes in transcription of corresponding gene in nucleus. Both TGS and PTGS are used to regulate endogenous genes. Interestingly, mechanisms of gene silencing also protect the organism's genome from transposons, viruses. In this paper, molecular aspects and mechanisms of gene silencing in plant were reviewed. Finally, its applications were discussed.

Keywords: Transcriptional gene silencing, post-transcriptional gene silencing, miRNA, siRNA.

INTRODUCTION

TGS is the result of histone modifications, creating an environment of heterochromatin around a gene that makes it inaccessible to transcriptional machinery (RNA polymerase, transcription factors, etc.). The primary cause of TGS is thought to be cytosine methylation, in 5-CG-3 sequences of eukaryotic DNA, of promoter sequences, but the exact role of promoter methylation in TGS remains unclear. Occasionally, 5'-CNG-3' sequences are also methylated. Methylation presumably inactivates the promoter by blocking its proper interactions with transcription factors

* Corresponding author e-mail: ershad110@yahoo.com

or by attracting chromatin-remodeling proteins, which could lead to the heterochromatinization of the promoter sequence. The methyl groups project into the major groove of the DNA and thus hinder the binding of most transcription factors. In addition, methylated CG sequences are recognized by methylcytosine binding proteins (MeCPs). Bound MeCPs are, in turn, recognized by other proteins that remove acetyl groups from the histones (especially H4). This results in the condensation of the DNA to form heterochromatin that is no longer accessible for transcription. The genes in such regions are said to be "silenced" (Clark 2005, Wang and Waterhouse 2002, Khan 2007).

When plant scientists first began to create transgenic plants, it was tacitly assumed that more copies of a given gene or mRNA would result in higher amounts of protein product. This assumption, however, proved to be false, as it became clear that poor expression of transgenes was frequently associated with multiple copies of particular sequences. Several studies revealed an inverse correlation between the probability of observing gene silencing and number of transgene copies integrated at a given locus. Distinct silencing effects were observed depending on whether promoters or protein coding regions were repeated. In some studies, multiple copies of the same promoter had been used to drive expression of distinct coding regions in different transgene constructs. When these different transgene constructs were combined in the same plant genome, TGS and increased methylation of the homologous promoters occasionally resulted. It can be divided into two groups; *cis*-TGS and *trans*-TGS (Vaucheret and Fagard 2001, McDonald 2000, Trent 2005, Choudhuri and Carlson 2009).

cis-TGS

It is assumed that TGS occurs in *cis* as a result of pairing between closely linked copies that leads to the formation of secondary DNA structures that attract methylation and heterochromatin components. However, these analyses did not determine whether hypermethylation is a cause or a consequence of TGS. Occasionally, single copies of a transgene are subject to TGS. It is assumed that this results from large discrepancies between the GC content of the transgene and that of the surrounding genomic sequences (Vaucheret and Fagard 2001, Verdel and Moazed 2005).

trans-TGS

Although transcriptional silencing is occasionally observed with single copies of transgene, most of systems analyzed in detail involve tras-inactivation effects in which a silencing transgene locus induce methylation and silencing of an unlinked target transgene locus with which it shares

DNA sequence identity in promoter regions. The mechanism by which *trans*-TGS occurs in these cases is not known. It was suggested that interaction of homologous sequences (DNA–DNA pairing) leads to the transfer of a silent chromatin state from one locus to the other. Alternatively, the involvement of a silencing RNA produced by one locus has also been invoked. Evidences for an RNA-DNA association as a trigger for *de novo* DNA methylation was obtained first with a viroid system in plants. Viroids are plant pathogens that consist solely of a non-coding, highly base paired rod-shaped RNA several hundred nucleotides in length. Viroid cDNA copies integrated into tobacco nuclear DNA became methylated *de novo* only when plants are infected with replication-competent viroids. Viroid RNA replication apparently induced methylation of homologous cDNA copies in tobacco genome, presumably via a direct RNA-DNA interaction. On the other hand, trans-silencing can also occur when an RNA is transcribed from an inverted DNA repeat. The resulting double-stranded RNA (dsRNA) can diffuse to the homologous target locus and trigger its methylation through an RNA-DNA interaction. Finally, RNA degradation step of PTGS takes place in the cytoplasm but some works have indicated that RNAs produced as a consequence of PTGS can re-enter the nucleus and trigger methylation of homologous DNA sequences. Cytoplasmic events can thus be imprinted at the chromosomal level. RNAs made either in the nucleus or the cytoplasm can provoke DNA methylation, thereby completing a cellular circuit of homology-based controls that are involved in regulating gene expression at multiple levels (Carrington et al. 2001, Jeong et al. 2002, Vaucheret and Fagard 2001).

MOLECULAR ASPECTS OF DNA METHYLATION AND TGS IN PLANTS

How is DNA methylation involved in TGS in plants? In general, plant genomes are more extensively methylated than genomes of other eukaryotic organisms. This is certainly due to the fact that in animal systems methylation of cytosines is almost restricted to CpG dinucleotides whereas in plants, 5-methyl cytosine (m^5C) is found at CpG, CpXpG and asymmetric sites. Despite the obvious significance of DNA methylation in plants, little information about the mechanism(s) of methylation-mediated TGS is available. Wu and co-workers (2000) characterized two *Arabidopsis thaliana* genes, AtRPD3A and AtRPD3B, which are homologous to the yeast histone deacetylase, RPD3. A Gal4–AtRPD3A, Gal-4 is an activator, fusion protein repressed transcription when directed to a promoter driving a reporter gene and inhibition of endogenous AtRPD3A activity led to delayed flowering. In parallel, the *AtRPD3A* cDNA was also isolated by Tian and Chen (2001) and named *AtHD1*. Tian and Chen also down-regulated the endogenous *AtHD1* gene by using the antisense technique. In addition to the delayed flowering phenotype seen by Wu et al., Tian and Chen described various developmental abnormalities, including early senescence, ectopic expression of silenced genes, suppression of apical dominance, homeotic changes, heterochronic shift toward juvenility, flower defects and male and female sterility in their transgenic *A. thaliana* lines. These data demonstrated that also in plants, histone deacetylation followed by chromatin remodeling (Meyer 2000) is involved in transcriptional repression. They have also shown that a reporter transgene was targeted for deacetylation but they did not point to a connection between DNA methylation and TGS. A first clue for this connection was documented by the observation that in plants, trichostatin (TSA), a drug which causes inhibition of histone deacetylase activity, can substitute for 5-azacytidine to depress silent endogenous rRNA genes (Chen and Pikaard 1997, Meyer1999). In other words, if transcriptional repression of the rRNA genes is relieved by both TSA and 5-azacytidine the methylated rRNA gene promoters most likely recruit a methyl-CpG-binding repressor complex (Pikaard 1999). Two methyl-CpG-binding proteins, dcMBP1 and dcMBP2, were recently detected in *Daucus carota* (carrot) plants (Pitto et al. 2000). Interestingly, dcMBP2 showed high affinity for methylated nonconventional CpXpX and CpXpG sites and lower affinity for conventional CpGs. This indicates that in plants, other or additional repressor complexes are present when compared with animal systems. The fact that plants have two classes of HDACs that are not present in animals may further indicate that chromatin remodeling plays an important role in plant gene regulation (Jeon 2002, Meyer 2000, Lusser et al. 2001).

PTGS

Although discovered independently, it seems that PTGS in plants, quelling in *Neurospora* and RNAi in animals share a conserved mechanism. This system is also found in protozoa such as *Paramecium* and trypanosomes. In both PTGS and RNAi the presence of dsRNA triggers the Dicer/RISCsystem for RNA degradation. RNAi works through a number of RNA species including (1) siRNA (small interfering RNA)—small, double-stranded RNA (long dsRNA) that degrade mRNA; (2) miRNA (micro RNA)-small, double-stranded RNA (hairpin dsRNA) that interfere with translation by imperfect base pairing with mRNA. Both siRNA and miRNA share common intermediates including DICER, an RNase III endoclease and RNA-induced silencing complex (RISC). Small amounts of dsRNAs have been shown to silence a vast excess of target mRNA (Paddison and Vogt 2008).

dsRNA may originate from bidirectional transcription of repetitive DNA elements, or transcription of RNA molecules that can base pair internally to form dsRNA segments. The former mechanism reliant upon recognition of an abundant sense RNA species and subsequent synthesis of a complementary antisense strand by the plant encoded genes *sde1/ sgs2, sde3, ago1* and *sgs3*. Transcription of aberrant RNAs that may lack proper processing signals may trigger dsRNA by RNA-dependent RNA polymerase (RdRP). The dsRNAs are translocated to the cytoplasm where they become target for the Dicer enzyme. Viral replication of many RNA viruses also includes a dsRNA replicative form that can be targeted by Dicer. In plants, two types of Dicer proteins generate short (21 nucleotide) and long (25 nucleotide) siRNA and miRNA. These RNAs are double-stranded with two nucleotide 3' overhangs and hydroxyl termini. They direct RNA interference in plants. The short ones are associated with local silencing and the long ones with systematic silencing (Allis et al. 2007, Munroe and Zhu 2006).

MECHANISMS OF RNAi

A dsRNA complementary to an endogenous mRNA is processed by Dicer into 21–25 nucleotide siRNA. Dicer also processes miRNAs, which function as negative regulators of specific target mRNAs. Although recombinant Dicer is active as a dsRNA-specific endo-nuclease *in vitro*, in cells it generally functions in association with other proteins as a component of multiprotein complexes. In *C. elegans*, the RDE-4 dsRNA RNA binding protein interacts with Dicer, the argonaute-like protein RDE-1 and a DExH-box helicase to regulate RNAi. Hiraguri et al. reported that HYL1/DRB interacts with Dicer in plants. The 21 to 25-nt siRNAs are recognized by the RISC. A helocase within the RISC unwinds the siRNA duplex, enabling the complementary antisense strand to guide target recognition. Finally, an endonuclease(s) within the RISC cleaves the mRNA to induce silencing. Thus, whether produced naturally or introduced by transfection, siRNA are involved in multiple protein interactions and participate in a minimum of four major steps: (1) initial duplex recognition by pre-RISC complex, (2) ATP-dependent RISC activation, (3) target recognition and (4) target cleavage.

Alternatively, the siRNA may act as a primer for replication by a dsRNA-dependent RNA polymerase (RdRP), resulting in a new dsRNA molecule that is then subsequently cleaved by Dicer (Appasani 2005, Engelke and Rossi 2005).

microRNA GENE STRUCTURE AND EXPRESSION IN PLANTS

miRNAs are single-stranded non-coding RNAs (ncRNA) that silence gene expression through translational repression. Genes for miRNA are found in all chromosomes. They are found in introns of coding and non-coding genes as well as in exons of non-coding genes.miRNA gene expression is under control of either a discrete promoter or the the promoter of the host gene. In both plants and animals, miRNA genes are transcribed by the same DNA-dependent RNA polymerase II (pol II) machinery as the coding genes. The primary miRNA transcripts (pri-miRNAs) contain 5′ m′G cap and 3′ poly (A) tail and they may be hundred s to thousands of nucleotides long with hairpin (stem-loop) structures. As with other miRNAs, the plant miRNA genes encode longer stem-loop precursor transcripts that are subsequently processed to form the mature, active miRNA sequence of 20–22 nt in length. It appeared until recently that the stem-loop precursor encodes just a single mature miRNA in plants which is contained in the stem of molecule. This is in contrast to animals where some miRNAs are known to form gene clusters and appear to be transcribed together as a polycistron. In plants, a similar organization has been discovered, for example the CA764701 transcript harbouring three mRNAs (Zhang et al. 2006). Recent studies of plants miRNA primary precursor from a wide range of species indicate that the vast majority of these plant miRNA primary precursors are even larger than animal miRNA primary precursors. Whereas most animal miRNA primary precursors fall into the 70 to 80 nt size range, plant miRNA primary precursors are far more variable in size ranging from 60–509 nt. It is possible that the longer precursors may contain regions that are functionally important for example regulating mature miRNA biogenesis. However, the basic stem-loop structure is conserved amongst these different molecules (Tost 2008).

Whilst plants animals start with the same basic stem-loop miRNA precursors in the nucleus, the subsequent processing system differs in important details. In plants, nuclear processing events are particularly important. The stem-loop miRNA primary precursor molecules are bound by the dsRNA binding protein HYPONASTIC LEAVES 1 (HYL1), which itself physically interacts with the RNase III protein DICER-LIKE 1 (DCL1; Kurihara et al. 2006). Two other dsRNA-binding proteins, DRB2 and DRB5, may also interact with DCL1 (Hiraguri et al. 2005). Another protein that interacts with HYL1 is SERRATE, a nuclear localized C2H2 zinc finger protein (Lobbes et al. 2006, Yang et al. 2006).

In plants, DICER-LIKE 1 makes two cuts that generate a double- stranded mRNAi precursor from the stem structure. These duplexes remain in the nucleus and, unlike their animal counterparts, are subjected to further chemical modifications at their 3′ ends. The nuclear HUA ENHANCER 1 (HEN1) protein is duplex-specific RNA methyltransferase enzyme that adds methyl groups to the 3′ ends (Yu et al. 2005). Methylation may increase miRNA stability and/or act as a tag to direct the miRNA in the processing pathway (Li et al. 2006).

Both animal and plant miRNA duplexes interact with Argonaute family proteins at the next stage. Argonautes are conserved slicer proteins characterized by their conserved PIWI domain-which adopts an RNase-like fold and is the predicted region of nuclease activity—and the conserved PAZ domain which is proposed to be the domain that binds small RNAs (Parker et al. 2004). It is not clear where the next stage of miRNA processing occurs in plants. Plants do have a protein with similarity to Exportin-5, called HASTY, but the exact function of this protein is unknown and it remains to be determined whether HASTY exports miRNA duplexs or mature miRNAs from the nucleus to the cytoplasm. If HASTY is the sole export system for miRNAs from the nucleus to cytoplasm, a more dramatic outcome might be expected. Indeed, there is some evidence that all the steps of miRNA biogenesis and activity could take place within the nucleus, making cytoplasmic processing unnecessary (Xie et al. 2003; Park et al. 2005). Although the precise location of these events remains unknown, it is now clear that the ARGONAUTE 1 (AGO1) protein of plants plays a major role in subsequent events. AGO1 may act alone in recruiting miRNAs and subsequent interactions with miRNA-bound mRNA targets and probably does not need to form a complex with other proteins to function (Baumberger and Baulcombe, 2005). A significant difference between plants and animals is readily distinguished in the miRNA-mediated processing of the target. In plants, the great majority of miRNAs so far explored directs the AGO1-catalysed cleavage of the mRNA. This is in contrast to animals, where miRNAs mostly act as translational repressors by inhibiting the passage of translational machinery (Bartel 2004). The difference in the mechanism may in part be due to general differences seen between plants and animals in the homology of the miRNAs to their target mRNAs. Whereas plant miRNAs are perfectly or almost perfectly complementary to the target mRNA, animal miRNAs tend to display less complementarity.

In plants, the miRNA always directs cleavage of the target mRNA at 10th or 11th nucleotide from the 5′ end of the miRNA to give cleavage products with3′ hydroxyl and 5′ phosphate groups. Both cleavage products enter further degradation pathways, with the 3′ products degraded by the cytoplasmic EXONUCLEASE 4 (Souret et al. 2004) whilst the 5′ products are the target of 3′ uridine addition and accelerated degradation (Shen and Goodman 2004).

APPLICATIONS OF GENE SILENCING IN PLANTS

(1) Defence against Viruses and Viroids

Several lines of evidences suggest that gene silencing is an important defense mechanism against invasive nucleic acid parasites. DNA viruses, RNA viruses and viroids are all initiators of gene silencing in plants. Accordingly, gene silencing mediates viral recovery and cross protection, whereby plants display a reduction in disease symptoms and develop *de novo* resistance to infection by viruses closely relate to the primary inoculums. Further, loss of gene silencing function in *sde1/sgs2, sde3, ago1 and sgs3* mutants results in hypersensivity to infection by certain viruses.

(2) Genome Stability

Indirect evidence suggests that gene silencing may help to maintain genome integrity by suppressing the activity of transposable elements. Transposons in plant genomes are generally located in heterochromatic regions that are highly methylated and transcriptionally inactive. The cloning and sequencing of siRNAs from tobacco has revealed a population of siRNAs homologous to integrated transposon sequences.

(3) Regulation of Endogenous Gene Expression

It is tempting to speculate that an ancient and highly conserved pathway such as PTGS could play a broader role in the control of endogenous gene expression. At present there is no clear data to verify this hypothesis in plants, but several lines of evidence support the possibility. Mutant of some known PTGS pathway components, notably *ago1*, display highly abnormal development. Thus it is possible that PTGS plays a role in development.

Naturally occurring antisense RNA transcripts have been described for several endogenous genes in plants, but the functional importance of these antisense RNAs has not been determined. The only example of PTGS-like control of endogenous gene expression is the small temporal RNA (stRNA) system in *C. elegans*. The *let-7* and *lin-4* genes of *C. elegans* produce short, self-complementary mRNAs which form stem-loop structures *in vivo*. The duplex RNA sequence is digested into ~21 nt fragments by Dicer and these stRNAs suppress expression of various homologous mRNA targets which are involved in developmental timing. However, stRNA is thought to operate by hindering translation of mRNA rather than by the RNA degradation mechanism associated with siRNA and PTGS (Barciszewski and Erdmann 2003).

REFERENCES

Allis, C.D., T. Jenuwein and D. Reinberg. 2007. Epigenetics. John Inglis. 502 p.

Appasani K. 2005. RNA interference technology: from basic science to drug development. Cambridge, 510 p.

Baumberger, N. and D.C. Baulcombe. 2005. Arabidopsis ARGONAUTE1 is an RNA Slicer that selectively recruits mRNAs and siRNAs. *PNAS*. **102:**11928–11933.

Barciszewski, J. and V.A. Erdmann. 2003. Noncoding RNAs: molecular biology and molecular medicine. Kluwer Academic/Plenum Publishers.

Bartel, D. 2004. MicroRNAs Genomics, Biogenesis, Mechanism and Function. *Cell.* **116(2):**281–297.

Carrington, J.C., K.D. Kasschau and L.K. Johansen. 2001. Activation and Suppression of RNA Silencing by Plant Viruses. *Virology*. **281:**1–5.

Chen, Z.J. and C.S. Pikaard. 1997. Epigenetic silencing of RNA polymerase I transcription: a role for DNA methylation and histone modification in nucleolar dominance. *Genes and Development*. **11:**2124–2136.

Choudhuri, S. and D.B. Carlson. 2009. Genomics: Fundamentals and Applications. *Informa healthcare*. 424 p.

Clark, D. 2005. Molecular Biology. Elsevier Inc., 784 p.

Engelke, D.R. and J.J. Rossi. 2005. Methods in enzymology. Elsevier Academic Press, 453 p.

Hiraguri, A., R. Itoh, N. Kondo, Y. Nomura, D. Aizawa, Y. Murai H. Koiwa, M. Seki, K. Shinozaki and T. Fukuhara. 2005. Specific interactions between Dicer-like proteins and HYL1/DRB-family dsRNA-binding proteins in *Arabidopsis thaliana*. *Plant Mol. Biol*. **57(2):**173–88.

Jeon, K.W. 2002. International Review of Cytology. Academic Press, Vol. 219, 273 p.

Jeong, B.R. and D. Wu-Scharf, C. Zhang and H. Cerutti. 2002. Suppressors of transcriptional transgenic silencing in *Chlamydomonas* are sensitive to DNA-damaging agents and reactivate transposable elements. *PNAS*. **99:**1076–1081.

Khan, I. 2007. Citrus genetics, breeding and biotechnology. *CABI*. 370 p.

Kurihara and Y. Takashi and Y. Watanabe. 2006. The interaction between DCL1 and HYL1 is important for efficient and precise processing of pri-miRNA in plant microRNA biogenesis. *RNA*. **12:**206–212.

Li, H., T. Rauch, Z.X. Chen, P.E. Szabo, A.D. Riggs and G.P. Pfeifer. 2006. The histone methyltransferase SETDB1 and the DNA methyltransferase DNMT3A interact directly and localize to promoters silenced in cancer cells. *J. Biol. Chem*. **281:**19480–19500.

Lobbes, D., G. Rallapalli, D.D. Schmidt, C. Martin and J. Clarke. 2006. SERRATE: new player on plant microRNA scene. *EMBO Rep*. **7:**1052–1058.

Lusser, A., D. Kolle and P. Loidle. 2001. Histone acetylation: Lessons from the plant kingdom. *Trends in Plant Science*. **6:**59–65.

McDonald, J.F. 2000. Transposable elements and genome evolution. Kluwer Academic Publishers, 299 p.

Meyer, P. 1999. The role of chromatin remodeling in transgene silencing and plant development. *In vitro Cell. Dev. Biol. Plant.* **35**:29–36.

Meyer, P. 2000. Transcriptional transgene silencing and chromatin components. *Plant Mol. Biol.* **43**:221–234.

Munroe, S.H. and J. Zhu. 2006. Overlapping transcripts, double-stranded RNA and antisense regulation: A genomic perspective. *Cellular and Molecular Life Sciences.* **63**:2102–2118.

Paddison, P.J. and P.K. Vogt. 2008. RNA interference. Springer, 273 p.

Park, Y.J. J.V. Chodaparambil, Y. Bao, S.J. McBryant and K. Luger. 2005. Nucleosome assembly protein1 exchange histone H2A-H2B dimmers and assists nucleosome sliding. *J. Biol. Chem.* **280**:1817–1825.

Parker, J.S., S.M. Roe and D. Barford. 2004. Crystal structure of a PIWI protein suggests mechanisms for siRNA recognition and slicer activity. *EMBO J.* **23(24)**:4727–4737.

Pikaard, C.S. 1999. Nucleolar dominance and silencing of transcription. *Trends in Plant Science.* **4**:478–483.

Pitto, L., F. Cernilogar, M. Evangelista, L. Lombardi, C. Miarelli and P. Rocchi. 2000. Characterization of carrot nuclear proteins that exhibit specific binding affinity towards conventional and non-conventional DNA methylation. *Plant Mol. Biol.* **44**:659–673.

Tian, L. and J. Chen J. 2001. Blocking histone deacetylase in Arabidopsis induces pleiotropic effects on plant gene regulation. *Proc. Natl. Acad. Sci. USA.* **98**:200–205.

Tost, J. 2008. Epigenetics. Caister Academic Press. 407 p.

Trent, R.J. 2005. Molecular medicine: an introductory text. *Elsevier Academic Journal.* 310 p.

Vaucheret, H. and M. Fagard. 2001. Transcriptional gene silencing in plants: targets, inducers and regulators. *Trends in Genet.* **17(1)**:29–35.

Verdel, A. and D. Moazed. 2005. RNAi-directed assembly of heterochromatin in fission yeast. *FEBS Letters.* **579(26)**:5872–5878.

Wang, M.B. and P.M. Waterhouse. 2002. Application of gene silencing in plants. *Current Opinion in Plant Biology.* **5(2)**:146–150.

Wu, K., K. Malik, L. Tian, D. Brown and B. Miki. 2000. Functional analysis of a RPD3 histone deacetlyase homologue in *Arabidopsis thaliana. Plant Mol. Biol.* **44**:167–176.

Xie, Z., K.D. Kasschau and J.C. Carrington. 2003. Negative feedback regulation of DICER-LIKE1 In Arabidopsis by miRNA-guided mRNA degradation. *Curr. Biol.* **13**:784–789.

Yang, L., Z. Liu, F. Lu. Feng and A. Dong. 2006. Huang H. SERRATE is a novel nuclear regulator in primary microRNA processing in Arabidopsis. *The Plant Journal.* **47(6)**:841–850.

Yu, B., Z. Yang, L. Li, S. Minakhina, M. Yang, R.W. Padgett, R. Steward and X. Chen. 2005. Methylation as a Crucial Step in Plant microRNA Biogenesis. *Science.* **307(5711):**932–935.

Zhang, X., Y.R. Yuan, Y. Pei, S.S. Lin, T. Tuschl, D.J. Patel and N.H. Chua. 2006. *Cucumber mosaic virus*-encoded 2b suppressor inhibits *Arabidopsis* Argonaute1 cleavage activity to counter plant defense. *Genes and Dev.* **20:**3255–3268.

CHAPTER 13

Use of RNAi Technology for Control of Rice Viruses

Indranil Dasgupta

ABSTRACT

Rice is the most important cereal food crop worldwide and provides dietary calories to more people than any other single crop. Rice Tungro is a viral disease of rice in South and Southeast Asia, responsible for significant losses during the production of this important food crop. Rice tungro disease is caused by the simultaneous infection of the rice plant by two viruses, *Rice tungro bacilliform virus* (RTBV) and *Rice tungro spherical virus* (RTSV). RNA-interference (RNAi) is a natural defence mechanism present in organisms against intracellular pathogens, such as viruses. RNAi pathway can be artificially stimulated, with the help of transgenes to enhance the defence pathways to obtain resistance against RTBV in rice. This chapter describes the use of RNAi technology to control RTBV in rice. A small DNA fragment from RTBV was cloned in a manner, which produced a sense and an anti-sense transcript simultaneously. This gave rise to triggering of RNAi, which takes place in the presence of such aberrant RNAs, against the transcript, giving rise to the degradation products, the small-interfering RNAs. Upon challenge inoculation with RTBV, its levels were much lower than the control untransformed plants and the transgenic plants developed only mild tungro symptoms.

Keywords: Rice, tungro, caulimovirus, RNA resistance, siRNA.

INTRODUCTION

Rice is the most important food crop in the world and acts as the primary source of dietary calories for a very large proportion of the world population living in tropical regions. It is thus very important that efforts are made to

* *Corresponding author e-mail*: indranil58@yahoo.co.in; indasgup@south.du.ac.in

increase its yield in the light of increase in population pressures, especially in the areas of the world where rice is the primary food. Of the several factors limiting rice production, those caused by biotic stress are most important and hence deserve serious study. Several biotic factors play an important role in reducing rice yields, which include mainly bacteria, fungi and viruses. Of the three, viruses are the only pathogenic agents, which operate in an intracellular manner, the rest operating largely in the space outside the cell, at least for a large part of their life cycle. Viruses infecting plants have small genomes, which code for only a few proteins, but are able to modify successfully the cell machinery to produce profound biochemical changes within the cell, resulting in pathogenesis. Thus, the mechanisms used by viruses leading to pathogenesis is generally simpler than the same for non-viral pathogens, hence understanding them leads to a better understanding of the disease process and its eventual control measures.

VIRUSES OF RICE

A number of viruses are known to affect rice worldwide, but the severity and the occurrence of viral diseases is very significant in South and Southeast Asia. The Tungro disease, the most common viral disease in this region (Azzam and Chancellor, 2002, Manwan et al. 1985, Herdt 1991) is caused by the joint infection with two dissimilar viruses; a DNA virus, the *Rice tungro bacilliform virus* (RTBV), shaped like a bullet and RNA virus, *Rice tungro spherical virus* (RTSV) (Hibino et al., 1978, Jones et al. 1991), having an icosahedral or spherical shape. Artificially produced single infections have indicated that RTBV causes most of the symptoms exhibited by plants affected by tungro, the typical yellow-orange discolouration and severe stunting, whereas RTSV infection produces only mild stunting (Rivera and Ou, 1967, Hibino, 1989, Dasgupta et al. 1991). The tungro viral complex (RTBV and RTSV) is transmitted exclusively by the insect Green leafhopper (*Nephotettix virescens*), GLH, which acts as an efficient vector (Hibino and Cabauatan 1987). However, interestingly, RTBV is unable to be transmitted by GLH in the absence of RTSV (Hibino, 1983, Cabauatan and Hibino, 1985). On the other hand, RTSV is efficiently transmitted by GLH, indicating that RTSV is more adapted to the GLH-rice system than RTBV, which might exist in some weed hosts in parallel to rice and may use rice only as an additional host. However, these possibilities have not been investigated as yet. The molecular nature of the transmission helper function of RTSV in the vectoring of RTBV by GLH, although investigated, has not been revealed.

RTBV is closely related to several viruses of the Genus Badnavirus (abbreviated from Bacilliform DNA-containing) whose members infect a variety of plants growing in the tropical regions of the world (Lockhart, 1990). Badnaviruses can infect plants as varied as rice, sugarcane, yam, cacao, citrus, kalanchoe, etc. All badnaviruses have a circular DNA molecule of approximately 7.5–8.0 kb and have a gene arrangement resembling retroviruses, i.e., a "gag-pol-env" plan. In this, structural proteins such as Coat protein and enzymatic proteins such as protease and replicase/polymerase occur adjacent to each other and are transcribed together from a single promoter. All badnaviruses show a gene arrangement suggesting that they produce a poly-protein as the primary translation product, which consists of at least three genes; Coat protein (CP), protease and reverse transcriptase (RT). Presence of RT is a common feature between retroviruses (animal viruses with an RNA genome which convert to DNA) and badnaviruses. In badnaviruses, a single, more than full-length transcript is reverse-transcribed to form the genomic DNA. This feature is present not only in badnaviruses, but also in the members of the related Genus Caulimovirus, a prime example being *Cauliflower mosaic virus*. Badnaviruses and caulimoviruses are also known as pararetroviruses because of their relationship with retroviruses. In retroviruses, the viral genome is an RNA, which gets converted into DNA, whereas in pararetroviruses, the viral genomic DNA is transcribed into RNA, which is then converted back to DNA (Hull, 2002).

In RTBV, the genome consists of four Open reading frames (ORFs, Figure 1). Of the four, ORF I, II and IV potentially encode proteins of molecular weights of 12, 24 and 46 kDa, whereas ORF III encodes a potential protein of molecular weight 194 kDa, P194 (Qu et al. 1991, Hay et al. 1991). Only the 194 kDa protein shows homology in the derived amino acid sequence to other proteins, mainly from retroviruses and caulimoviruses (Hay et al. 1991, Covey, 1986, Qu et al. 1991, Kano et al. 1992). The analysis of the amino acid sequence of purified CP has shown that it is encoded by the P194. ORF IV of RTBV is absent in other badnaviruses and exists only in RTBV (Qu et al. 1991).

RNA INTERFERENCE AND VIRAL RESISTANCE

The common method of generating disease resistance in plants is to take varieties or wild relatives having resistance against the disease in the form of Resistance genes and transfer them in cultivated crop varieties. However, extensive searches in the available rice germplasm have failed to find resistance sources against RTBV, although there were several against RTSV (Hibino et al. 1990, Azzam and Chancellor, 2002). Hence, to achieve RTBV and tungro resistance in rice, it is important to adopt new strategies, namely the transfer of genes from outside the rice genome, participating in pathways which might help build resistance against viral

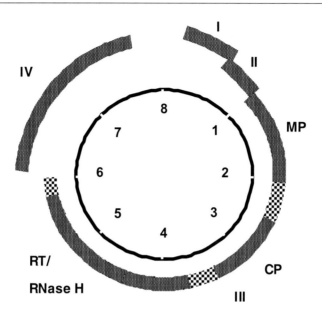

Figure 1: Genome organisation of RTBV. The inner circle represents the double-stranded 8 kb DNA. The arcs represent the positions of the four open reading frames: I, II, III and IV. The position of genes encoding proteins of the ORF III polypotein have been marked. MP: movement protein; CP: coat protein; PR: protease and RT/RNase H: reverse transcriptase/ribonuclease H.

infections and thus to confer tungro resistance in rice. The use of RNA-interference (RNAi) as a technology, which has been used previously against several plant viruses to develop resistance, thus gains importance in this context.

RNAi was discovered first in plants, during experiments to darken the purple hue of petunias following expression of a gene responsible for purple flower colour, resulted in many lines showing altered patterns of flower colour (Napoli et al. 1990). Further analysis indicated a low expression of the heterologous gene, despite being present in several copies. This indicated that the gene expression has been compromised in such plants, leading to a phenomenon known as gene silencing. RNAi has now been recognized as a universal gene regulatory mechanism. RNAi involves both modifications at the DNA as well as RNA levels, affecting the methylation patterns and the stability and translatability of mRNA. Depending upon the system, one or the other phenomena is important. In plants, as in contrast to animals, RNAi mainly determines the RNA stability. RNAs having a strong secondary structure or double-stranded form (aberrant RNA) trigger the action of proteins known as Dicers, which recognize and degrade the aberrant RNA forms to 21–24 polyribonucleotide products known as small-interfering (si)-RNA. A multi-protein complex known as the RNA-induced silencing complex (RISC), consisting of Argonaute (a ribonuclease), recognizes the siRNA and incorporates one strand giving sequence-specificity to RISC, which can then seek and degrade mRNA having homology to the incorporated siRNA fragments. The hallmark of RNAi in plants is the drastic fall in the levels of the target mRNA of the gene and the rise in the levels of siRNA (Tijsterman et al. 2002, Susi et al. 2004, Chen, 2009).

Double-stranded (ds) or hairpin (hp)-RNAs constitute the most common forms of aberrant RNA, which trigger RNAi. Most plant viruses, in fact produce dsRNA as intermediates during their replication process. That is the reason why RNAi is considered to be a natural defence mechanism in plants against viruses, because it can target viral RNAs for degradation. Upon viral infection, the plant host mounts a defence reaction against the invading virus by inducing RNAi (Voinnet, 2005, Ding and Voinnet, 2007). The strategy of strengthening this pre-existing defence mechanism to generate resistance against viruses is thus attractive. Generating ds- or hp-RNA derived from viral genes in plants has produced a number of instances of viral resistance in plants (Waterhouse et al, 1998, Pooggin et al. 2003; Vanitharani et al. 2003; Tenllado et al. 2003; Di Nicola-Negri et al. 2005; Lennefors et al. 2006; Abhary et al. 2006). Here we describe the use of RNAi in generating resistance against Rice tungro disease in rice.

DESIGN AND CONSTRUCTION OF RNAi VECTORS

For developing RNAi against RTBV in transgenic rice plants, ORF IV of RTBV was cloned in both sense and anti-sense orientation under a strong constitutively expressed promoter, the CaMV35S (Figure 2). Cloned DNA fragments, representing full length RTBV genomic DNA were generated from tungro affected plants, collected from the field by first obtaining full-length viral cloned DNA, following a partial purification step (Nath et al., 2002). DNA fragments representing ORF IV were generated using appropriate restriction enzymes. Each fragment was cloned under the control of a separate Cauliflower mosaic virus 35S (CaMV35S) promoter and transcription termination signals. The RTBV ORF IV fragments were cloned into the binary vector pCAMBIA 1380, having the hygromycin resistance selection marker (encoding hygromycin phosphotransferase) for selection of transformed rice tissues as calli.

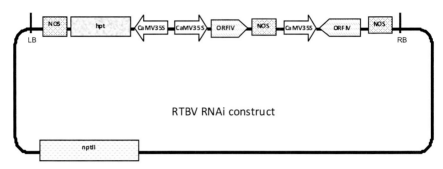

Figure 2: Schematic representation of the RTBV RNAi plasmid construct. Nos: nopaline synthase transcription terminator signal, hpt: hygromycin phosphotransferase gene, CaMV35S: promoter of Cauliflower mosaic virus 35S transcript, ORFIV: RTBV ORFIV gene, LB: Left border region of Ti Plasmid, RB: Right border region of Ti plasmid.

RICE TRANSFORMATION AND ANALYSIS OF TRANSFORMED PLANTS

Rice transformation was performed using the widely used method of co-cultivation (Wang et al. 1997) using *Agrobacterium tumefaciens* super-virulent strains suited for this purpose (Holsters et al. 1978) and carrying the appropriate plasmids described above. Rice calli were generated from embryos in seeds and were allowed to grow in the presence of hygromycin till roots and shoots developed. The plantlets were then transferred to soil and allowed to mature. Seeds (T_1) obtained from the above plants were used for molecular analysis. Copy numbers of transgenes were determined by Southern analysis, using a probe encoding RTBV ORF IV. For this purpose, the rice DNA was digested with restriction endonucleases capable of recognizing sites present outside the RTBV DNA. Total DNA was isolated from young leaves of T_1 plants using standard method as described by Dellaporta et al. (1983) and analyzed as described above. Accumulation of transcripts was determined by Northern analysis, in which total RNA was electrophoresed under denaturing conditions, followed by capillary transfer onto appropriate membranes. The accumulation of transcripts was visualized by using labeled probes. Accumulation of siRNA was investigated using a fraction of RNA enriched for small RNA molecules. The electrophoresis was performed using polyacrylamide gels and blotted onto membranes as earlier.

To challenge plants with RTBV, GLH were fed to tungro-infected rice plants and used to inoculate test plants at 10–15 day stage. The plants were kept with the viruliferous GLH for 24 hours. Viral titers were assayed in the plants by DNA dot-blot hybridization. The autoradiograms were developed using similar conditions to allow comparison of the viral DNA titers on the basis of hybridization signals and the signal intensities were used as a measure of the viral DNA titers. The values of the viral titers of the control plants (untransformed) were taken as 100% and those of the transgenic plants were expressed as percent fractions of the values of the controls.

Transgenic rice plants, selected for growth on a medium containing hygromycin, showed normal growth and development when transferred to soil and bore the normal quantities of seeds. The plants contained either one or two copies of the transgene, as suggested by the results of Southern analysis. The plants were also analyzed to see whether transcripts arising from the RTBV transgene, cloned in a sense and anti-sense orientation, accumulated in the plants or not. This was done using Northern hybridization and using RTBV ORF IV as a probe. The RNA samples from transgenic plants did not show hybridizing bands with labeled probe (ORF IV) in any of the three lines tested. To make sure that the quality of RNA was satisfactory, the same RNA preparation was probed with a labeled DNA encoding hygromycin phosphotransferase, the marker gene included in the T-DNA, which showed a prominent band in all samples tested from transgenic plants. This indicated that the quality of RNA was satisfactory and transcripts specific to hygromycin phosphotransferase gene accumulated in the plants, but the ones specific to ORF IV did not (Tyagi et al. 2008).

Most of the transgenic lines analyzed displayed accumulation of siRNA between 21–23 bases specific to ORF IV transgene. This indicated that the transgene is getting transcribed but is being degraded by the activated RNAi machinery in the plant.

VIRAL RESISTANCE IN THE TRANSFORMED PLANTS

Control plants (untransformed) showed a very rapid buildup of RTBV levels by 10 days post-inoculation (dpi), as assessed by DNA dot-blot of the inoculated plants. In the period till 60 dpi, the RTBV levels came down to about 60% of the maximal levels at about 30 dpi, followed by a rise again by 45 dpi. So, even in the untransformed plants, the RTBV levels were not constant. The transgenic lines could be divided into two classes depending upon their response pattern to RTBV buildup; in the Class I, plants showed a much slower buildup of RTBV DNA levels. At 30 dpi, the RTBV levels in transgenic lines were only 20% of the control levels, rising to 60% at 40 dpi and almost the same levels as control plants by 50 dpi. In the Class II, the transgenic plants showed a rapid buildup of RTBV DNA levels similar to the control plants at 10 dpi, but then showed a sharp decline in the RTBV DNA levels, becoming undetectable by 20 dpi

(Tyagi et al. 2008). In general, both the types of transgenic plants showed lower buildup of viral DNA, but the timings of the buildup were different.

Transgenic plants belonging to the Class I showed mild tungro symptoms. The height reduction was only about 50% as compared to that observed in control plants. Class I plants flowered and bore seeds whereas the control plants showed no flowering and usually succumbed to secondary infections in the glass house. Transgenic plants belonging to the Class II were however equally symptomatic as the control plants.

RNAi, if fully functional, should rapidly degrade RTBV transcripts and lead to a reduction in the DNA levels upon infection. The initial degradation is the function of the RISC complex. There is a lack of information on the exact sequence of events taking place upon infection with RTBV. The viral DNA probably enters the nucleus, forms a mini-chromosome and is transcribed using the cellular transcriptional machinery. The transcripts enter the cytoplasm, where they are translated into the viral proteins, including the Reverse transcriptase-RNAseH. This results in the conversion of further RTBV RNAs as templates for the production of RTBV DNA, which spreads throughout the plants using the plant vasculature. RTBV transcripts, if targeted by RISC will block the cycle of viral DNA accumulation and spread.

However, there could be several reasons why the block in the RTBV replication cycle could be delayed or disrupted. RISC is a multi-enzyme complex, which puts limits in its ability to target its substrate. The targets of RISC may not be readily accessible or the viral DNA formation and spread could be too fast for the RISC to keep pace and degrade the viral transcript. Lastly, there could be a specific suppression of RNAi by viral proteins (Ding and Voinnet, 2007). All the above could result in several variations in the response of the transgenic plants to RTBV infection, as was observed in this study.

Many viral proteins have been shown to have RNAi suppressing activity, which can result from the protein binding and sequestering siRNAs or interfering with the formation of siRNAs by binding or interacting with the DCL proteins, required for siRNA biogenesis (Ding and Voinnet, 2007). RTBV proteins have not been tested for RNAi suppressor properties and there is a possibility that some RTBV proteins do in fact have such activities. It may actually depend upon the dynamics of the activated RISC in the RNAi triggered rice plants for degrading the viral transcript and the activity of RNAi suppressor proteins from RTBV, if any, which might emerge from the RTBV DNA formed from any viral transcripts escaping the RISC activity.

The development of tungro symptoms is the resultant of several steps in the infection process. The rising levels of RTBV DNA within infected rice cells brings about a battery of changes, both at the molecular and the cellular levels, the identities of such changes are yet unknown. In this study, it was generally observed that the rapid accumulation of RTBV DNA in the plant within the first two weeks of infection decided the pattern of development of symptoms. Transgenic plants belonging to Class I showed very low levels of RTBV DNA during the early period of plant growth, which maybe an important reason why they escaped the tungro symptoms. On the other hand, transgenic plants belonging to Class II showed high accumulation of RTBV DNA in the early period and thus developed tungro symptoms, although the viral DNA levels declined rapidly later on during infection. Work by Dai et al. (2008) suggests the depletion of certain rice transcription factors, necessary for the proper growth and development of the plant, to be an important reason for the development of tungro symptoms. These transcription factors are sequestered by the accumulating RTBV DNA in the infected rice cells and prevented from performing their usual transcriptional activities. It is quite possible that such interference by the RTBV depends upon the stage of plant development and hence the observations on the RNAi transgenic plants as to the RTBV resistance levels can be explained.

CONCLUSION

RNAi is a very promising technology for the generation of virus resistant plants. DNA fragments derived from the virus can be cloned so as to produce a double-stranded or a hairpin-loop transcript and used to generate transgenic plants. The plants are primed for RNAi against the virus and generally show resistance when challenged by the virus. The degree and pattern of virus resistance may vary depending upon the virus pathology and the RNAi suppressors carried by the virus, if any. Rice tungro disease, caused by the joint infection with RTBV and RTSV, two unrelated viruses, has been also targeted using RNAi. Although the titer of RTBV in the inoculated transgenic plants was not very low, especially towards late stages of infection, there was a significant amelioration of symptoms. More understanding would be required of the process of pathogenesis, especially the interplay between the primer RNAi machinery and the viral suppressors if RNAi to analyze critically the development of virus resistance using RNAi approaches.

ACKNOWLEDGEMENTS

The research work is supported by grants from Department of Biotechnology, Government of India and University of Delhi.

REFERENCES

Abhary, M.K., G.H. Anfoka, M.K. Nakhla and D.P. Maxwell. 2006. Post-transcriptional gene silencing in controlling viruses of the Tomato yellow leaf curl virus complex. *Arch. Virol.* **151**:2349–2363.

Azzam, O and T.C.B. Chancellor. 2002. The Biology, Epidemiology and Management of Rice Tungro Disease in Asia. *Plant Dis.* **86**:88–100.

Cabauatan, P.Q. and H. Hibino. 1985. Transmission of rice tungro bacilliform and spherical viruses by Nephotettix virescens Distant. *Philippine Phytopathol.* **21**:103–109.

Chen, X. 2009. Small RNAs and Their Roles in Plant Development. *Annu Rev. Cell. Dev. Biol.* **35**:21–44.

Covey, S.N. 1986. Amino acid sequence homology in gag region of reverse transcribing elements and the coat protein gene of cauliflower mosaic virus. *Nucleic Acids Res.* **14**:623–633.

Dai, S., X. Wei, A.A. Alfonso, L. Pei, U.G. Duque, Z. Zhang, G.M. Babb and R.N. Beachy. 2008. Transgenic rice plants which overexpress transcription factors RF2a and RF2b are tolerant to rice tungro virus replication and disease. *Proc. Natl. Acad. Sci. USA.* **105**:21012–21016.

Dasgupta, I., R. Hull, S. Eastop, C. Poggi-pollini, M. Blakebrough, M.I. Boulton and J.W. Davies. 1991. Rice tungro bacilliform virus DNA independently infects rice after Agrobacterium- mediated transfer. *J. Gen. Virol.* **72**:1215–1221.

Dellaporta, S.L., J. Wood and J.B. Hicks. 1983. A plant DNA minipreparation: Version II. *Plant Mol. Biol. Rep.* **1**:19–21.

Di Nicola-Negri, E., A. Brunetti, M. Tavazza and V. Ilardi. 2005. Hairpin RNA-mediated silencing of Plum pox virus P1 and HC-Pro genes for efficient and predictable resistance to the virus. *Transgenic Res.* **14**:989–994.

Ding, S.W. and O. Voinnet. 2007. Antiviral immunity directed by small RNAs. *Cell.* **130**:413–26.

Hay, J.M., M.C. Jones, M.L. Blakebrough, I. Dasgupta, J.W. Davies and R. Hull. 1991. An analysis of the sequence of an infectious clone of rice tungro bacilliform virus, a plant pararetrovirus. *Nucleic Acids Res.* **19**:2615–2621.

Herdt, R.W. 1991. Research priorities for biotechnology. In: G.S. Khush, G.H. Toennissen (eds.). Rice biotechnology. CAB International, Wallingford, UK, pp. 19–54.

Hibino, H. 1983. Relations of rice tungro bacilliform and rice tungro spherical viruses and their vector Nephotettix virescens. *Ann. Phytopathol. Soc. Jpn.* **49**:545–553.

Hibino, H. 1989. Insect-borne viruses of rice. In: Advances in Disease Vector Research, Vol. 6. K.F. Harris, ed. Springer-Verlag, New York. pp. 209–241.

Hibino, H. and P.Q. Cabauatan. 1987. Infectivity neutralization of rice tungro-associated viruses acquired by vector leafhoppers. *Phytopath.* **77**:473–476.

Hibino, H., M. Roechan and S. Sudarisman. 1978. Association of two types of virus particles with penyakit habang (tungro disease) of rice in Indonesia. *Phytopath.* **68**:1412–1416.

Hibino, H., R.D. Daquioag, E.M. Mesina and V.M. Aguiero. 1990. Resistances in rice to tungro-associated viruses. *Plant Dis.* **74:**923–926.

Holsters, M., D. Waele, M. Depicker, E. Messens, M. Van Montagu and J. Schell. 1978. Transfection and transformation of *A. tumefaciens. Mol. Gen. Genet.* **168:**181–187.

Hull, R. 2002. Matthew's Plant Virology, Academic Press, New York.

Jones, M.C., K. Gough, I. Dasgupta, B.L. Subba Rao, J. Cliffe, R. Qu, P. Shen, M. Kaniewska, M. Blakebrough, J.W. Davies, R.N. Beachy and R. Hull. 1991. Rice tungro disease is caused by an RNA and a DNA virus. *J. Gen. Virol.* **72:**757–761.

Kano, H., M. Koizumi, H. Noda, H. Hibino, K. Ishikawa, T. Omura, P.Q. Cabauatan and H. Koganezawa. 1992. Nucleotide sequence of capsid protein gene of rice tungro bacilliform virus. *Arch. Virol.* **124:**157–163.

Lennefors, B-L., E.I. Savenkov, J. Bensefelt, E. Wremerth-Weich, P. van Roggen, S. Tuvesson, J.P.T. Valkonen and J. Gielen. 2006. dsRNA-mediated resistance to Beet necrotic yellow vein virus infections in sugar beet (*Beta vulgaris* L. ssp. vulgaris). *Mol. Breed.* **18:**313–325.

Lockhart, B.E.L. 1990. Evidence for a circular double-stranded genome in a second group of plant viruses. *Phytopath.* **80:**127–131.

Manwan, I., S. Sama and S.A. Rizvi. 1985. Use of varietal rotation in the management of tungro disease in Indonesia. *Indonesia Agric. Res. Dev. J.* **7:**43–48.

Napoli, C., C. Lemieux and R. Jorgensen. 1990. Introduction of a chimeric chalcone synthase gene into *Petunia* results in reversible co-suppression of homologous genes in trans. *Plant Cell.* **2:**279–289.

Nath, N., S. Mathur and I. Dasgupta. 2002. Molecular analysis of two complete Rice tungro bacilliform virus genomic sequences from India. *Arch. Virol.* **147:**1173–1187.

Pooggin, M., P. V. Sivaprasad, K. Veluthambi and T. Hohn. 2003. RNAi targeting of DNA virus in plants. *Nature Biotech.* **21:**131–132.

Qu, R.D., M. Bhattacharyya, G.S. Laco, A. De Kochko, B.L. Subba Rao, M.B. Kaniewska, J.S. Elmer, D.E. Rochester, C.E. Smith and R.N. Beachy. 1991. Characterization of the genome of rice tungro bacilliform virus: comparison with Commelina yellow mottle virus and caulimoviruses. *Virol.* **185:**354–364.

Rivera, C.T. and S.H. Ou. 1967. Transmission studies of the two strains of rice tungro virus. *Plant Dis. Rep.* **51:**877–881.

Susi, P., M. Hohkuri, T. Wahlroos and N.J. Kilby. 2004. Characteristics of RNA silencing in plants: similarities and differences across kingdoms. *Plant Mol. Biol.* **54:**157–174.

Tenllado, F., B. Martinez-Garcia, M. Vargas and J.R. Dias-Ruiz. 2003. Crude extracts of bacterially-expressed dsRNA can be used to protect plants against virus infections. *BMC Biotechnol.* **3:**3.

Tijsterman, M., R.F. Ketting and R.H.A. Plasterk. 2002. The Genetics of RNA Silencing. *Annu. Rev. Genet.* **36:**489–519.

Vanitharani, R., P. Chellappan and C.M. Fauquet. 2003. Short interfering RNA-mediated interference of gene expression and viral DNA accumulation in cultured plant cells. *Proc. Natl. Acad. Sci. USA.* **100:**9632–9636.

Voinnet, O. 2005. Induction and suppression of RNA silencing: insights from viral infections. *Nat. Rev. Gen.* **6:**206–221.

Wang M.B., N.M. Upadhyay, R.I.S. Brettel and P.M. Waterhouse. 1997. Intron-mediated improvement of a selectable marker gene for plant transformation using *Agrobacterium tumefaciens*. *J. Genet. Breed.* **51:**325–334.

Waterhouse, P.M., M.W. Graham and M.B. Wang. 1998. Virus resistance and gene silencing in plants can be induced by simultaneous expression of sense and antisense RNA. *Proc. Natl. Acad. Sci. USA.* **95:**13959–13964.

CHAPTER 14

Use of RNAi Technology in Medicinal Plants

Pravej Alam, Athar Ali, Mather Ali Khan and M.Z. Abdin*

ABSTRACT

Plant metabolic engineering has generally been proposed as a means for increasing the level of a valuable pathway end-product or removing an undesirable metabolite. RNAi has mainly been used as a readily available, rapid, reverse genetics tool to create plants with novel chemical phenotypes and to determine the phenotypes of genes responsible for the synthesis of many different secondary metabolites. These manipulations have also greatly facilitated the identification and improvement of specific plant-insect and plant-pathogen interactions and have set the stage for greater exploitation of plants to produce commercially-valuable, plant-derived drugs, flavoring agents, perfumes, etc. The powerful approach of gene knockdown combined with foreign gene overexpression as a means for creating new secondary-product-based phenotypes in medicinal plants.

Keywords: Secondary metabolite, RNAi, Medicinal plants, overexpression.

INTRODUCTION

Plant secondary metabolites represent an enormous value from economical point of view. First of all quite a few are used as specialty chemicals, such as drugs, flavours, fragrances, insecticides and dyes. Of all drugs used in western medicine about 25% is derived from plants, either as a pure compound or as derived from a natural synthon (Verpoorte 2000). Examples of the former are morphine, codeine, paclitaxel, vinblastine, vincristine, scopolamine, atropine, pilocarpine, physostigmine, digoxin and artemisinin. With a total world market for

* Corresponding author e-mail: mzabdin@rediffmail.com

medicines of about 250 billion US $ per year, it is obvious that natural products from plants are a valuable commodity. A much more difficult group to assess in terms of economical terms, at least in value of money, is that of the medicinal plants (Balick et al. 1996). It is estimated that about 80% of the world population depends on traditional medicinal plants for their primary health care. Though in most cases no scientific studies have been made to confirm their activity, of those studied quite a few showed activities related to their use. Traditional medicines are very important in primary health care, where they can be used instead of expensive western medicines. Their potential value is in the possibility that they may contain new biologically active compounds, which can be further developed into drugs for the international market. The other major group of economically important natural products is that of flavours and fragrances. This group comprises both pure chemical entities and mixtures of compounds (e.g. various essential oils). These compounds are on the market as such, but of course they are also of great importance for the quality of our food and spices. For example the bitter taste of beer is dependent on the bitter acids from hops. Moreover, food plants also contain all kinds of other compounds which are very much quality determining, such as caffeine. Presently there is much interest in health promoting effects of secondary metabolites in food. Anthocyanins, flavonoids and carotenoids are now well known examples, but certainly one may expect others that will be discovered in the coming years.

Many of the more valuable secondary metabolites are produced in only small amounts and, hence, are difficult to obtain in sufficient quantities. It is estimated that 70–80% of the people worldwide rely mainly on traditional, largely herbal, medicines to meet their primary healthcare needs. The global demand for herbal medicine is not only large, but also growing. Over the years, several technologies have been adopted for enhancing bioactive molecules in medicinal plants (Khan et al. 2009). Biotechnological tools are important for the multiplication and genetic enhancement of the medicinal plants by adopting techniques such as *in vitro* regeneration and genetic transformation (Liew and Yang 2008, Abdin and Kamaludin 2006). It could also be harnessed for the production of secondary metabolites using plants as bioreactors. Recent advances in the combined molecular biology and enzymology through post-transcriptional gene silencing (PTGS) of medicinal plants suggest that this system is a viable source of important secondary metabolites. Metabolic engineering using RNAi provides a promising means of overcoming some of these limitations. Medicinal plants are one of the most important sources of life saving drugs for majority of the world's population. Plant secondary metabolites are economically important as drugs, fragrances, pigments, food additives and pesticides. The biotechnological tools are important for selecting, multiplying, improving and analyzing medicinal plants (Khan et al. 2009). In the present write-up, few examples have been highlighted to indicate the powerful approach of gene knockdown as a means for enhancing secondary metabolite production in medicinal plants.

Mechanism of RNAi

RNAi has also been used to engineer metabolic pathways to overproduce secondary products with health, yield or environmental benefits. The ability of RNA silencing to silence several genes simultaneously should enhance metabolic ability to create novel traits in plants. RNA silencing and RNA interference RNA silencing is a homology-based process that is triggered by double-stranded RNA (dsRNA), which leads to the suppression of gene expression (Denli et al. 2003). It was initially discovered in plants and was thought to function as part of a defence mechanism against viruses. Subsequently it was shown to be a ubiquitous silencing mechanism that is present in all eukaryotes including protozoa, plants and animals (Figure 1). The term RNA interference (RNAi) was first coined for the phenomenon when it was observed in the nematode *Caenorhabditis elegans* (Hannon 2000). This showed that both sense and antisense RNA are able to silence gene expression and it was proposed that dsRNA is involved in the process of gene silencing (Fire et al. 1998). The dsRNA is a more efficient elicitor of RNAi than either sense or antisense RNA alone. Subsequently RNA silencing works on at least three different levels in plants: (i) cytoplasmic silencing by dsRNA results in cleavage of mRNA and is known as post-transcriptional gene silencing (PTGS) (ii) endogenous mRNAs are silenced by micro- RNAs (miRNAs), which negatively regulate gene expression by base-pairing to specific mRNAs, resulting in either RNA cleavage or arrest of protein translation; (iii) RNA silencing is associated with sequence-specific methylation of DNA and the consequent suppression of transcription (Hamilton and Baulcombe 1999).There are two small RNAs in the RNAi pathway: small interfering RNAs (siRNAs) and microRNAs (miRNAs) that are generated via processing of longer dsRNA and stem loop precursors (Bernstein et al. 2001; Hammond et al. 2000; Stevenson 2004). Dicer enzymes play a critical role in the formation of these two effectors of RNAi (Elbashir et al. 2001). They can cleave long dsRNAs and stem loop precursors into siRNAs and miRNAs in an ATP-dependent manner, respectively. A primary miRNA transcript (pri-miRNA) (Yin et al. 2002), which is frequently synthesized from intronic regions of protein-coding RNA polymerase II transcripts (Novina et al. 2002, Tijsterman et al. 2004), is first processed by a protein complex containing the double-strand specific ribonuclease Drosha in the nucleus to produce a hairpin intermediate of 70 nucleotide (nt) (Hamilton et al. 1999). This precursor miRNA (pre-miRNA) is subsequently transported by Exportin-5/Ran-

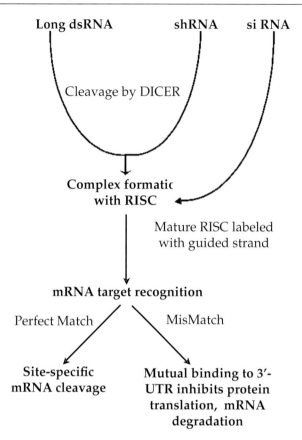

Figure 1. Molecular mechanism of gene silencing.

GTP (Hamilton et al. 1999, Mette et al. 2000) to the cytoplasm where it is cleaved by another dsRNA specific ribonuclease, Dicer, (Jones et al. 2001, Tang et al. 2007) into miRNA duplexes. After strand separation of the duplexes, the mature singlestranded miRNA is incorporated into an RNA-induced 78 silencing complex (RISC)-like ribonucleo-protein particle (miRNP) (Hamilton et al. 1999, Wassengger et al. 1994, Mette et al. 2000). This complex inhibits translation or, depending on the degree of Watson-Crick complementarily, induces degradation of target mRNAs (Jones et al. 2001, Tang et al. 2007).

APPLICATIONS OF RNAi TECHNOLOGY IN MEDICINAL PLANT RESEARCH

Metabolic engineering using RNAi provides a promising means of overcoming some of these limitations. The first real breakthrough for

RNAi-mediated metabolic engineering came in 2004 when RNAi was used to silence enzymes in the codeine reductase (COR) gene family in the *Papaver somniferum* (Allen et al. 2004) (Table 1). This was achieved using anhpRNAconstruct containing sequences from multiple cDNAs of genes in the pathway. A precursor, the nonnarcotic alkaloid (S)-reticuline, which occurs upstream of codeine in the pathway, accumulated at the expense of morphine, codeine, opium and thebaine in transgenic plants. The researchers involved in this study were the first to report metabolic engineering of the opium poppy using RNAi and the first to interfere with multiple steps in a complex biochemical pathway. RNAi has also been used to improve an edible crop. RNAi has also shown promise in the development of hypoallergenic grasses. The major cause of hay fever and seasonal allergic asthma, which affects 25% of the population in temperate climates, is ryegrass pollen (Lolium spp.). The main allergens are the pollen proteins Lol p1 and Lol p2: 90% of allergy sufferers are sensitive to these proteins. Levels of Lol p1 and Lol p2 can be downregulated by expressing antisense cDNA sequences under the control of a maize pollen-specific promoter. Transgenic Coffea spp. expressing hpRNAi constructs containing CaMXMT1 sequences, a cDNA encoding one of the genes involved in the caffeine biosynthetic pathway, showed theobromine and caffeine accumulation in the range of 30–50% of that normally found in these species. This finding showed that theobromine is involved in the major biosynthetic pathway of coffee and that the technology can be applied to engineer decaffeinated coffee plants.

Alkaloids

By definition alkaloids contain nitrogen which is usually derived from amino acids. Because of the presence of a nitrogen atom, alkaloids react mostly alkaline and are able to form soluble salts in aqueous environments. In plants however, they can occur in the free state, as a salt or as an N-oxide and they are accumulated in the plant vacuole as reservoir or often coupled to phenolic acids like chlorogenic acid or caffeic acid. Alkaloids can be classified in terms of their biological activity, their chemical structure, or more accepted according their biosynthetic pathway.

Most efforts have been concentrated on mapping the early part of the pathway and on overexpression of early genes, aiming to increase the metabolic flux into the alkaloid pathway (Hashimoto et al. 2003). Park et al. (2000) have recently demonstrated antisense suppression of the berberine bridge enzyme in poppy cells, reducing the amount of benzophenanthridine alkaloids but increasing the levels of several amino acids. There have been many attempts to modulate the indole alkaloid biosynthetic pathway, which produces important

Table 1

S. No.	Gene	Plant	Product	References
1.	Theobromine synthase	*Coffee Arabica* *C. canephora*	Reduction of theobromine and caffeine content	Ogita et al. 2004
2.	Nicotine-N-demethylase	*Nicotina tobacum*	Suppresed nicotine to norniconine	Gavilano et al. 2006
3.	Codeinone reductase	*Papaner somniferum*	No deatable morphine, codeine	Allen et al. 2004
4.	Satutaridinol 7-o-acetyl transferase	*P. somniferum*	Narcolic reduction	Allen et al. 2008
5.	CBT-ol hydroxylase	*N. tobacum*	Moderate suppression of CBT-diol	Wang et al. 2003
6.	Cadinene synthase	*G. hirsutum*	Gossypol-reduction	Kumar et al. 2006
7.	Dammarenediol synthase	*Panax ginseng*	Reduction of ginsenoside	Han et al. 2006
8.	P450cyP79D1 p450cyP79D2	*Menihot esculenta* (*cassava*)	Reduced cyanogenic glucoside content	Jorgensen et al. 2005
9.	Cinnomoyl-COA reductase	*Solanum lycopersicum* L.	Lignin reducrion, increase phenolies	Vander et al. 2006
10.	Cinnamyl alcohol dehydrogenase	*Linum usitatissimum* L.	Lignine reduction improve the elastic properties	Wrobel et al. 2007
11.	Chalcone isomerase	*N. tobacum*	Novel colour, functional analysis, reduction of anthocyanins in petals; accumulation of high level of chalcone in pollen	Nishihara et al. 2005
12.	Anthocyanidin synthase	*Torenia hybrid*	Decrease the anthocyanidin and flavones	Nakamura et al. 2006
13.	Flavonol synthase and flavonone-3-hydroxylase and expression of hetrologous dihydroflavonol-4-reductase (DFR)	*Nicotiana tobacum*	High amounts of pelargonidin and reduced amounts of flavonons	Nakatsuka et al. 2007

14.	Endogenous dihydro-flavonol-4-reductase (DFR)	*Rosa hybrida*	Accumulation of delphinidinm in the petals with absence of cynidine and pelargonidine	Katsumoto et al. 2007
15.	Phenylacetaldehyde syn-thase PAAS	*Petunia hybrida*	Complete suppression of phenylacetaldehyde and 2-phenylethanol emission	Kaminga et al. 2006
16.	Coniferyl alcohol acetyltransferase phCFAT	*Petunia hybrida*	Decreased synthesis and emission of isoeugenol and 5 other volatiles, 5x and 10x increased coniferyl aldehyde and homovanillic acid in the petals	Dexter et al. 2007
17.	Chalcone synthase	*Fragaria x anassa cv. Elsanta*	Large increase in levels of (hydroxyl) cinnomoyl glucose esters	Hoffmann et al. 2006
18.	Isoflavone synthase	*Glycine max*	Reduction of isoflavone levels	Subramanian et al. 2005
19.	Chalcone synthase CHS	*Medicago truncatula*	No nodulation of due to the deficiency in flavonoids altered auxin transfer	Wasson et al. 2006
20.	cytochrome P450 (+)	*Mentha x piperata*	Menthofuran	Mahmoud and Croteau, 2001
21.	berberine bridge enzyme (BBE)	*Eschscholizia*	Benzophenanthridin Alkoloids	Park and Facchini, 2002
22.	Putrescine N-PMT	*Nicotiana tabacum*	Pyridine and tropane alkoloids	Chintapakorn and Hamill, 2003
23.	Codeinone reductase (COR)	*Papaver somniferum*	Codeine and morphine	Allen et al. 2004
24.	Limonene-3- hydroxylase gene	*Mentha x piperata*	Limonene	Mahmoud et al. 2004
25.	Berberine bridge enzymes (BBE) the N methyjococlaurine hydroxylase (CYP80B1)	*Papaver somniferum*	Morphine, codeine, sanguinarine	Frick et al. 2004
26.	De-etiolated1 DET1	*Solanum lycopersicum*	Carotenopid and contents of fruits	Davuluri et al. 2005
27.	Benzoic acid/salicylic acid methyltransferase	*Petunia x hybrida*	Methylnenzoate	Underwood et al. 2005

contd. ... Table

Table ... contd.

S. No.	Gene	Plant	Product	References
28.	R2R3 MYB-type transcription factor ODORANT1	*Petunia hybrida*	Fragrance	Verdonk, et al. 2005
29.	Glutatione S tranferase 1	*Torenia fournieri*	Anthocyanin	Nagira et al. 2006
30.	Dammarenediol synthase gene	*Pinax ginseng*	*Ginsenoside*	Han et al. 2006
31.	Phenylacetaldehyde synthase (PAAS)	*Petunia hybrida*	Ginsenoside Complete suppression of phenylacetaldehyde and 2-phenylethanolemission	Kaminaga et al. 2006
32.	Cinnamoyl-Co A reducatse	*Solanum lycopersicum*	Phenolics	Van der Rest et al. 2006
33.	Coniferyl alcohol acyltransferase *Ph*CFAT)	*Petunia x hybrida*	Benzylaldehyde	Orlova et al. 2006
34.	Cinnamyl alcohol dehydrogenase	*Linum usitatissimum*	Lignin reducatse	Wrobel-Kwiatkowska et al. 2007
35.	Tryptophan Decarboxylase	*Catharanthusroseus*	Tryptamine	Runguphan et al. 2009

pharmaceutical compounds such as the anticancer drugs vinblastine and vincristine (Verpoorte et al. 2002).

Allen et al. (2004) used RNAi technology to modify the alkaloid content of the latex in *P. somniferum*. The transformation construct comprised a hybrid double-stranded RNA composed of a sequence corresponding to conserved members of the gene family fused to a sequence corresponding to a divergent member. This strategy was successful in achieving a substantial reduction of COR enzyme activity (Figure 2). The transgenic poppies produced not morphine and codeine but reticuline. The precursor alkaloid (S)-reticuline which is seven enzymatic steps upstream of codeinone, accumulated in transgenic plants at the expense of morphine,

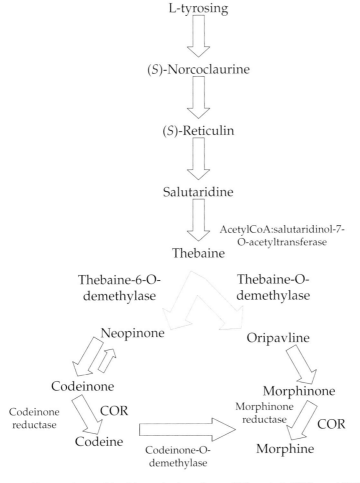

Figure 2: Proposed morphine biosynthetic pathway (Allen et al. 2004) used RNAi to block codeinone reductase (COR), which catalyzes the reduction of both codeinone and morphinone.

codeine, oripavine and thebaine. The surprising accumulation of (S)-reticuline suggests a feedback mechanism preventing intermediates from general benzylisoquinoline synthesis entering the morphine-specific branch. One would expect that a decrease in COR activity would lead to a reduction of morphine content and an accumulation of immediate precursors. Surprisingly, although morphine content was reduced, none of the morphine-type precursors normally present in the latex were detectable. Blocking a terminal biosynthesis step leads to complete loss of all morphinetype alkaloids from the poppy latex and to accumulation of reticuline and methylated reticuline derivatives, which normally do not occur at appreciable levels. RNA interference was also used to down-regulate bbe in *E. californica* cells (Fujii et al. 2007). Unlike the largely unsuccessful earlier attempts to detect changes in pathway intermediates when an antisense approach was used to silence bbe in transgenic roots and cell lines of *E. californica* (Park et al. 2002, 2003a), the RNAi-mediated approach resulted in accumulation of (S)-reticuline and substantial reduction in the level of sanguinarine. A methylated derivative of reticuline, laudanine was also produced in the silenced lines. Allen et al. (2007) demonstrated that the over-expression and suppression of the gene encoding the morphinan pathway enzyme salutaridinol 7-O-acetyltransferase (SalAT) in opium poppy affects the alkaloid products that accumulate. Over-expression of the gene in most of the transgenic events resulted in an increase in capsule morphine, codeine and thebaine on a dry-weight basis. The transgenic line with the highest alkaloid content had 41%, 37% and 42% greater total alkaloids than the control in three independent trials over 3 years. DNA-encoded hairpin RNA-mediated suppression of SalAT resulted in the novel accumulation of the alkaloid salutaridine at up to 23% of total alkaloid; this alkaloid is not detectable in the parental genotype.

Terpenoids

Terpenoids are by far the largest group of plant secondary metabolites. Following the recent discovery of the role of the 2-C-methyl -D-erythritol-4-phospate (MEP) pathway in the biosynthesis of plastidial terpenoids, such as the carotenoids and monoterpenes and diterpenes, several genes of this pathway have been cloned (Brown et al. 2001, Mahmoud et al. 2001, Lenge et al. 1999, 2000). Modifying the MEP (non-mevalonate pathway) pathway is potentially useful for a wide range of applications. A co-suppression and an antisense strategy used to knock out a cytochrome P450 enzyme in tobacco trichome glands, for example, conferred an increased resistance to aphids (Chen et al. 2000). There was a clear shift in the cembranoid spectrum, with a 19-fold increase of the diterpenoid cembratriene-ol and a decrease in its oxidation product cembratriene-diol. The application of RNAi to terpenoid metabolism has not been as extensive as its use in studying and manipulating phenylpropanoid and alkaloid metabolism. However, the antisense technology, has been successfully applied to manipulating, at least, monoterpenoid metabolism (Mahmoud and Croteau, 2003). In case of trichome diterpene e.g. CBT-ol hydroxylase from *Nicotiana tabaccum* moderate suppression to increase of CBT to produced aphid resistance (Lerkin et al. 2007, Leither et al 2006) Cadinene synthase, the first committed step in the biosynthesis of sesquiterpene gossypol was suppressed in cotton seed. Efficiency of knockdown in that study was about 6% (Kumar et al. 2006). The use of a seed specific promoter allowed seed-specific gossypol suppression, allowing retention of this pest resistance conferring compound in leaves and flowers. Han et al. (2006) demonstrate that oxidosqualene cyclase, the first committed step enzyme in the synthesis of ginsenosides (triterpenoid saponins) of ginseng was key in the formation of these pharmacologically-active compounds. RNAi silencing of this gene led to ~85% reduction in ginsenoside production in roots to assess the role of plastid lipid-associated protein LeCHRC in chromoplastogenesis and stress in tomato (Leitner et al. 2006). This protein is involved in the sequestration/stabilization of hyperaccumulated carotenoids in developing flower and fruit chromoplasts. Flowers of transgenic plants accumulated ~30% less carotenoids than controls. Suppressed plants were also more susceptible to *Botrytis cinerea* infection. Since certain plastid lipid associated proteins are induced and accumulated by plant stresses, it was suggested that increased *Botrytis* sensitivity was due to reduced plastid lipid-associated protein. RNAi was used to reduce the cyanogenic glucoside contents of cassava, an extremely important food crop, particularly in third world countries. The presence of potential-cyanide-generating cyanogenic glucosides in cassava requires careful processing during food preparation, which results in loss of nutritional value. The first committed steps in the biosynthesis of cassava cyanogenic glucosides are catalyzed by the P450 enzymes CYP79D1 and CYP79D2 (Jorgeuser et al. 2005). A fusion RNAi construct was prepared to simultaneously knock down both genes. The independent transgenic lines were found to have <1% of wild type cassava cyanogenic glucosides.

The mechanism of artemisinin biosynthesis has recently become much clearer (Liu et al. 2006, Bertea et al. 2005). It has been shown that artemisinin belongs to the isoprenoid group of compounds, which are derived from two common precursors, namely IPP (isopentenyl diphosphate) and its isomer DMAPP (dimethylallyl diphosphate). GPP (geranyl diphosphate) is formed by chain elongation from IPP and DMAPP when they react with a carbonium ion and GPP can then further react with IPP to produce

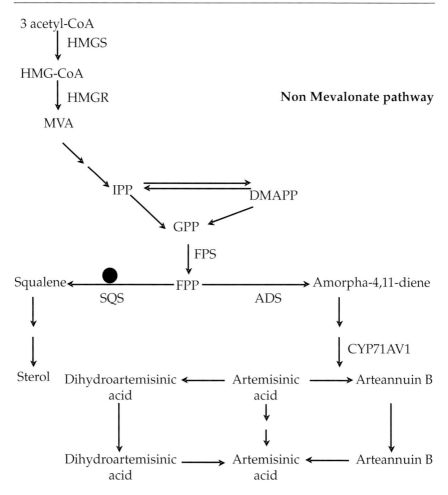

Figure 3: Proposed Artemisinin biosynthetic pathway. Zhang et al. (2009) used RNAi to block the SQS (squalene synthase) in *Artemisia annua* L. Enzyme abbrebations: DMAPP dimethylallyl diphosphate, GPP geranyl diphosphate, FPP farnesyl diphosphate, FPS farnesyl diphosphate synthase, HMGR 3-hydroxy- 3-methylglutaryl coenzyme A (HMG-CoA) reductase, HMGS HMG-CoA synthase, SQS squalene synthase, ADS amorpha 4,11-diene synthase, CYP71AV1 cytochrome P450 monooxygenase.

FPP (farnesyl diphosphate). FPP can be converted by enzymic catalysis to produce various isoprenoid final products such as artemisinin and sterol (Figure 3). On the other hand, SQS (squalene synthase) is the key enzyme catalysing the first step of the sterol biosynthetic pathway, a pathway in competition with that artemisinin biosynthesis. It has already been cloned from *A. annua* (Liu et al. 2003). Previous studies showed that the inhibition of the sterol biosynthetic pathway by chemical methods could improve the artemisinin content of *A. annua*. Woerdenbag et al. (1993) proved that artemisinin production was enhanced by the addition of naphtiphine, an inhibitor of the enzyme squalene epoxidase, to the medium. Kudakasseril et al. (1987) demonstrated that the application of many sterol inhibitors, including miconazole or chlorocholine, resulted in an increase in artemisinin in shoot cultures of *A. annua*. Recently, Ro et al. (2006) proved that down-regulation of *ERG9* (ergosterolbiosynthesis-pathway gene 9), a gene that encodes SQS in yeast, using a methionine-repressible promoter (PMET3), increased amorpha-4,11-diene (first rate limiting enzyme in artemisinin biosynthetic pathway. These studies imply that suppression of the sterol-biosynthetic pathway by genetic engineering may be an effective way to improve artemisinin content. Earlier by using hpRNA-mediated RNAi technology to suppressing the expression of SQS in *Artemisia annua* L. The hpSQS transgenic showed that suppressing SQS by 60% in the transgenic plants, which resulted in the sterol content being reduced to 37–58% of the control value, still did not alter plant growth significantly (Zhang et al. 2009).. The present study demonstrates that the genetic engineering strategy of suppressing sterol pathway using RNAi is an effective and suitable means of increasing the artemisinin content of plants. RNAi (RNA interference)-mediated by hpRNA (hairpin RNA) has been used in gene silencing in many species of plants. Liu et al. (2006) reported that hpRNA mediated downregulation of ghSAD-1 and ghFAD2-1, two key enzymes in the fatty-acid-biosynthesis pathway in cotton (*Gossypium hirsutum*), elevated the stearic acid content (44% compared with a normal level of 2%) and oleic acid content (77% compared with a normal level of 15%) in cottonseeds. It was also reported that suppression of one key enzyme, CaMXMT1, involved in the caffeine-biosynthetic pathway through hpRNA-mediated interference in coffee (*Coffea* spp.), decreased obromine and caffeine accumulation efficiently (30–50% of that normally found in the species.

Flavonoids

The flavonoids are phenolic compounds derived ultimately from phenylalanine and thus share a common metabolic source with lignins and lignans. The first committed step in flavonoid biosynthesis is the conversion of the precursor 4-coumaroyl-CoA into chalcone by the enzyme chalcone synthase (Figure 4). Chalcone is then derivatized in a series of enzymatic steps to produce a variety of molecules that act as pigments, defense chemicals (phytoalexins) and regulatory molecules (Winkel et al. 2001). Flavonoids have been a favorite target for metabolic engineering since the early 1990s, because the result

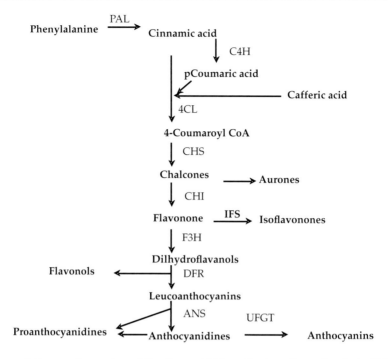

Figure 4: Proposed overview of the flavonoid biosynthesis pathway in plants (Schijlen et al. 2007). Enzyme abbreviations: PAL, phenylalanine ammonia-lyase; C4H, cinnamate 4-hydroxylase; 4CL, 4-coumaroyl:CoA ligase; CHS, chalcone synthase; CHI, chalcone isomerase; F3H, flavanone 3-hydroxylase; DFR, dihydroflavonol 4-reductase; ANS, anthocyanidin synthase; UFGT, UDP glucose-flavonoid 3-o-glucosyl transferase wave Blue has two CHS genes and that the transcription of the double-strand RNA of the 3 non-coding sequence of one of them could specifically down-regulate that molecular species. The exclusive production of delphinidin and the most significant color change toward blue were achieved by the replacement of the function of rose DFR genes with an iris DFR gene in transgenic roses by using RNAi. RNAi gave more efficient gene suppression than antisense or sense suppression (Nakamura et al. 2006). The replacement was efficient; a higher frequency and less deviated delphinidin accumulation and color change were observed.

of different modifications can be determined through alterations in flower color (Forkmann and Martens 2001). As for other pathways, the number of experiments involving multiple gene transfer strategies is beginning to increase. A recent example is the stacking of dihydroflavonol 4-reductase and anthocyanidin synthase transgenes in forsythia by sequential transformation (Rosati et al. 2003).

Flower color comes mainly from anthocyanins, a color class of flavonoids. All structural genes leading to anthocyanidins in the pathway have been cloned from many plants (Tanaka et al. 2005). Some of these genes have been used to generate white flowers from color flowers through antisense or sense suppression. Antisense and sense suppression were described in petunia using the chalcone synthase (*CHS*) gene (van der Krol, 1988 and Napoli et al. 1990). Since then, flower color biosynthesis has been down regulated through antisense or sense suppression in many ornamental plants, such as rose, carnation and chrysanthemum (Gutterson 1995, Tanaka and Mason 2003). For example, transgenic petunia plants whose endogenous genes had been suppressed through sense or antisense suppression lost their original phenotypes after a few years (Tsuda et al. 2005). Aida et al. (2000) generated transgenic *T. fourunieri* plants that had lighter flower colors than those of the host through sense and antisense suppression of the CHS or dihydroflavonol 4-reductase (DFR) gene and reported that sense and antisense genes gave different phenotypes. Fukuzaki et al. (2004) reported that *T. hybrida* cv. Summer wave Blue has two CHS genes and that the transcription of the double-strand RNA of the 3 non-coding sequence of one of them could specifically down-regulate that molecular species. The exclusive production of delphinidin and the most significant color change toward blue were achieved by the replacement of the function of rose DFR genes with an iris DFR gene in transgenic roses by using RNAi. RNAi gave more efficient gene suppression than antisense or sense suppression (Nakamura et al. 2006). The replacement was efficient; a higher frequency and less deviated delphinidin accumulation and color change were observed.

Increasing the levels of flavonoids in food plants can provide health benefits, as these molecules often have antioxidant activity. This has been demonstrated in the case of tomatoes by overexpressing the petunia gene for chalcone isomerase, leading to an 80-fold increase in the flavonoid content of the tomato peel and a corresponding 20-fold increase in the flavonoid level in tomato paste (Muir et al. 2001). In addition, chalcone synthase and flavonol synthase transgenes were found to act synergistically to upregulate flavonol biosynthesis significantly in the flesh of tomato fruits (Verhoeyen et al. 2002).

FUTURE PROSPECTS

RNAi is a potentially powerful tool for a wide variety of gene silencing applications. Since its initial discovery in 1998 by Fire et al. RNAi has taken the scientific community by storm. Despite many rapid advances, RNAi is still in its infancy in medicinal plant research. As described, the RNAi and gene disruption methods differ in principle and therefore, have their own strengths and limitations. The drawbacks of RNAi, such as incomplete repression and possible unintended targets, are often described for pharmacological studies. In this regard, RNAi seems to be somewhere between gene disruption and pharmacological approaches. RNAi might prove to be useful for the studies towards production of important biomedical products by medicinal plants, which in turn can provide novel and rapid applications. Compounds like Ginsenoside, morphinan alkaloid and (S)-reticuline may be produced from the RNAi incorporated medicinal plants in near future. To make RNAi a better tool to increase the production of compound in a medicinal plant, the next challenges are to understand the extent of complete effects in medicinal plant cells and to develop an inducible RNAi system with a combination of a strictly controlled promoter and a convenient inducer applicable to a wide range of medicinal plant. Finally, the generation of additional RNAi protocols for genome-wide screening might assist in the rapid identification of genes involved in novel compound production. Thus, though the promise of RNAi is yet to be fulfilled, its potential is begun to be realized.

REFERENCES

Abdin, M.Z., Kamaludin. 2006. Improving quality of medicinal herbs through physio-chemical and molecular approaches. In: Abdin M.Z., Abdol Y.P. (eds.) Traditional systems of medicine, Narosa Publishing House Pvt Ltd., India, pp. 30–39.

Aida, R., S. Kishimoto, Y. Tanaka and M. Shibata. 2000. Modification of flower colour in torenia (*Torenia fournieri* Lind.) by genetic transformation. *Plant Sci.* **153**:33–42.

Allen, R.S., J.A. Miller and J.A. Chitty et al. 2008. Metabolic engineering of morphinan alkaloids by over-expression and RNAi suppression of salutaridinol 7-O-acetyltransferase in opium poppy. *Plant Biotechnol. J.* **56**:22–30.

Allen, R.S. et al. 2004. RNAi-mediated replacement of morphine with the nonnarcotic alkaloid reticuline in opium poppy. *Nat. Biotechnol.* **22**:1559–1566.

Balick, M.J., E. Elisabetsky and S.A. Laird (eds.). 1996. Medicinal resources of the tropical forest. Columbia University Press, New York.

Bernstein, E., A.A. Caudy, S.M. Hammond and G.J. Hannon. 2001. Role for a bidentate ribonuclease in the initiation step RNA interference. *Nature.* **409**:363–366.

Bertea, C.M., J.R. Freije, H. van der Woude, F.W. Verstappen, L. Perk and V. Marquez et al. 2005. Identification of intermediates and enzymes involved in the early steps of artemisinin biosynthesis in Artemisia annua. *Planta Med.* **71**:40–47.

Broun, P. and C. Somerville. 2001. Progress in plant metabolic engineering. *Proc. Natl. Acad. Sci. USA.* **98**:8925–8927.

Chen, D.H., H.C. Ye and G.F. Li. 2000. Expression of a chimeric farnesyl diphosphate synthase gene in *Artemisia annua* L. transgenic plants via *Agrobacterium tumefaciens*-mediated transformation. *Plant Sci.* **155**:179–185.

Davuluri, G.R., A. van Tuinen and P.D. Fraser et al. 2005. Fruit-specific RNAi-mediated suppression of DET1 enhanced carotenoid and flavonoid content in tomatoes. *Nat. Biotechnol.* **23:**890–895.

Denli, A.M. and G.J. Hannon. 2003. RNAi: an ever-growing puzzle. *Trends Biochem. Sci.* **28:**196–201.

Dexter, R., A. Qualley and C.M. Kish. 2007. Characterization of a petunia acetyltransferase involved in the biosynthesis of the floral volatile isoeugenol. *Plant J.* **49:**265–275

Elbashir, S.M., J. Martinez, A. Patkaniowska, W. Lendeckel and T. Tuschl. 2001. Functional anatomy of siRNAs for mediating efficient RNAi in Drosophila melanogaster embryo lysate. *EMBO J.* **20:**6877–6888.

Fire, A. et al. 1998. Potent and specific genetic interference by double-stranded RNA in Caenorhabditis elegans. *Nature.* **391:**806–811.

Forkmann, G. and S. Martens. 2001. Metabolic engineering and application of flavonoids. *Curr. Opin. Biotechnol.* **12:**155–160.

Frick, S., J.A. Chitty, R. Kramell, J. Schmidt, R.S. Allen, P.J. Larkin and T.M. Kutchan. 2004. Transformation of opium poppy (*Papaver somniferum* L.) with antisense berberine bridge enzyme gene (anti-bbe) via somatic embryogenesis results in an altered ratio of alkaloids in latex but not in roots. *Transgenic Res.* **13:**607–613.

Fujii, N., T. Inui, K. Iwasa, T. Morishige and F. Sato. 2007. Knockdown of berberine bridge enzyme by RNAi accumulates (S)-reticuline and activates a silent pathway in cultured California poppy cells. *Transgenic Res.* **16:**363–375.

Gavilano, L., N. Coleman and E. Burnley et al. 2006. Genetic engineering of *Nicotiana tabacum* for reduced nornicotine content. *J. Agric. Food Chem.* **54:**9071–9078

Hamilton, A.J. and D.C. Baulcoumbe. 1999. A species of small antisense RNA in posttranscriptional gene silencing in plants. *Science.* **286(5441):**950–952.

Hammond, S.M., E. Bernstein, D. Beach and G.J. Hannon. 2000. An RNA-directed nuclease mediates post-transcriptional gene silencing in Drosophila cells. *Nature.* **404:**293–296.

Han, J.Y., Y.S. Kwon, D.C. Yang, Y.R. Jung and Y.E. Choi. 2006. Expression and RNA interference-induced silencing of the dammarenediol synthase gene in *Panax ginseng. Plant Cell. Physiol.* **47(12):**1653–1662.

Hannon, G.J. 2002. RNA interference. *Nature.* **418:**244–251.

Hashimoto, T. and Y. Yamada. 2003. New genes in alkaloid metabolism and transport. *Curr. Opin. Biotechnol.* **14:**163–168.

Hoffmann, T., G. Kalinowski and W. Schwab.2006. RNAi-induced silencing of gene expression in strawberry fruit (*Fragaria x ananassa*) by agroinfiltration: a rapid assay for gene function analysis. *Plant J.* **48:**818–826.

Jorgensen, K., S. Bak and P.K. Busk et al. 2005. Cassava plants with a depleted cyanogenic glucoside content in leaves and tubers. Distribution of cyanogenic glucosides, their site of synthesis and transport and blockage of the biosynthesis by RNA interference technology. *Plant Physiol.* **139:**363–374.

Kaminaga, Y., J. Schnepp and G. Peel et al. 2006. Plant phenylacetaldehyde synthase is a bifunctional homotetrameric enzyme that catalyzed phenylalanine decarboxylation and oxidation. *J. Biol. Chem.* **281:**23357–23366

Katsumoto, Y., M. Fukuchi-Mizutani and Y. Fukui et al. 2007. Engineering of the rose flavonoid biosynthetic pathway successfully generated blue-hued flowers accumulating delphinidin. *Plant Cell. Physiol.* **48**:1589–1600.

Khan, M.Y., S. Aliabbas, V. Kumar and S. Rajkumar. 2009. Recent advances in medicinal plant biotechnology. *Indian J. Biotech.* **8**:9–22.

Kinghorn, A.D. and M.F. Balandrin (eds.). 1993. Human medicinal agents from Plants. ACS Symposium Series 534, Washington.

Kudakasseril, G.J., E.J. Lukem and E.J. Stabam.1987. Effect of sterol inhibitors on incorporation of 14C-isopentenyl pyrophosphate into artemisinin by a cell free system from *Artemisia annua* tissue culture and plants. *Planta Med.* **53**:280–284.

Lange, B.M. and R. Croteau. 1999. Genetic engineering of essential oil production in mint. *Curr. Opin. Plant Biol.* **2**:139–144.

Larkin, P.J., J.A.C. Miller, R.S. Allen, J.A. Chitty, W.L. Gerlach, S.F.T.M. Kutchan and A.J. Fist. 2007. Increasing morphinan alkaloid production by over- expressing codeinone reductase in transgenic *Papaver somniferum*. *Plant Biotechnol. J.* **5**:26–37.

Leitner-Dagan, Y., M. Ovadis and E. Shklarnam et al. 2006, Expression and functional analyses of the plastid lipid-associated protein CHRC suggest its role in chromoplastogenesis and stress. *Plant Physiol.* **142**:233–244.

Liew, S.T. and L.X. Yang. 2008. Design, synthesis and development of novel camptothecin drugs. *Curr. Pharm. Des.* **14**:1078–1097.

Liu, C.Z., Y. Zhao and Y.C.Wang. 2006. Artemisinin: current state and perspectives for biotechnological production of an antimalarial drug. *Appl. Microbiol. Biotechnol.* **71**:11–20.

Liu, Y., H.C. Ye, H. Wang and G.F. Li. 2003. Molecular cloning, *Escherichia coli* expression and genomic organization of squalene synthase gene from *Artemisia annua*. *Acta Bot. Sin.* **45**:608–613.

Mahmoud, S.S., M. Williams and R. Croteau. 2004. Cosuppression of limonene-3-hydroxylase in peppermint promotes accumulation of limonene in the essential oil. *Phytochemistry.* **65**:547–554.

Mahmoud, S.S. and R.B. Croteau. 2001. Metabolic engineering of essential oil yield and composition in mint by altering expression of deoxyxylulose phosphate reductoisomerase and menthofuran synthase. *Proc. Natl. Acad. Sci. USA.* **98**:8915–8920.

Mette, M.F., W. Aufsatz, J. Vander Winden, A. Matzke and A.J. Matzke. 2000. Transcriptional silencing and promoter methylation triggered by double-stranded RNA. *EMBO J.* **19**:5194–5201.

Muir, S.R., G.J. Collins, S. Robinson, S. Hughes, A. Bovy, C.H.R. De Vos, A.J. Van Tunen and M.E. Verhoeyen. 2001. Overexpression of petunia chalcone isomerase in tomato results in fruits containing increased levels of flavonoids. *Nat. Biotechnol.* **19**:470–474.

Nagira, Y., K. Shimamura and S. Hirai et al. 2006. Identification and characterization of genes induced for anthocyanin synthesis and chlorophyll degradation in regenerated torenia shoots using suppression subtractive hybridization, cDNA microarrays and RNAi techniques. *J. Plant Res.* **119**:217–230.

Nakamura, N., M. Fukuchi-Mizutani, K. Suzuki, K. Miyazaki and Y. Tanaka. 2006. RNAi suppression of the anthocyanidin synthase gene in *Torenia hybrida* yields white flowers with higher frequency and better stability than antisense and sense suppression. *Plant Biotechnol.* **23**:13–17.

Nakatsuka, T., Y. Abe and Y. Kakizaki et al. 2007. Production of red-flowered plants by genetic engineering of multiple flavonoid biosynthetic genes. *Plant Cell. Rep.* doi:10.1007/s00299-007–0401–0.

Napoli, C., C. Lemieux and R. Jorgensen. 1990. Introduction of a Chimeric Chalcone Synthase Gene into Petunia Results in Reversible Co-Suppression of Homologous Genes *In* trans *The Plant Cell.* **2**:279–289.

Nishihara, M., T. Nakatsuka and S. Yamamura. 2005. Flavonoid components and flower color change in transgenic tobacco plants by suppression of chalcone isomerase gene. *FEBS Lett.* **579**:6074–6078.

Ogita, S., H. Uefuji and M. Morimoto et al. 2004. Application of RNAi to confirm theobromine as the major intermediate for caffeine biosynthesis of coffee plants with potential for construction of decaffeinated varieties. *Plant Mol. Biol.* **54**:931–941.

Orlova, I., A. Marshall-Colon and J. Schnepp et al. 2006. Reduction of benzenoid synthesis in petunia flowers reveals multiple pathways to benzoic acid and enhancement in auxin transport. *Plant Cell.* **18**:3458–4375

Park, S.U., M. Yu and P.J. Facchini. 2002. Antisense RNA-mediated suppression of benzophenanthridine alkaloid biosynthesis in transgenic cell cultures of California poppy. *Plant Physiol.* **128**:696–706.

Ro, D.K., E.M. Paradise, M. Ouellet, K.J. Fisher, K.L. Newman, J.M. Ndungu, K.A. Ho, R.A. Eachus, T.S. Ham, J. Kirby, M.C.Y. Chang, S.T. Withers, Y. Shiba, R. Sarpong and J.D. Keasling. 2006. Production of the antimalarial drug precursor artemisinic acid in engineered yeast. *Nature.* **440**:940–943.

Rosati, C., P. Simoneau, D. Treutter, P. Poupard, Y. Cadot, A.M. Cadic and M. Duron. 2003. Engineering of flower color in forsythia by expression of two independently-transformed dihydroflavonol 4-reductase and anthocyanidin synthase genes of flavonoid pathway. *Mol. Breed.* **12**:197–208.

Runguphan, W., J.J. Maresh and S.E. O'Connor. 2009. Silencing of tryptamine biosynthesis for production of nonnatural alkaloids in plant culture. *PNAS.* **106(33)**:13673–13678.

Stevenson, M. 2004. Therapeutic Potential of RNA Interference. *The new England journal of medicine.* **351**:1772–1777.

Subramanian, S., M.Y. Graham and O. Yu et al. 2005. RNA interference of soybean isoflavone synthase genes leads to silencing I in tissues distal to the transformation site and to enhanced susceptibility to *Phytophthora sojae*. *Plant Physiol.* **137**:1345–1353.

Sunilkumar, G., L.M. Campbell and L. Puckhaber et al. 2006. Engineering cottonseed for use in human nutrition by tissue-specific reduction of toxic gossypol. *Proc. Natl. Acad. Sci. USA.* **103**:18054–18059.

Tanaka, Y., Y. Katsumoto and F. Brugliera et al. 2005. Genetic engineering in floriculture. *Plant Cell. Tissue Org. Cult.* **80**:1–24.

Tang, G., G. Galili and X. Zhuang. 2007. RNAi and microRNA: breakthrough technologies for the improvement of plant nutritional value and metabolic engineering. *Metabolomics.* **3**:357–369.

Tsuda, S., Y. Fukui, N. Nakamura, Y. Katsumoto, K. Yonekura Sakakibara, M. Fukuchi-Mizutani, K. Ohira, Y. Ueyama, H. Ohkawa, T.A. Holton, T. Kusumi and Y. Tanaka. 2004. Flower color modification of *Petunia hybrida* commercial varieties by metabolic engineering. *Plant Biotechnol.* **21**:377–386.

Underwood, R.B., D.M. Tieman and K. Shibuya et al. 2005. Ethylene-regulated floral volatile synthesis in petunia corollas. *Plant Physiol.* **138**:255–266.

Van der Krol, A.R. L.A. Mur, M. Beld, J.N.M. Mol and A.R. Stuitje. 1990. Flavonoid genes in petunia: addition of a limited number of gene copies may lead to a suppression of gene expression. *Plant Cell.* **2**:291–299.

Van der Rest, B., S. Danoun and A.M. Boudet. 2006. Down-regulation of cinnamoyl-CoA reductase in tomato (*Solanum lycopersicum* L.) induces dramatic changes in soluble phenolic pools. *J. Exp. Bot.* **57**:1399–1411.

Verdonk, J.C., M.A. Haring and A.J. van Tunen et al. 2005. ODORANT1 regulates fragrance biosynthesis in petunia flowers. *Plant Cell.* **17**:1612–1624.

Verhoeyen, M.E., A. Bovy, G. Collins, S. Muir, S. Robinson, C.H.R. de Vos and Colliver. 2002. Increasing antioxidant levels in tomatoes through modification of the flavonoid biosynthetic pathway. *J. Exp. Bot.* **53**:2099–2106.

Verpoorte, R. and A.W. Alfermann. 2000. *Metabolic Engineering of Plant Secondary Metabolism.* Dordrecht: Kluwer Academic Publishers.

Verpoorte, R. and J. Memelink. 2002. Engineering secondary metabolite production in plants. *Curr. Opin. Biotechnol.* **13**:181–187.

Wang, E. and G.J. Wagner. 2003. Elucidation of the functions of genes central to diterpene metabolism in tobacco trichomes using PTGS. *Planta.* **216**:686–691.

Wasson, A.P., F.I. Pellerone and U. Mathesius. 2006. Silencing the flavonoid pathway in *Mediago truncatula* inhibits root nodule formation and prevents auxin transport regulation by Rhizobia. *Plant Cell.* **18**:1617–1629.

Winkel-Shirley, B. 2001. Flavonoid biosynthesis. A colorful model for genetics, biochemistry, cell biology and biotechnology. *Plant Physiol.* **126**:485–493.

Woerdenbag, H.J., J.F.J. Lüers, W. Van Uden, N. Pras, Th. Malingŕe and A.W. Alfermann. 1993. Production of the new antimalarial drug in shoot cultures of *Artemisia annua* L. *Plant Cell. Tiss. Org. Cult.* **32**:247–257.

Wrobel-Kwiatkowska, M., M. Starzycki and J. Zebrowski et al. 2007. Lignin deficiency in transgenic flax resulted in plants with improved mechanical properties. *J. Biotechnol.* **128**:919–934.

Yin J.Q. and Y. Wan. 2002. RNA-mediated gene regulation system: now and the future. *Int. J Mol. Med.* **10(4)**:355–365.

Zhang, L., F. Jing, F. Li, M. Li, Y. Wang, G. Wang, X. Sun and K. Tang. 2009. Development of transgenic Artemisia annua (Chinese wormwood) plants with an enhanced content of artemisinin, an effective anti-malarial drug, by hairpin-RNA-mediated gene silencing. *Biotechnol. Appl. Biochem.* **52**:199–207.

CHAPTER 15

RNA Interference and Wheat Functional Genomics

Pradeep Sharma* and Rajender Singh

ABSTRACT

Over the course of 20 years, the scientific understanding of RNAi has developed from the initial observation of unexpected expression patterns to a sophisticated understanding of a multi-faceted, evolutionarily conserved network of mechanisms that regulate gene expression in many organisms. RNA interference (RNAi) is a post-transcriptional gene-silencing phenomenon induced by double-stranded RNA. It has also been developed as a genetic tool that can be exploited in a wide range of species. However, its use in polyploid species is still in the early stage. Nonetheless, it can possibly be used to silence multigene families and homoeologous genes in polyploids. This is of great importance for functional studies in hexaploid wheat (*Triticum aestivum*), where most of the genes are present in at least three homoeologous copies and conventional insertional mutagenesis is not effective. To date, RNAi has been used to target a number of genes like those coding for transcription factors, enzymes necessary for starch synthesis and signal and storage proteins. Wheat varieties with accelerated flowering time, reduced amylase content and delayed senescence have been previously produced by RNAi. RNAi response has been documented in different tissues and developmental stages. It has also been shown to be stably inherited and sequence specific. One of the limitations of RNAi in wheat is targeting genes with conserved domains and duplications in the genome, as there is a high chance of silencing unwanted genes.

Keywords: RNAi, *T. aestivum*, HMV, VRN, siRNA.

INTRODUCTION

The cereal crops are essential components to the human and animal food supply. Solutions to many of the problems challenging cereal production will require identification of gene responsible for particular traits. Wheat (*Triticum aestivum* L. Em. Thell.) is one of the most important food crops

* Corresponding author e-mail: neprads@yahoo.com

in the world and understanding its genetics and genome organization using molecular markers is of great value for genetic and plant breeding purposes. It is an allohexaploid (2n = 6x = 42) with the three genomes A, B and D and has extremely large genome 16×10^9 bp/IC (Bennett and Smith, 1976) with more than 80% repetitive DNA.

Unfortunately, the process of identifying gene function is very slow and complex in crop plants. In wheat, this process is made very difficult by the very large size and complexity of their genomes and the difficulty with which these crops can be genetically transformed. Additionally, the polyploidy of wheat greatly complicates any approach based on mutational analysis because functional, homeologous gene often mask genetic mutations. Therefore, RNAi is an important new tool that overcomes many of these obstacles and problems to greatly facilitate the assessment of gene function. Most of the techniques currently available to study gene function in plants involve knockouts of target genes followed by the phenotypic characterization of the mutant plants. In Arabidopsis and rice, large gene knockout collections have been assembled from chemical mutagenesis and T-DNA or transposon insertional mutagenesis (Jeon et al. 2000, Alonso et al. 2003, Sallaud et al. 2004). In wheat, a powerful reverse genetics approach was recently implemented through the combination of EMS-mediated mutagenesis and TILLING technology (Slade et al. 2005). Because of the tolerance of polyploid wheat to high mutation rates, this method is very efficient in identifying mutations in the target genes. However, the effect of single-gene knockouts can be masked by the functional redundancy of homoeologous genes present in the other wheat genomes (Lawrence and Pikaard 2003). This limitation can be overcome by generating double and triple mutants, although this process is cumbersome and time consuming. Therefore, faster alternatives are required for functional gene analysis in polyploid wheat. In this regard, Virus induced gene silencing (VIGS) system based on barley stripe mosaic virus (BSMV) has been developed for use in wheat and barley. The BSMV-VIGS system allows researchers to switch off or knockdown the expression of chosen genes so that the genes function may be inferred based on knockout phenotype.

Although RNA interference (RNAi) was first discovered in worms, related phenomena such as post-transcriptional gene silencing and coat protein mediated protection from viral infection had been observed in plants prior to this. In plants, RNAi is often achieved through transgenes that produce hairpin RNA. For genetic improvement of crop plants, RNAi has advantages over antisense-mediated gene silencing and co-suppression, in terms of its efficiency and stability. It also offers advantages over mutation-based reverse genetics in its ability to suppress transgene expression in multigene families in a regulated manner. RNAi is a highly coordinated gene regulatory mechanism that appears to be highly conserved across all metazoans studied thus far. Several biochemical and genetic investigations have focused on elucidating the regulatory mechanism for RNAi. These studies have revealed that this phenomenon plays a variety of cellular roles, including protection against harmful mobile genetic elements such as viruses or transposons, regulation of developmental events and elimination of unwanted run-on mRNA transcripts. RNAi is the result of well-coordinated RNA-to-protein interactions, a key participant in this mechanism is the RNA-induced silencing complex (RISC), which plays a role in the binding of siRNA and its target mRNA to effect eventual mRNA cleavage and resulting gene suppression. RNAi-induced gene silencing is now commonly used by scientists as a tool to characterize the individual biological roles of specific genes and to illuminate their participation in important pathways or mechanisms.

The discovery of RNA mediated gene silencing creates a viable alternative strategy for gene functional analysis through the simultaneous knockdown of expression of multiple related gene copies. This phenomenon was discovered in the late nineties in *Caenorhabditis elegans* (Fire et al. 1998) and plants (Jorgensen et al. 1996) and involves formation of double stranded RNA (dsRNA) that subsequently leads to the cleavage of homologous mRNAs at the post-transcriptional stage (RNA interference (RNAi) or to the blockage of these mRNAs at the transcriptional stage. Even though there are a limited number of RNAi studies in wheat, some general trends are emerging from its first applications that have successfully modified important agronomic and quality traits.

In bread wheat, in particular, the technology provides an additional advantage of silencing all genes of a multigene family including homoeoloci for individual genes, which are often simultaneously expressed, leading to a high degree of functional gene redundancy. It has been shown that delivery of specific dsRNA into single epidermal cells in wheat transiently interfered with gene function. Yan et al. (2004) and Loukoianov et al. (2005) used RNAi for stable transformation and to demonstrate that RNAi mediated reduction of *VRN2* and *VRN1* transcript levels, respectively, accelerated and delayed flowering initiation in winter wheat. Similarly, Regina et al. (2006) used RNAi to generate high-amylose wheat. However, none of the above studies reported long-term phenotypic stability of RNAi mediated gene silencing over several generations, neither did they report any molecular details on silencing of homoeologous genes. However, Travella et al. (2006) showed RNAi results in stably inherited phenotypes suggesting that RNAi can be used as an efficient tool for functional genomics studies in polyploid wheat. They introduced dsRNA-expressing constructs containing fragments of genes encoding *Phytoene Desaturase*

(*PDS*) or the signal transducer of ethylene, *Ethylene Insensitive 2* (*EIN2*) and showed stably inherited phenotypes of transformed wheat plants that were similar to mutant phenotypes of the two genes in diploid model plants. Synthetic microRNA constructs can also be used as an alternative to large RNA fragments for gene silencing, as has been demonstrated for the first time in wheat by Yao et al. (2008) by discovering and predicting targets for miRNAs, which are monocot specific. This study will serve as a foundation for the future functional genomics studies.

HISTORY OF RNA INTERFERENCE

The discovery of RNAi phenomenon came accidently when Guo and Kemphues (1995) injected the anti-sense strand to block expression of the *par*-1 gene in the nematode *C. elegans*. The expression was disrupted, however, upon performing their controls they found that the sense strand also reduced the expression of that gene. Even earlier biologists had unknowingly witnessed the process of RNA interference when performing experiments on petunias and found that when they introduced a pigment-producing gene under the control of a promoter into the flowers they did not get expected results. Instead of getting the expected deep purple color, the flowers were variegated or they were completely white (Napoli et al.1990). So what was the reason of these unusual results? It was in 1998 when Fire and Mello first injected double stranded RNA into *C. elegans* and were rewarded with a much more efficient gene silencing effect (Zamore 2002). Now the mystery was unrevealed and it was found that the initiator of this post-transcriptional gene silencing (PTGS) was dsRNA, but how it happens, was still a question.

THE MECHANISM OF RNA INTERFERENCE

The dsRNA can be introduced into the cell in a number of ways. In simple organisms such as *C. elegans* and *Paramecium*, the dsRNA can be delivered by feeding the organisms with bacteria engineered to express the dsRNA of choice (Galvani and Linda 2002). In other cells, the dsRNA may be injected directly (Shuey et al. 2002). The process of RNAi is triggered by dsRNA precursors which are processed into siRNA in the presence of ATPs. Once the dsRNA is in the cell, it is the target for an enzyme named DICER. This enzyme is a dsRNA specific endonuclease that cuts it into smaller fragments, specifically into 21–23 nucleotides. These siRNAs are then incorporated to RNA-induced silencing complex (RISC) which contains several proteins besides siRNAs. Some well known proteins are AGO2, FMRP and P100 (Hutvagner and Zamore 2002, Caudy et al. 2003, Doench et al. 2003, Zeng et al. 2003). Now RISC is activated which is ATP dependent process and unwinds the double stranded siRNAs.

It binds to the targeted mRNA using the siRNA as a guide to find the target sequence and an endoribonuclease cleaves the mRNA which is then degraded by exo-ribonucleases resulting in a loss of expression of the gene (Zamore et al. 2000, Meister and Tuschl 2004) (Figure 1). Some of the double stranded siRNAs may be used as primers by an RNA-dependent RNA polymerase (RdRp) resulting in the formation of another long strand of dsRNA that can continue through the RNAi pathway. This may enhance the efficiency of the gene silencing by dsRNA (Shuey et al. 2002).

Since the RNAi pathway was first discovered in cells as a natural process, the question has arisen as to what its purpose is in the cell. Two answers came from researchers (1) to inhibit transposon mobilization (Tabara et al. 1998) and (2) to act as an antiviral mechanism in plants (Kasschau and Carrington, 1998). It has also been discovered that a disruption in the genes required for RNAi to take place often leads to developmental defects in the organism. This observation has

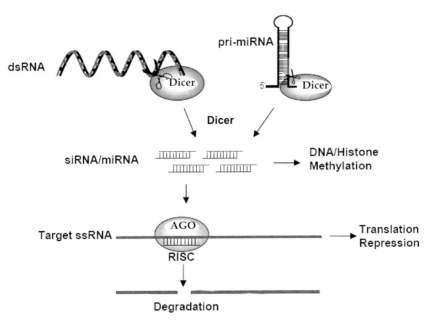

Figure 1: Schematic diagram of plant mediated RNAi in plants. PTGS involves the generation of dsRNA by RdRP. miRNA is an endogenous siRNA-like RNA known to be involved in the developmental regulation of gene expression. Its pre-cursor (pre-miRNA) is a small hpRNA with bulges in its stem region. All dsRNA. hpRNA and pre-miRNA are processed by Dicer into 21 nt RNA complexes and the unwound ssRNA is then incorporated into RISC. In plants, dsRNA and pre-miRNA can be processed by distinct Dicer like proteins (DCLs).

(Color image of this figure appears in the color plate section at the end of the book.)

suggested that the process of RNA interference is involved in at least one developmental pathway.

DIVERSE RNA SILENCING PATHWAYS

There are at least three diverse RNA silencing pathways for silencing specific genes in plants. In these pathways, silencing signals can be amplified and transmitted between cells and may even be self regulated by feedback mechanisms. Three natural pathways of RNA silencing (cytoplasmic siRNA silencing, silencing of endogenous messenger RNAs by siRNAs and DNA methylation and suppression of transcription) in plants have been revealed by genetic and molecular analysis (Baulcombe, 2004). Several lines of evidence have established that small RNAs play an important role in plant development by triggering RNA degradation, translational inhibition or chromatin modification (Vaucheret 2006). First, small RNAs were detected in transgenic plants undergoing cosuppression. Second, plants undergoing silencing of the transgenes β-*glucuronidase* or *green fluorescent protein* (*GFP*), which have no homology to endogenous genes, accumulated small RNAs (sRNA) with homology to these foreign gene sequences. Finally, the finding that plants infected with potato virus X (PVX, genus *Potexvirus*) accumulated siRNAs homologous to PVX demonstrated that plant viruses can be targets of PTGS (Hamilton and Baulcombe, 1999). Characterization of small RNAs revealed two major types, which are referred to as siRNA and miRNA, based on their mode of biogenesis and function. Furthermore, numerous types of siRNAs have been recognized, including trans-acting siRNA (ta-siRNA), natural-anti-sense siRNA (nat-siRNA) and repeat-associated siRNA (ra-siRNA) (Ramachandran and Chen. 2008). Over the last decade, 20–30 nucleotide RNA molecules have emerged as critical regulators in the expression and function of eukaryotic genomes. Two primary categories of these small RNAs-short interfering RNAs (siRNAs) and microRNAs (miRNAs) act in both somatic and germ line lineages in a broad range of eukaryotic species to regulate endogenous genes and to defend the genome from invasive nucleic acids. Recent advances have revealed unexpected diversity in their biogenesis pathways and the regulatory mechanisms that they access. Our understanding of siRNA- and miRNA-based regulation has direct implications for fundamental biology as well as disease etiology and treatment.

miRNA

In plants, miRNAs are a small, non-coding regulatory RNAs that regulate gene expression by guiding mRNA cleavage or translational inhibition. So far, identification of miRNAs has been limited to a few model plant species, such as *Arabidopsis,* rice and populous, whose genomes have been sequenced. Wheat is one of the most important cereal crops worldwide. These miRNAs play a role in regulation of hormone responses, development, control of cell division and flowering. This class of small RNAs is also involved in the response to environmental stress and pathogens (Kurihara et al. 2006). The database of miRNA Release 15, April 2010, contains 14,197 entries representing hairpin precursor miRNAs, expressing 15,632 mature miRNA products, in 133 species. The data are freely available to all through the web interface at http://www.mirbase.org/ and in fiatfile form from ftp://mirbase.org/pub/mirbase/.

A total of more than 3000 new hairpin sequences and more than 4000 new mature sequences been deposited in the current edition of miRNA registry. These miRNAs include 199 from *Arabidopsis*, 477 from rice, 148 from sorghum, 230 from *Phycomiterella*, 246 from populus, 375 from *Medicago traunculata*, 85 from soybean, 30 from tomato and 170 from maize. The *MIR* genes are involved in biogenesis of long single-stranded(ss) precursor miRNAs (pri-miRNAs), which are transcribed by a DNA-dependent RNA polymerase II and include a hairpin structure. The pri-miRNA transcripts are then processed by DCL1 to produce the shorter precursor miRNAs (pre-miRNAs) (Jones-Rhoades et al. 2006). The proteins HYPONASTIC LEAVES 1 (HYL1) and SERRATE (SE) interact with DCL1 to form a complex for processing of pri- and pre-miRNAs (Dong et al. 2008). The 21-nt mature miRNA is generated from the pre-miRNAs by DCL1 and is methylated by the HUA ENHANCER 1 (HEN1) (Yang et al. 2006). Components of the miRNA processing pathway have been localized to nuclear dicing bodies, indicating that these bodies may be the miRNA processing center (Fang and Spector 2007). After DCL1-mediated cleavage and HEN1-mediated methylation, mature ss-miRNAs are exported from the nucleus to the cytoplasm. Here, the miRNAs bind an AGO protein, most commonly AGO1, to form the RISC (Brodersen et al. 2008). The AGO-RISC complex recognizes and cleaves target mRNAs. Cleavage of the target mRNA occurs in the middle of the sequence complementary to the miRNA. In total, 58 wheat miRNAs were identified which comprising 43 miRNA families, 20 of these families conserved and 23 are novel in wheat.

ta-siRNA

ta-siRNA is a form of siRNA that represses gene expression via PTGS (Xie et al. 2005). The ta-siRNA is derived from action of *TRANS-ACTING siRNA* (*TAS*) genes. Transcripts targeted by TAS are cleaved via a *TAS*-targeted miRNA-programmed RISC containing AGO1 or AGO7 (Allen et al. 2005). As is the case for miRNAs, the AGO1/AGO7 in the ta-siRNA-RISC complex facilitates cleavage of target mRNAs. DCL1, AGO1/AGO7, HEN1 and HYL1 are all involved in the ta-siRNA pathway. Transcripts cleaved by TAS are converted to dsRNA by RDR6 with the help of SGS3 and the dsRNA is processed into 21-nt ta-siRNA by DCL4 in association with DRB4 (Peragine et al. 2004).

nat-siRNA

More than two million small RNAs from seedlings and inflourescence of *Arabidopsis* has been sequenced and among these sequences, more than half represent lower abundance siRNA that match repetitive sequences, intergenic regions and genes. Natural antisense transcripts (NATs) are formed either by antisense transcription at a specific genomic locus, or from transcription at a different locus, in which case the NAT is referred to as a trans-NAT. The NATs are induced by abiotic or biotic stress and the dsRNA formed by the two complementary (sense and anti-sense) transcripts is processed by DCL2 (Borsani et al. 2005). The nat-siRNAs that are subsequently generated are 21-nt or 24-nt. Some of the initial steps of this process involve RDR6, SGS3 and the atypical DNA-dependent RNA polymerase-like subunit NUCLEAR RNA POLYMERASE D 1A (NRPD1a). The nat-siRNA targets dsRNA formed by hybridization of the sense and anti-sense transcripts for cleavage. The cleaved RNA product is converted to dsRNA by RDR6, in association with SGS3 and the 21-nt nat-siRNA is generated by DCL1. Accumulation of the nat-siRNA also requires NRPD1a. Both the 21-nt and 24-nt nat-siRNAs are hypothesized to facilitate sequence-specific mRNA degradation in an AGO-dependent manner.

Recenlty, Yao et al. (2010) have reported that some of the wheat small RNAs exhibit developmental stage-dependent and stress-responsive expression patterns and five putative wheat nat-siRNAs are also identified. They further predicted 4,249 *trans* targets for the 1,106 small RNAs and these predicted target genes include not only transcription factors implicated in development but also other genes involved in broad range of physiological processes.

ra-siRNA and RNA-directed DNA methylation

In *Arabidopsis*, other forms of endogenous siRNAs are generated from regions of the genome with direct or inverted repeat loci. The siRNAs associated with repeated sequences are 24-nt and are called ra-siRNA. The ra-siRNA plays an important role in transcriptional gene silencing (TGS), which is an epigenetic mechanism that results in the silencing of an endogenous gene through DNA methylation (Chan, 2008). At some loci, methylation-induced aberrant RNA is transcribed by DNA-dependent RNA polymerases (e.g., Pol II, Pol III, or Pol IVa). Pol IVa also synthesizes additional aberrant RNAs (Xie et al. 2004). This aberrant RNA moves to the nucleolus where it is converted into dsRNA by RDR2. This dsRNA is processed by DCL3, in association with HEN1, to generate the 24-nt ra-siRNA. The mature ra-siRNA becomes incorporated into AGO4-RISC complexes. AGO4 also interacts with NRPD1b, which is the large subunit of Pol IVb. The 24-nt ra-siRNAs, NRPD1b, RDR2 and DCL3 have been co-localized to nuclear Cajal bodies (Pontes et al. 2006), indicating a possible site of ra-siRNA synthesis. AGO6 has a partially redundant function with AGO4 in ra-siRNA accumulation. AGO4-ra-siRNA-RISC complexes are required for DNA methylation at the target site loci. In addition, Pol IVb, DOMAINS REARRANGED METHYLTRANSFERASE 2 (DRM2) and a SNF2-like protein, DEFECTIVE IN RNA-DIRECTED DNA METHYLATION 1 (DRD1), are required downstream of ra-siRNA formation to induce RNA-directed DNA methylation of target loci by an unknown mechanism.

DCLs

RNase III family members are among the few nucleases that show specificity for dsRNAs and cleave them with 3' overhangs of 2 to 3 nucleotides and 5'-phosphate and 3'-hydroxyl termini. Bernstein et al. (2001) identified an RNase III-like enzyme in *Drosophila* extract which was shown to have the ability to produce fragments of 22 nucleotides, similar to the size produced during RNAi. These authors showed that this enzyme is involved in the initiation of RNAi. Owing to its ability to digest dsRNA into uniformly sized small RNAs (siRNA), this enzyme was named Dicer (DCL). These nucleases are evolutionarily conserved in worms, flies, fungi, plants and mammals. Dicer has four distinct domains: an aminoterminal helicase domain, dual RNase III motifs, a dsRNA binding domain and a PAZ domain (a 110-amino-acid domain present in proteins like Piwi, Argo and Zwille/Pinhead), which it shares with the RDE1/QDE2/Argonaute family of proteins that has been genetically linked to RNAi by independent studies (Huang, 1999). Cleavage by Dicer is thought to be catalyzed by its tandem RNase III domains. Some DCR proteins, including the one from *D. melanogaster*, contain an ATP-binding motif along with the DEAD box RNA helicase domain.

Overall, the responses of DCLs to drought, cold and salt are quite different, indicating that plants might have specialized regulatory mechanism in response to different abiotic stresses. Further analysis of the promoter regions reveals a few of *cis*-elements that are hormone- and stress-responsive and developmental-related. However, gain and loss of cis-elements are frequent during evolution and not only paralogous but also orthologous DCLs have dissimilar *cis*-element organization.

Genetic and biochemical evidence has demonstrated that small RNAs such as miRNAs and siRNAs in eukaryotic organisms play important roles in developmental regulation, epigenetic modifications, tumorigenesis and biotic and abiotic stress responses (Kidner and Martienssen 2005, Vaucheret 2006, Llave 2004). The two kinds of non-coding RNAs, miRNAs and siRNAs, are produced from different types of precursors. Dicer or Dicer-like proteins are key components in the miRNA and siRNA biogenesis pathways in processing long double-stranded RNAs into mature small RNAs (Millar and Waterhouse 2005, Großhans and Filipowicz 2008).

Relative to animals and fungi, the notable expansion of DCL family members in monocot and dicot plants may reflect the deployment of RNA silencing approach in antiviral defense (Deleris et al. 2006). In *A. thaliana*, four Dicer-like proteins (DCL1–DCL4) with different roles were found (Dunoyer et al. 2005, Moissiard et al. 2007, Mlotshwa et al. 2008): DCL1 not only is associated with miRNA production but also has a role in the production of small RNAs from endogenous inverted repeats. The other three DCLs are siRNA-generating enzymes. DCL2 generates siRNAs from natural cis-acting antisense transcripts and functions in viral resistance. DCL3 generates siRNAs for a guide of chromatin modification, while DCL4 is associated with ta-siRNA metabolism and acts during posttranscriptional silencing. The functions of DCL1 and DCL3 overlap to promote Arabidopsis flowering. Overlaps in function are also found for DCL2 and DCL4 with respect to antiviral defense and for DCL2, DCL3 and DCL4 in siRNA and ta-siRNA production and in the establishment and maintenance of DNA methylation (Henderson et al. 2006).

The expression profiles of DCL genes in response to stresses such as drought, cold and salt were examined using rice datasets. Compared to the biological control, the expression of rice OsDCLs is slightly repressed under such above stress treatments (Liu et al. 2009). The expression of OsDCL3 reduces significantly under the condition of drought or salt treatment ($p<0.05$), as is the case for OsDCL4 under drought treatment.

To further confirm the responsiveness of DCLs to stresses, an examination of Arabidopsis AtDCLs expression in roots and shoots under the treatments of drought, cold and salt confirmed that AtDCL1 expression decreases extensively. Similar to expression in roots, AtDCL1 expression in shoots showed the lowest expression at 1.0 h and then increased to a high level at 24.0 h after drought treatment. There has been no obvious change in expression for AtDCL3 under different drought treatment conditions. Both AtDCL2 and AtDCL4 increase their expression at 6.0 h. However, the former gene shows the highest expression at 6.0 h, while the latter reaches its expression peak at 12.0 h after treatment and then declined at 24.0 h. After cold treatment, the expression of AtDCL1 continues to increase and shows its highest level at 24.0 h in roots and AtDCL4 decreased from 6.0 to 24.0 h. similar patterns were observed in shoots where the expression of AtDCL1 has increased extensively after long time of cold treatment, whereas other AtDCLs show an inverse tendency. More complicated expression patterns of AtDCLs were revealed after salt treatment (Yan et al. 2009).

PLANT RNAi VECTORS

The construction of plant RNAi vectors is time-consuming and laborious work because sense and antisense target sequences and a spacer should be incorporated into a single vector. As a rapid and easy method for construction of RNAi vectors, Wesley et al. (2001) made a pHELLSGATE vector system that allows us to construct plant RNAi vectors by using Gateway technology (Invitrogen). In this system, PCR fragments with sense and antisense orientations are amplified by using primers having the recombinase recognition sites, *att*B1 and *att*B2 sequences. These PCR fragments are cloned into a plasmid containing *att*P1 and *att*P2 sites by a BP clonase. pHELLSGATE vector consists of a CaMV 35S promoter and a catalase intron as a spacer. pHELLSGATE can be directly used for *Agrobacterium*-mediated transformation of plants and facilitates high-throughput application of RNAi studies. Miki and Shimamoto (2004) developed another high-throughput RNAi vector system, the pANDA vector, in which a maize ubiquitin promoter is used for high expression of an RNAi cassette and a DNA fragment originated from the β-glucuronidase (*GUS*) gene was used as a spacer. To easily clone target DNA fragments, the Gateway technology is used in the pANDA vector system. A target PCR fragment containing CACC sequences at the 5′ end of forward primer is cloned into the pENTR/D-TOPO vector, resulting that these PCR fragments are flanked with two recombination sites, *att*L1 and *att*L2. A LR clonase recombines this PCR fragment into two recombination sites (*att*R1 and *att*R2) of pANDA vector in opposite directions. pANDA vector is also ready for use in *Agrobacterium*-mediated transformation of monocots. Rest of the RNAi vectors developed by using Gateway technology are summarized in Table 1.

The sequence specificity in RNAi is determined by hybridization of siRNAs to the corresponding target mRNAs. Knockdown of a single gene among the several paralog genes or simultaneous silencing of multiple genes has been demonstrated, in which the choice of target sequences is important.

Table 1. RNAi vector using Gateway technology.

vector	Promoter	Spacer	Selection marker	Transformation	References
pHELLSGATE	CaMV 35S	Pdk interon	NPTII	Agrobacteruim infection	Wesley et al. (2001)
pHELLSGATE12	CaMV 35S	Pdk interon and Cat intron	NPTII	Agrobacteruim infection	Helliwell and Waterhouse (2003)
pAGRIKOLA	CaMV 35S	Pdk interon and Cat intron	Bar	Agrobacteruim infection	Hilson et al. (2004)
P*7GWIWG2	CaMV 35S	Intron	NPTII, HPT, Bar	Agrobacteruim infection	Karimi et al. (2002)
pANDA	Maize Ubiquitin	GUS	NPTII, HPT	Agrobacteruim infection	Miki and Shimamoto (2004)
pANDA-mini	Maize Ubiquitin	GUD	No marker	Particle bombardment	Miki and Shimamoto (2004)
pIPKTA30N	CaMV 35S	RGA2 interon	No marker	Particle bombardment	Douchkov et al. (2005)
pOpOff1	DEX-inducible promoter	Pdk interon and Cat intron	HPT	Agrobacteruim infection (2005)	Wielopolska et al.
pIPKb007, pIPKb0010	Maize Ubiquitin, Rice Actin Enhanced CaMV 35S, wheat Gst A1	RGA2 interon	HPT	Agrobacteruim infection	Himmelbach et al. (2007)

RNAi-MEDIATED STARCH METABOLISM

RNAi technology is now widely used and representative examples of its application to plant molecular breeding as shown in Table 2. RNAi is being used as a tool of genetic manipulation of primary metabolites, including starch and oil, level control of the target proteins and modulation of secondary metabolisms, in crop plants is not well documented. Starch consists of two types of α-glucan polymer that are called amylose and amylopectin. Amylose is a linear α-1,4 glucan and is synthesized in amyloplast by granule-bound starch synthase (GBSS). Amylopectin, the main constituent of starch, is a highly branched glucan in which the glucose unit is joined by α-1,6 linkage. The starch-branching enzyme (SBE) is responsible for formation of α-1,6 linkage in amylopectin. Starch is a source of energy in diets and has been used as a renewable raw material (Jobling, 2004). Physicochemical properties of starch can be altered through chemical and enzyme modifications and such altered properties of starches are followed by development of unique applications. Thus, designing starch by genetic engineering of starch-synthesizing enzymes has been an important breeding subject.

Amylose-free (waxy) starch has been produced by a corn *waxy* mutant. Corn waxy starch easily gelatinizes and yields clear pastes without gel formation. Heilersig et al. (2006) reported an inhibition of potato (*Solanum tuberosum* L.) granule-bound starch synthase I (GBSSI) activity by introduction of the *GBSSI* RNAi constructs. Waxy potato starch is expected to have clarity and stability, especially stability against the freeze-thaw treatment. These properties have been desired in the food industry and in paper manufacture (Jobling 2004). Heilersig et al. (2006) made eight different RNAi constructs that contained different regions, lengths and orientation of potato *GBSSI* cDNA fragment and transcription of inverted repeats was controlled by the *GBSSI*'s own promoter. Otani et al. (2007) showed the effective elimination of the sweet potato (*Ipomoea batatas* L.) GBSSI protein by RNAi. Over 70% of the transgenic plants with an RNAi construct targeting the 351-bp-first exon of the *GBSSI* gene produced amlyose-free storage roots. This amylose-free starch showed altered physicochemical properties. They used a part of *GBSSI* cDNA as a spacer. The results showed that sense–antisense or antisense–sense orientations in the inverted repeat had nearly the same effects on silencing efficiency. High silencing efficiencies were observed when the constructs using 5' region or middle region, but not 3' region, of *GBSSI* cDNA were introduced into potato plants. The constructs with a stem size of 500~600 bp and a spacer of about 150 bp in length more efficiently silenced than the larger stems with a larger spacer. Although strong reduction in

Table 2. Summary of transgene induced-dsRNA silencing studies in wheat.

| Silencing target | Targetgene | RNAi cassette | | Homoeologues | Phenotype | Ref. |
		Target region	Promoter			
Zinc finger, domain protein	VRN2	Coding region	35S + Adh1 Intron, B	A,B, D genomes (not specified)[c]	Accelerated flowering time	Yan et al. CCT (2004)
MADS, K-box domain protein	VRN1	Coding region	35S + Adh1 Intron	A, B, D genomes (not specified)[c]	Delayed flowering time	B. Loukoianov et al. (2005)
Granule bound starch synthase enzyme	GBSSI	Coding region	Maize Ubi , A	A, B, D genomes (not specified)[c]	Reduced amylose content	Li et al. (2005)
Starch branching enzyme	SBE-IIa	Coding region	HMWG, A	A, B, D genomes (not specified)[c]	Reduced amylopectin content	Regina et al. (2006)
Starch branching enzyme	SBE-IIb	Coding region	HMWG, A	A, B, D genomes (not specified)[c]	No detectable phenotype	Regina et al. (2006)
Transmembrane protein	EIN2	Coding region	Maize Ubi, B	A, B, D genomes	Ethylene insensitivity	Travella et al. (2006)
Phytoene desaturase enzyme	PDS	Coding region	Maize Ubi	A, B, D genomes	Photobleaching	Travella et al. (2006)
NAM, NAC transcription factor	GPC	Coding region	35S + Adh1 intron B	A, D genomes & B, D paralogues	Delayed senescence, reduced grain protein, Zn and Fe	Uauy et al. (2006b)
Seed storage protein	1Dx5	Coding region	Maize Ubi	A, B, D genomes & B, D paralogues	Reduced gluten and mixing quality	Yue et al. (2007)

[c]Studies did not analyze homoeologous genomes separately.

amylose core can be obtained, a complete reduction of *GBSSI* mRNA was never observed in the resulting transgenic potato plants.

High-amylose corn starch has been commercially produced, where naturally occurring mutants of maize (*Zea mays*) are cultivated. In monocots, three major isoforms of the starch branching enzyme have been identified, namely *SBEI*, *SBEIIa* and *SBEIIb*. In maize, a high-amylose phenotype is caused by a mutation in the *SBEIIb* gene (Kim et al. 1998). Regina et al. (2006) generated high amylase in wheat (*Triticum aestivum*) by RNAi-based technology. They showed that *SBEIIa* hairpin RNAs reduced the expression of both *SBEIIa* and *SBEIIb* genes, although the target region of *SBEIIa* is only 70% identical to the corresponding region of *SBEIIb*. There is one region in which a 21-nt long identical continuous stretch exists between *SBEIIa* and *SBEIIb* genes. In contrast, a hairpin construct against the *SBEIIb* gene did not inactivate the *SBEIIa* gene. The similarity between the *SBEIIb* target region and the corresponding *SBEIIa* gene sequences is about 70% and one 21-nt long stretch is found. The precise reason was unclear why the RNAi construct against the *SBEIIa* gene can reduce the expression of *SBEIIb* while the opposite cannot. Transgenic wheat plants produced starch with >70% amylase content. The analysis of total starch content in the endosperm revealed a slight reduction in the transgenic wheat plants (43%) compared with 52% in the control plants. The high amylose wheat grain improved the digestion of rats, indicating that it served as a source of resistant starch. High amylose starch is unique in its film forming, gelling properties and resistance against digestion in the human gastrointestinal tract. This 'resistant starch' has nutritional benefits as dietary fibers (Topping et al. 2003). Until now, no other major crops with the high amylose starch have been found. Andersson et al. (2006) have succeeded in producing of high amylose potato plants by simultaneous inactivation of two genes encoding starch branching enzymes (*SBE1* and *SBE2*). The 200-bp fragments of both *SBE* genes were fused and a single inverted repeat cassette was prepared. This RNAi cassette was inserted behind a potato *GBSS* promoter. The resulting transgenic potato plants showed a high-amylose phenotype with amylose contents ranging from 38% to 87% whereas the amylose content of parental plants was about 20%.

RNAi based manipulation of Storage Proteins

Wheat flour has been used in a wide range of distinct foods and gluten proteins containing high molecular weight glutenin subunit (HMW-GS) are the most important component affecting cooking quality (Anjum et al. 2007). The contents and relative proportion of HMW-GS are closely involved in determination of dough performance, namely its strength and elasticity. Hexaploid wheat has five HMW-GS genes called 1Ax2, 1Dx5, 1Bx7, 1By9 and 1Dy10 on the homoeologous chromosomes 1A, 1B and 1D. Yue et al. (2008) reported the RNAi inhibition of 1Dx5 expression. The RNAi construct comprised of inverted repeat sequences harboring a 200-bp fragment of the 1Dx5 gene, spacer sequences from the intron 4 of wheat *waxy* gene and the maize ubiquitin promoter. Target sequences of the 1Dx5 gene showed 81–91% identity to the corresponding sequences of 1Ax2, 1Bx7, 1By9 and 1Dy10 genes. There was no >20-nt long perfect identity on target regions. Wheat transformation was conducted by particle bombardment. Analysis of HMG-GS expression in the resultant transgenic plants showed that the 1Dx5 expression was completely silenced. Furthermore, expression of the 1Bx7 gene was reduced in spite of no difference in the protein level of 1Ax2, 1By9 and 1Dy10 compared to the control plants. These results suggest that the presence of continuous nucleotide stretches of over 21-nt and high sequence identity with the trigger sequence may not always be sufficient to trigger the RNAi process. Analysis of flour quality showed that wheat flour from transgenic plants could not produce dough necessary to make bread because of a reduction of gluten development. Therefore, the genetic control of gluten components is possible by RNAi (for undesired HMG-GS gene) and overexpression (for desirable HMW-GS gene), which may develop new quality and taste of wheat flour.

Zeins, major maize seed storage proteins, are deficient in lysine content. Lysine is a limiting essential amino acid for animal nutrition. From analyses of naturally occurring maize mutants with high lysine contents, reduction of the zein content is considered to be a potent approach for increasing lysine content. A decrease of zein proteins should be associated with an increase of non-zein, lysine-rich proteins. Segal et al. (2003) made two RNAi constructs to reduce the 22-kD α-zein. One construct had an inverted repeat harboring the target zein cDNA fragment under the control of the endosperm-specific 27-kD γ-zein promoter. In this construct, sense and antisense fragments were directly linked. Another one had a green fluorescent protein (*GFP*) spacer sequence and its inverted repeat cassette was transcribed by a strong endosperm-specific promoter of the zp22/6 zein gene. Both resulting transgenic maize plants showed reduced accumulation of 22-kD γ-zein and the lysine content of kernels markedly increased. Huang et al. (2006b) produced transgenic maize plants with reduced accumulation of both 19- and 22-kD α-zeins. Two distinct RNAi constructs (pMON73566 and pMON73567) were prepared, in which the inverted repeat cassette was transcribed by the 27-kD γ-zein promoter. pMON73566 has an inverted repeat against the 19-kD zein gene and antisense 22-kD zein sequences are included as a loop of the resulting hairpin RNAs. Introduction of pMON73566 resulted in low silencing frequency of both 19- and 22-kD zein genes. pMON73567 is comprised of an inverted repeat that contains tandem fused sequences with the 19- and 22-kD zein genes and also harbors a spacer consisting of an unpaired 22-kD zein gene fragment. pMON73567 transformants showed an efficient reduction in the expression of both 19- and 22-kD zein genes. Total amino acid analysis demonstrated that two times higher levels of lysine and tryptophan were found in the pMON73567 kernels than those of parental plants. As an alternative approach to breeding of high-lysine corn could be RNAi-mediated suppression of a maize lysine degradation enzyme, lysine-ketoglutarate reductase/saccharopine dehydrogenase. The hairpin RNAs were specifically expressed in endosperm and lysine content in transgenic maize kernels significantly increased.

RNAi based γ–gliadins silencing

The wheat grain is widely consumed and therefore major research efforts have focused on modifying grain composition to improve its quality for processing and for the nutrition of humans and livestock. The major protein fraction of wheat grain is gluten which is largely responsible for the functional properties of dough. However, gluten is not a single protein but a complex mixture of components with the monomeric gliadins accounting for about 50% of the gluten proteins. The gliadins are classified into three groups (α/β, γ and ω) on the basis of their electrophoretic mobility in acidic polyacrylamide gel electrophoresis (A-PAGE) (Metakovsky et al. 1984). It is generally accepted that gliadins contribute mainly to the extensibility and viscosity of gluten and dough, with the polymeric glutenins being responsible for elasticity. However, because the gliadins are encoded by large multigene families and inherited in blocks, the roles of individual components are not well understood (Fido et al. 1997). Gliadins are also associated with the development of celiac disease, a food-sensitive enteropathy caused by the ingestion of gluten proteins. There is also clear evidence that α- and γ-gliadins contain clusters of epitopes that are active in coeliac disease (Arentz-Hansen et al.2000, Shan et al. 2002). The down regulation of specific individual gliadins or groups of gliadins may therefore be of interest in relation to determining their role in both grain processing properties and in triggering coeliac disease. The aim of this work was therefore to silence the expression of specific gliadins by hpRNA, to demonstrate the feasibility of systematically silencing specific groups of gluten proteins. RNAi technology can be used to down-regulate groups of proteins encoded by multigene families, such as gliadins and glutenins. Such material is ideal for elucidating the roles of specific groups of proteins in determining the functional properties of flours and their relative activities in triggering coeliac disease.

VIGS

A faster alternative to stable transformation is transient RNAi silencing by virus induced gene silencing (VIGS). The biological principles on which VIGS is based were uncovered as molecular biologists studied the consequences of virus infection in plants (Lindbo and Dougherty, 1992). Through this work, it was discovered that many RNA viruses activate response, which targets the RNA produced by infecting viruses for sequence-specific degradation (Ratcliff et al. 1997). This RNA based plant defense response is triggered when dsRNA accumulates within cells. This response is transient since it depends on the virus replicating within the plant. This strategy has been demonstrated to be effective in monocots using either Wheat streak mosaic virus (WSMV) or Barley strip mosaic virus (BSMV) as a viral vector for post transcriptional gene silencing (Choi et al. 2000, Holzberg et al.2002). Microarray studies have generated large lists of differentially expressed genes during compatible and incompatible interactions. The great challenge now is to sort through these lists to determine which genes have causal roles in the outcome of the plant-pathogen interactions. Several recent studies have employed BSMV-VIGS to test if genes implicated in disease resistance by introduction of transcription during resistance may have bonafide functions in resistance. VIGS has effective in determining gene function for *Lr1* (Cloutier et al. 2007), *Lr21* (Scofiled et al. 2005), phytoene desaturase (Holzberg et al.2002, Scofield et al. 2005) and recepetor-like kinases (TaRLKs) in stripe rust resistance (Zhou et al. 2007). Although VIGS offer a means for obtaining the rapid determination of gene function, the assay were generally transient and may be limited to silencing of genes to specific tissue where the virus infects. Oikawa et al. (2007) employed BSMV-VIGS to characterize the function of the P23k gene that is unique to monocots. Silencing P23k in barley resulted in asymmetrically shaped leaves with frequent cracks along the margins. These results together with the bundles, supports their assertion that this gene in involved in the synthesis of cell wall polysaccharides and secondary wall formation. The studies conducted by Scofield et al. (2005) and Hein et al. (2005) showed that *Lr21-* and *Mla13-* mediated resistance were abrogated when RAR1, SGT1 and HSP90 were silenced, indicating their essential role in these resistance pathway.

ACKNOWLEDGEMENTS

We would like to thank PI (CI), DWR, Karnal and Prof. H.L. Sharma, Palampur and Prof. N. Rishi, Director Amity University, Delhi for their valuable comments and encouragement during preparation of this

Chapter. Due to space constraint, we could not add up all the references. We also acknowledge partly support by Indian Council of Agricultural Research (ICAR) Young Scientist's scheme.

REFERENCES

Adenot, X., T. Elmayan, D. Lauressergues, S. Boutet, N. Bouche, V. Gasciolliet and H. Vaucheret. 2006. DRB4-dependent *TAS3 trans*-acting siRNAs control leaf morphology through AGO7. *Curr. Biol.* **16**:927–32.

Allen, E.l., Z.l. Xie, A.M.l. Gustafson and J.C. Carrington. 2005. microRNA-directed phasing during *trans*-acting siRNA biogenesis in plants. *Cell.* **121**:207–221.

Andersson, M., M. Melander, P. Pojmark, H. Larrson, L. Bulow and P. Hofvander. 2006. Targeted gene suppression by RNA interference: an efficient method for production of high-amylose potato lines. *J. Biotechnol.* **123**:137–148.

Anjum, F.M., M.R. Khan, A. Din, M. Saeed, I. Pasha and M.U. Arshad. 2007. Wheat gluten: high molecular weight glutenin subunits structure. genetics. and relation to dough elasticity. *J. Food Sci.* **72**:R56–R63.

Arentz-Hansen, E.H., S.N. McAdam, O. Molberg, C. Kristiansen and LM.L. Sollid. 2000. Production of a panel of recombinant gliadins for the characterisation of T cell reactivity in coeliac disease. *Gut.* **461**:46–51.

Baulcombe, D. 2004. RNA silencing in plants. *Nature.* **431**:356–363

Bernstein, E., A.A. Caudy, S.M. Hammond and G.J. Hannon. 2001. Role for a bidentate ribonuclease in the initiation step of RNA interference. *Nature.* **409**:363–366.

Borsani, O., J.l. Zhu, P.E.l. Verslues, R.l. Sunkar and J.K. Zhu. 2005. Endogenous siRNAs derived from a pair of natural *cis*-antisense transcripts regulate salt tolerance in *Arabidopsis*. *Cell.* **123**:1279–1291.

Brodersen, P.l., L.L. Sakvarelidze-Achard, M.l. Bruun-Rasmussen, P.L. Dunoyer, Y.Y.L. Yamamoto, L.L. Sieburth and O. Oliverl. 2008. Widespread translational inhibition by plant miRNAs and siRNAs. *Science.* **320**:1185–90.

Caudy, A.A., R.F. Ketting, S.M. Hammond, A.M. Denli, A.M. Bathoorn and B.B. Tops. 2003. A micrococcal nuclease homologue in RNAi effector complexes. *Nature.* **425**:411–414.

Chan, S.W.L. 2008. Inputs and outputs for chromatin-targeted RNAi. *Trends Plant Sci.* **13**:383-389.

Cloutier, S., B.D. McCallum, C. Loutre, T.W. Banks, T. Wicker, C. Feuillet, B. Keller and M.C. Jordan. 2007. Leaf rust resistance gene *Lr1*, isolated from bread wheat (*Triticum aestivum* L.) is a member of the large psr567 gene family. *Plant Mol. Biol.* **65**:93–106.

Deleris, A.l., J.l. Gallego-Bartolome, J.L. Bao, K.D.l. Kasschau, J.C.L. Carrington and O. Voinnet. 2006. Hierarchical action and inhibition of plant Dicer-like proteins in antiviral defense. *Science.* **313**:68–71.

Doench, J.G., C.P. Petersen and P.A. Sharp. 2003. siRNAs can function as miRNAs. *Genes Dev.* **17**:438–442.

Dong, Z.L., M.l. Han and N. Fedoroff. 2008. The RNA-binding proteins HYL1 and SE promote accurate *in vitro* processing of pri-miRNA by DCL1. *PNAS USA*. **105**:9970–9975.

Douchkov, D. and D. Nowara. 2005. A high-throughput gene-silencing system for the functional assessment of defense-related genes in barley epidermal cells. *Mol. Plant Microbe Interact*. **18**:755–761.

Dunoyer, P.l., C.L. Himber and O. Voinnet. 2005. DICER-LIKE 4 is required for RNA interference and produces the 21-nucleotide small interfering RNA component of the plant cell-to-cell silencing signal. *Nat. Genet*. **37**:1356–1360.

Fang, Y. and D.L. Spector. 2007. Identification of nuclear dicing bodies containing proteins for microRNA biogenesis in living *Arabidopsis* plants. *Curr. Biol*. **17**:818–823.

Fido, R.J.L., F. Bekes, P.W. Gras and A.S. Tatham. 1997. Effects of C. D. I and Y-gliadins on the dough mixing properties of wheat flour. *J. Cereal Sci*. **26**:271–277.

Fire, A., S. Xu, M.K. Montgomery, S.A. Kostas, S.E. Driver and C.C. Mello. 1998. Potent and specific genetic interference by double stranded RNA in *Caenorhabditis elegans*. *Nature*. **97**:1165–11654.

Galvani, A. and S. Linda. 2002. RNA Interference by feeding in *Paramecuim*. *Trends in Genetics*. **18**:11–12.

Guo, S., K.J. Kemphues. 1995. *Par-1* a gene required for establishing polarity in *C. elegans* embryos, encodes a putative expression Ser/Thr kinase that is asymmetrically distributed. *Cell*. **81**:611–620.

Hamilton, A.J. and D.C. Baulcombe. 1999. A species of small antisense RNA in posttranscriptional gene silencing in plants. *Science*. **286**:950–2.

Heilersig, H.J.B., A. Loonen, M. Bergervoet, A.M.A. Wolters and R.G.F. Visser. 2006. Post-transcriptional gene silencing of GBSSI in potato: effects of size and sequence of the inverted repeats. *Plant Mol. Biol*. **60**:647–662.

Hein, I.M., K. Barciszewska-Pacak, S. Hrubikova, M. Williamson, I.E. Dinesen, S. Soenderby, A. Sundar, K. Jarmolowski, Shirasu and C. Lacomme. 2005. Virus-induced gene silencing– based functional characterization of genes associated with powdery mildew resistance in barley. *Plant Physiol*. **138**:2155–2164.

Henderson, I.R., X. Zhang, C. Lu, L. Johnson, B.C. Meyers and P.J. Green and S.E. Jacobsen. 2006. Dissecting Arabidopsis thaliana DICER function in small RNA processing. gene silencing and DNA methylation patterning. *Nat. Genet*. **38**:721–725.

Henz, S.R., J.S. Cumbie, K.D. Kasschau, J.U. Lohmann, J.C. Carrington, D. Weigel and M. Schmid. 2007. Distinct expression patterns of natural antisense transcripts in *Arabidopsis*. *Plant Physiol*. **144**:1247–55.

Hilson, P., J. Allemeersch, T. Altmann, S. Aubourg, A. Avon, J. Beynon, R. Bhalerao, F. Bitton, M. Caboche, B. Cannoot, V. Chardakov, C. Cognet-Holliger, V. Colot, M. Crowe, C. Darimont, S. Durinck, H. Eickhoff, A.F. de Longevialle, E.E. Farmer, M. Grant, M.T. Kuiper, H. Lehrach, C. Léon, A. Leyva, J. Lundeberg, C. Lurin, Y. Moreau, W. Nietfeld, J. Paz-Ares, P. Reymond, P. Rouzé, G. Sandberg, M.D. Segura, C. Serizet, A. Tabrett, L. Taconnat, V. Thareau, P. Van Hummelen, S. Vercruysse, M. Vuylsteke, M. Weingartner, P.J. Weisbeek, V. Wirta, F.R.

Wittink, M. Zabeau and I. Small I. 2004. Versatile gene-specific sequence tags for *Arabidopsis* functional genomics: transcript profiling and reverse genetics applications. Genome Res **14**:2176–2189.

Himmelbach, A., U. Zierold, G. Hensel, J. Riechen, D. Douchkov, P. Schweiser and J. Kumlehn. 2007. A set of modular binary vectors for transformation of cereals. *Plant Physiol.* **145**:1192–1200.

Holzberg, S., P. Brosio, C. Gross and G.P. Pogue. 2002. Barley stripe mosaic virus–induced gene silencing in a monocot plant. *Plant J.* **30**:315–327.

Houmard, N.M., J.L. Mainville, C.P. Bonin, S. Huang, M.H. Luethy and T.M. Malvar. 2007. High-lysine corn generated by endosperm specific suppression of lysine catabolism using RNAi. *Plant Biotechnol J.* **5**:605–614.

Howell, M.D., N. Fahlgren, E.J. Chapman, J.S. Cumbie, C.M. Sullivan, S.A. Givan, K.D. Kasschau, J.C. Carrington. 2007. Genome-wide analysis of the RNA-DEPENDENT RNA POLYMERASE6/DICER-LIKE4 pathway in *Arabidopsis* reveals dependency on miRNA- and tasiRNA-directed targeting. *Plant Cell.* **19**:926–942.

Huang, G., R. Allen, E. Davis, T.J. Baum and R.S. Hussey. 2006. Engineering broad root-knot resistance in transgenic plants by RNAi silencing of a conserved and essential root-knot nematode parasitism gene. *Proc. Natl. Acad. Sci. USA.* **103**:14302–14306.

Hutvagner, G. and P.D. Zamore. 2002. A microRNA in a multiple turnover RNAi enzyme complex. *Science.* **297**:2056–2060.

Jobling, S. 2004. Improved starch for food and industrial applications. *Curr. Opin. Plant Biol.* **7**:210–218.

Jones-Rhoades, M.W., D.P. Bartel and B. Bartel. 2006. MicroRNAs and their regulatory roles in plants. *Ann. Rev. Plant Biol.* **57**:19–53.

Karimi M., D. Inzé and A. Depicker. 2002. GATEWAY vectors for *Agrobacterium*-mediated plant transformation. *Trends Plant Sci.* **7**:193–195.

Kasschau, K.D. and J.C. Carrington. 1998. A counterdefense strategy of plant viruses: suppression of posttranscriptional analysis of *Caenorhabditis elegans* genome using RNAi. *Nature.* **421**:231–237.

Kidner, C.A. and R.A. Martienssen. 2005. The developmental role of microRNA in plants. *Curr. Opin. Plant Biol.* **8**:38–44.

Kim, K.N., D.K. Fisher, M. Gao and M.J. Guiltinan. 1998. Molecular cloning and characterization of the *Amylose-Extender* gene encoding starch branching enzyme IIB in maize. *Plant Mol. Biol.* **38**:945–956.

Kurihara, Y., Y. Takashi and Y. Watanabe. 2006. The interaction between DCL1 and HYL1 is important for efficient and precise processing of pri-miRNA in plant microRNA biogenesis. *RNA.* **12**:206-12.

Lindbo, J.A. and W.G. Dougherty. 1992. Untranslatable transcripts of the tobacco etch virus coat protein gene sequence can interfere with tobacco etch virus replication in transgenic plants and protoplasts. *Virology.* **189**:725–733.

Liu, Q., Y. Feng and Z. Zhu. 2009. Dicer-like (DCL) proteins in plants. *Funct. Integr. Genomics.* **9**:277–286.

Llave, C. 2004. MicroRNAs: more than a role in plant development. *Mol. Plant Pathol.* **5:**361–366.

Matzke, M.A., T. Kanno, B. Huettel, L. Daxinger and A.J.M. Matzke. 2007. Targets of RNA-directed DNA methylation. *Curr. Opinion Plant Biol.* **10:**512–519.

Meister, G. and T. Tuschl. 2004. Mechanisms of gene silencing by double stranded RNA. *Nature.* **431:**343–349.

Metakovsky, E.V., A.Y. Novoselskaya and A.A. Sozinov. 1984. Genetic analysis of gliadin components in winter wheat using two-dimensional polyacrylamide gel electrophoresis. *Theor. App. Gen.* **69:**31–37.

Miki, D. and K. Shimamoto. 2004. Simple RNAi vectors for stable and transient suppression of gene function in rice. *Plant Cell. Physiol.* **45:**490–495

Millar, A.A. and P.M. Waterhouse. 2005. Plant and animal microRNAs: similarities and differences. *Funct. Integr. Genomics.* **5:**129–135.

Mittal, V. 2004. Improving the efficacy of RNA interference in mammals. *Nat. Rev. Genet.* **5:**355-365.

Mlotshwa, S., G.J. Pruss, A. Peragine, M.W. Endres, J. Li, X. Chen, R.S. Poethig, L.H. Bowman and V. Vance. 2008. DICER-LIKE2 plays a primary role in transitive silencing of transgenes in Arabidopsis. *PLOS One.* **3:**e1755.

Moissiard G., E.A. Parizotto, C. Himber and O. Voinnet. 2007. Transitivity in Arabidopsis can be primed. requires the redundant action of the antiviral Dicer-like 4 and Dicer-like 2 and is compromised by viral-encoded suppressor proteins. *RNA.* **13:**1268–1278.

Napoli, C., C. Lemieux and R. Jorgensen. 1990. Introduction ofna chemeric chalcone synthase gene into petunia results in reversible co-suppression of homologous genes in trans. *Cell.* **2:**279–289.

Oikawa, A., A. Rahman, T. Yamashita, H. Taira and S. Kidou. 2007. Virus-induced gene silencing of P23k in barley leaf reveals morphological changes involved in secondary wall formation. *J. Exp. Bot.* **58:**2617–2625.

Onodera, Y., J.R. Haag, T. Ream, P.C. Nunes, O. Pontes and C.S. Pikaard. 2005. Plant nuclear RNA polymerase IV mediates siRNA and DNA methylation-dependent heterochromatin formation. *Cell.* **120:**613–622.

Otani, M., T. Hamada, K. Katayama, K. Kitahara, S.H. Kim, Y. Takahata, T. Saganuma and T. Shimada. 2007. Inhibition of the gene expression for granule-bound starch synthase I by RNA interference in sweet potato plants. *Plant Cell. Rep.* **26:**1801–1807.

Peragine, A., M. Yoshikawa, G. Wu, H.L. Albrecht and R.S. Poethig. 2004. *SGS3* and *SGS2/SDE1/RDR6* are required for juvenile development and the production of *trans*-acting siRNAs in *Arabidopsis. Genes Develop.* **18:**2368–79.

Pontes, O., C.F. Li, P.C. Nunes, J. Haag, T. Ream, A. Vitins, T. Lagrange, S. Craig,

C.S. Pikaard and J.E. Jacobsen. 2006. The *Arabidopsis* chromatin-modifying nuclear siRNA pathway involves a nucleolar RNA processing center. *Cell.* **126:**79–92.

Qi, Y., A.M. Denli and G.J. Hannon. 2005. Biochemical specialization within *Arabidopsis* RNA silencing pathways. *Molecular Cell.* **19:**421–428.

Ramachandran, V. and X. Chen. 2008. Small RNA metabolism in *Arabidopsis. Trends Plant Sci.* **13:**368–384.

Ratcliff, F., B.D. Harrison and D.C. Baulcombe. 1997. A similarity between viral defense and gene silencing in plants. *Science.* **276**:1558–1560.

Regina, A., A. Bird, D. Topping, S. Bowden, J. Freeman, T. Barsby, B. Kosar-Hashemi, Z. Li, S. Rahman and M. Morell. 2006. High-amylose wheat generated by RNA interference improves indices of large-bowel health in rats. *Proc. Natl. Acad. Sci. USA.* **103**:3546–3551.

Scofield, S.R., L. Huang, A.S. Brandt and B.S. Gill. 2005. Development of a virus-induced gene-silencing system for hexaploid wheat and its use in functional analysis of the Lr21-mediated leaf rust resistance pathway. *Plant Physiol.* **138**:2165–2173.

Segal, G., R. Song and J. Messing. 2003. A new opaque variant of maize by a single dominant RNAinterference- inducing transgene. *Genetics.* **165**:387–397.

Shan, L., O. Molberg, I. Parrot, F. Hausch, F. Filiz, G.M. Gray, L.M. Sollid and C. Khosla. 2002. Structural basis for gluten intolerance in celiac sprue. *Science.* **297**:2275–2279.

Shuey, D., D. McCallus and G. Tony. 2002. RNAi: gene silencing in therapeutic intervention. *Drug Discov. Today.* **7**:1040–1046.

Song, L., M.H. Han, J. Lesicka and N. Fedoroff. 2007. *Arabidopsis* primary microRNA processing proteins HYL1 and DCL1 define a nuclear body distinct from the Cajal body. *PNAS USA.* **104**:5437–5442.

Tabara, H., A. Grishok and C.C. Mello. 1998. RNAi in *C. elegans*: soaking in the genome sequence. *Science.* **282**:430–431.

Topping, D.L., M. Fukushima and A.R. Bird. 2003. Resistant starch as a prebiotic and synbiotic: state of the art. *Proc. Nut. Soc.* **62**:171–176.

Travella, S., T.E. Klimm and B. Keller. 2006. RNA interference based gene silencing as an efficient tool for functional genomics in hexaploid bread wheat. *Plant Physiol.* **142(1)**:6–20.

Uauy, C., A. Distelfeld, T. Fahima, A. Blechl and J.A. Dubcovsky. 2006. A NAC Gene regulating senescence improves grain protein, zinc and iron content in wheat. *Science.* **314(5803)**:1298–301.

Vaucheret, H. 2006. Post-transcriptional small RNA pathways in plants: mechanisms and regulations. *Genes Dev.* **20**:759–771.

Wesley, S.V., C.A. Helliwell, N.A. Smith, M.B. Wang, D.T. Rouse, Q. Liu, P.S. Gooding and S.P. Singh. 2001. Construct designs for efficient effective and high-throughput gene silencing in plants. *Plant J.* **27**:581–590.

Wielopolska, A., H. Townley, I. Moore, P. Waterhouse and C. Helliwell. 2005. A high-throughput inducible RNAi vector for plants. *Plant Biotechnol. J.* **3**:583–590.

Xie, Z., E. Allen, A. Wilken and J.C. Carrington. 2005. DICER-LIKE 4 functions in trans-acting small interfering RNA biogenesis and vegetative phase change in *Arabidopsis thaliana. PNAS USA.* **102**:12984–12989.

Xie, Z., L.K. Johansen, A.M. Gustafson, K.D. Kasschau, A.D. Lellis, D. Zilberman, S.E. Jackobson and J.C. Carrington. 2004. Genetic and functional diversification of small RNA pathways in plants. *PLoS. Biol.* **2**:e104.

Yan, F., J. Peng, Y. Lu, L. Lin, H. Zheng, H. Chen, J. Chen and M.J. Adams. 2009. Molecular cloning and characterization of *Dicer-like* e gene from *Brassica rapa. Mol. Biol. Rep.* **36**:1283–1289.

Yang, Z., Y.W. Ebright, B. Yu, X. Chen. 2006. HEN1 recognizes 21–24 nt small RNA duplexes and deposits a methyl group onto the 2′ OH of the 3′ terminal nucleotide. *Nucleic Acids Res.* **34:**667–675.

Yue, S.J., H. Li and Y.W. Li. 2008. Generation of transgenic wheat lines with altered expression levels of 1Dx5 high-molecular weight glutenin subunit by RNA interference. *J. Cereal Sci.* **47:**153–161.

Zamore, P.D. 2002. Ancient pathways programmed by small RNAs. *Science.* **296:**1265–1269.

Zamore, P.D., T. Tuschl, P.A. Sharp and D.P. Bartel. 2000. Double-stranded RNA directs the APT-dependent cleavage of mRNA at 21–23 nucleotide intervals. *Cell.* **101:**23–33.

Zeng, Y., R. Yi and B.R. Cullen. 2003. MicroRNAs and small interfering RNAs can inhibit mRNA expression by similar mechanisms. *Proc. Natl. Acad. Sci. USA.* **100:**9779–9784.

Zhang, X., I.R. Henderson, C. Lu, P.J. Green and S.E. Jacobsen. 2007. Role of RNA polymerase IV in plant small RNA metabolism. *PNAS USA.* **104:**4536–4541.

Zhou, H., S. Li, Z. Deng, X. Wang, T. Chen, J. Zhang, S. Chen, H. Ling, A. Zhang, D. Wang and X. Zhang. 2007. Molecular analysis of three new receptor-like kinase genes from hexaploid wheat and evidence for their participation in the wheat hypersensitive response to stripe rust fungus infection. *Plant J.* **52:**420–434.

CHAPTER 16

Biogenesis, Processing and Function of miRNA in Plants and Animals

Sakshi Issar and R.K. Gaur*

ABSTRACT

miRNA are single stranded molecules, ~21–23 nucleotide long that play important roles through regulating gene expression in many biological and metabolic processes of plants and animals like—tissue identity, developmental timing, response to environmental stress, plant-microbe interaction, apoptosis, fat metabolism, cell differentiation etc. It has been found that miRNA plays very efficient role in the regulation of nutritional balance in plants as well as also take an effectual part in rhizobia-legume symbiosis.

Keywords: miRNA, biogenesis, ribonuclease, drosha, DICER.

INTRODUCTION

MicroRNAs (miRNAs), initially discovered in *C. elegans* (Lee et al. 2001), are a large group of small endogenous RNAs (Ambros 2004, Bartel 2004, Carrington et al. 2003), that widely exist in animals (Tongwen et al. 2007), plants (Yi et al. 2003) and in some viruses (Cullen 2006). Increasing evidence demonstrates that miRNAs play an important function in many biological and metabolic processes, including tissue identity, developmental timing and response to environmental stress (Jones-Rhodes et al. 2006, Zhang et al. 2006).

MicroRNAs (miRNAs) are single strand RNA molecules ~21–23 nucleotides long that play important roles in many biological processes

---

* Corresponding author *e-mail*: gaurrajarshi@hotmail.com

through regulating gene expression (Ambros 2004). Each miRNA is thought to regulate multiple genes and since hundreds of miRNA genes are predicted to be present in higher eukaryotes (Lim et al. 2003b) Several research groups have provided evidence that miRNAs may act as key regulators of processes as diverse as early development (Reinhart et al. 2000), cell proliferation and cell death (Brennecke et al. 2003), apoptosis and fat metabolism (Xu et al. 2003) and cell differentiation (Chen 2003, Dostie et al. 2003). The biogenesis and functional mechanism of miRNAs are similar in both animals and plants. A majority of the miRNA genes exists as independent transcriptional units and are transcribed by RNA polymerase II into long primary transcripts (termed pri-miRNAs) (V.N. Kim 2005). Then, the miRNA precursor is processed by ribonuclease III enzymes (Drosha and Dicer in animals and DICER-LIKE1 in Arabidopsis) to release the mature miRNA that is located in a hairpin structure within the primary transcript. The miRNA is subsequently incorporated into an RNA induced silencing complex (RISC), where it guides the cleavage or translational repression of its target mRNA by base-pairing with the target (Bartel 2004). Small interfering RNAs (siRNAs), which originate from transcripts from transgenes, endogenous repeat sequences or transposons, are another type of small RNAs. They are similar in structure, biogenesis and functional mechanism to miRNAs (Bartel 2004, Hamilton et al. 1999). The main difference is that miRNAs come from endogenous transcripts that can form local hairpin structures, while siRNAs come from perfect pairing, long exogenous or endogenous dsRNA transcripts (Bartel 2004, Hamilton et al. 1999).

miRNAs were first discovered as components of the heterochronic pathway in *Caenorhabditis elegans* (Lee et al. 1993) Although miRNA sequences are not conserved between animals and plants, they are highly conserved within each kingdom and this evolutionary conservation is one of their defining characteristics (Ambros et al. 2003). Many miRNAs in animals were identified to be involved in a wide variety of roles such as developmental timing, cell death, cell proliferation, haematopoiesis, patterning of the nervous system and even response to environmental stimuli (V. Ambros 2004). Plant miRNAs have been predicted or confirmed to regulate a wider variety of developmental and physiological processes than animal miRNAs (Dugas et al. 2004).

Since the first discovery of miRNAs in Arabidopsis in 2002 (Reinhart et al. 2002, Tongwen et al. 2007), more than 700 plant miRNAs have been identified using three primary strategies: direct cloning from small RNA libraries with different tissues and conditions; bioinformatic prediction from existing genomic database, based on sequence conservation of plant miRNAs and on the ability of the surrounding region to adopt a fold-back structure; and traditional mutant screen method that led to the

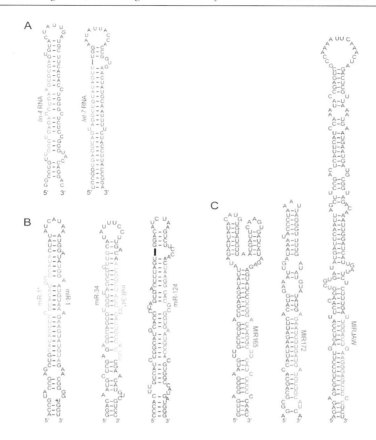

Figure 1: Examples of Metazoan miRNAs Shown are predicted stem loops involving the mature miRNAs (red) and flanking sequence. The miRNAs* (blue) are also shown in cases where they have been experimentally identified (Lim et al. 2003a).

(A) Predicted stem loops of the founding miRNAs, *lin-4* and *let-7* RNAs (Lai et al. 2003, Olsen et al. 1999). The precise sequences of the mature miRNAs were defined by cloning (Lau et al. 2001). Shown are the *C. elegans* stem loops, but close homologs of both have been found in flies and mammals (Lagos-Quintana et al. 2002, Lagos-Quintana et al. 2001, Pasquinelli et al. 2000).

(B) Examples of miRNAs from other metazoan genes, *mir-1*, *mir-34* and *mir-124*. Shown are the *C. elegans* stem loops, but close homologs of these miRNAs have been found in flies and mammals (Quintana et al. 2002, Lagos-Quintana et al. 2001, Lau et al. 2001, Lee et al. 2001).

(C) Examples of miRNAs from plant genes, *MIR165a*, *MIR172a2* and *JAW*. Shown are *Arabidopsis* stem loops, but close homologs of these miRNAs have been found in rice and other plants (Palatnik et al. 2003, Park et al. 2002, Reinhart et al. 2002). (Tongwen et al. 2007).

(Color image of this figure appears in the color plate section at the end of the book.)

first identification of small RNAs in *C. elegans* (Lee et al. 1993). A useful database of miRNAs has been constructed through the Sanger Institute and miRNA annotated sequences are available in the Rfam miRNA registry (http:// www.sanger.ac.uk/cgi-bin/Rfam/mirna/browse.pl) (Axtell et al. 2005). According to the miRNA Registry Database (Release 8.0, February 2006), 731 miRNA genes have been identified in various plants, including 117 from Arabidopsis, 178 from rice (*Oryza sativa*), 97 from Zea may and the rest from *Populus trichocarpa, Saccharum officinarum, Sorghum bicolor, Medicago truncatula, Glycine max*, respectively (http://www. sanger. ac.uk/cgi-bin/Rfam/mirna/browse.pl) (Griffiths-Jones 2004, Tongwen et al. 2007).

Though similar to animal miRNAs in general, plant miRNAs show several substantial differences. First, plant pre-miRNAs have larger and more variable stem-loop structures. Second, mature plant miRNAs pair their target sites with near-perfect complementarity. Third, in animals, miRNAs usually recognize several target sequences in the 30 UTR of mRNAs and cause translation inhibition, while plant miRNAs often recognize a single target site in the coding region and guide the mRNA to cleavage. The specificity of plant miRNA targeting in coding sequences with fewer mismatches suggests that plant miRNAs may act more like siRNAs than do animal miRNAs (Bartel 2004, Tongwen et al. 2007).

BIOGENESIS OF miRNA

According to the current convention, a miRNA is defined as 19–25 nucleotides in length, which is generated by the RNase-III-type enzyme Dicer from an endogenous transcript that contains a local hairpin structure (Lim et al. 2005). This evidence brings to attend the transcription miRNA genes are mediated by the RNA polymerase II (poly II). The primary transcripts of miRNAs (pri-miRNAs) contain a 5′ 7-methyl guanosine cap and a 3′ poly-A tail, modifications that are trademarks of the poly II transcription (Aravin et al. 2003, Bartel 2004, He et al. 2004). The association of miRNA with poly II, the miRNAs are under elaborate control of various regulation factors in different development stages and tissues. It is becoming evident that miRNA expression profiles are indeed complicated. It should bear in mind, though, that there is still possibility that a small portion of miRNAs are transcribed by poly III. And so far the DNA elements that are common to most poly II promoters such as that TATA box and the TFIIB recognition elements have not been identified for miRNAs (http://www.ambion.com/ techlib/resources/miRNA/mirna_fun.html). Overviews of the model of miRNA biogenesis (He et al. 2004) (shown in figure 2), miRNA genes are transcribed by RNA polymerase II (poly II) to generate the primary transcript (pri-miRNAs). The initiation step ('cropping') is mediated by the Drosha complex (also known as the Microprocessor complex). Drosha is located mainly in the nucleus. The product of this unclear processing step is a ~70-nucleotide (nt) pre-miRNA, which possesses a short stem plays a ~2-nucleotide 3' overhang. This structure might serve as signature motif that is recognized by the nuclear export factor exportin-5. Pre-miRNA constitutes a transport complex together with exportin-5 and it cofactor Ran (the GTP-bound form). Following export, the cytoplasmic RNase III Dicer participates in the stand is selected as the mature miRNA, whereas the other strand is degraded. Pre-miRNAs are transported to the cytoplasm by exportin 5 and are processed into miRNA:miRNA duplexes by Dicer. Dicer also processes long dsRNA molecules into small interfering RNA (siRNA) duplexes. Only one strand of the miRNA: miRNA duplex or the siRNA duplex is preferentially assembled into the RNA-induced silencing complex (RISC), which subsequently acts on its

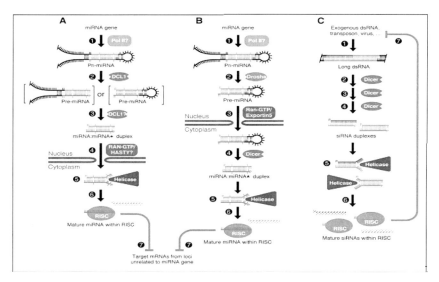

Figure 2: The Biogenesis of miRNAs and siRNAs (Bartel 2004).

(A) The biogenesis of a plant miRNA (steps 1–6) and its hetero-silencing of loci unrelated to that from which it originated (step 7). The pre-miRNA intermediates (bracketed), thought to be very short-lived, have not been isolated in plants. The miRNA (red) is incorporated into the RISC (step 6), whereas the miRNA* (blue) is degraded (hatched segment). A monophosphate (P) marks the 5_ terminus of each fragment.

(B) The biogenesis of a metazoan miRNA (steps 1–6) and it's hetero-silencing of loci unrelated to that from which it originated (step 7).

(C) The biogenesis of animal siRNAs (steps 1–6; see text for details) and their auto-silencing of the same (or similar) loci from which they originated (step 7). (Bartel 2004).

(Color image of this figure appears in the color plate section at the end of the book.)

target by translational repression or mRNA cleavage, depending, at least in part, on the level of complementarily between the small RNA and its target.

PROCESSING OF miRNA

A number of miRNAs are encoded in introns of primary mRNA transcripts. Typically they are encoded in the same orientation as the parent transcript, indicating that transcription of this class of miRNA gene is driven by an mRNA promoter (Aravin et al. 2003, Lagos-Quintana et al. 2002, Lai et al. 2003, Lim et al. 2003a).

The excision and activation of active single-stranded miRNAs from precursor transcript occurs through a multi-step process described below (Figure 3).

1. Transcription

miRNAs are initially expressed as part of transcripts termed primary miRNAs (pri-miRNAs) (Lee et al. 2002). They are apparently transcribed by RNA Polymerase II and include 5′ caps and 3′ poly (A) tails (Cai et al. 2004, Smalheiser et al. 2003). The miRNA portion of the

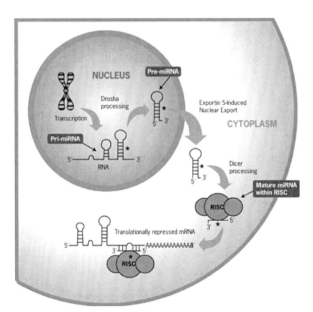

Figure 3: miRNA Processing and Activity (http://www.ambion.com/techlib/resources/miRNA/mirna_pro.html).

(Color image of this figure appears in the color plate section at the end of the book.)

pri-miRNA transcript likely forms a hairpin with signals for dsRNA-specific nuclease cleavage.

2. Hairpin release in the nucleus

The dsRNA-specific ribonuclease Drosha digests the pri-miRNA in the nucleus to release hairpin, precursor miRNA (pre-miRNA) (Lee et al. 2002). Pre-miRNAs appear to be approximately 70 nt RNAs with 1–4 nt 3′ overhangs, 25–30 bp stems and relatively small loops. Drosha also generates either the 5′ or 3′ end of the mature miRNA, depending on which strand of the pre-miRNA is selected by RISC (Lee et al. 2002, Yi et al. 2003).

3. Export to the cytoplasm

Exportin-5 (Exp5) seems to be responsible for export of pre-miRNAs from the nucleus to the cytoplasm. Exp5 has been shown to bind directly and specifically to correctly processed pre-miRNAs. It is required for miRNA biogenesis, with a probable role in coordination of nuclear and cytoplasmic processing steps (Liu et al. 2009, Yi et al. 2003).

4. Dicer processing

Dicer is a member of the RNase III superfamily of bidentate nucleases that has been implicated in RNA interference in nematodes, insects and plants. Once in the cytoplasm, Dicer cleaves the pre-miRNA approximately 19 bp from the Drosha cut site (Lee et al. 2002, Xu et al. 2003). The resulting double-stranded RNA has 1–4 nt 3′ overhangs at either end (Lund et al. 2003). Only one of the two strands is the mature miRNA; some mature miRNAs derive from the leading strand of the pri-miRNA transcript and with other miRNAs the lagging strand is the mature miRNA.

5. Strand selection by RISC

To control the translation of target mRNAs, the double-stranded RNA produced by Dicer must strand separate and the single-stranded mature miRNA must associate with the RISC. Selection of the active strand from the dsRNA appears to be based primarily on the stability of the termini of the two ends of the dsRNA (Khvorova et al. 2003, Schwarz et al. 2003). The strand with lower stability base pairing of the 2–4 nt at the 5′ end of the duplex preferentially associates with RISC and thus becomes the active miRNA (Schwarz et al. 2003).

MODE OF ACTION OF miRNA

Virtually all of the miRNAs that have been studied in animals reduce steady state protein levels for the targeted gene(s) without impacting the corresponding levels of mRNA (Olsan et al. 1999). The mechanism by which miRNAs reduce protein levels is not fully understood, but one study involving the *C. elegans* lin-4 miRNA/lin-14 mRNA pair indicates that lin-4 miRNA does not affect the poly (A) tail length, transport to the cytoplasm, nor entry into polysomes of the lin-14 mRNA (Olsan et al. 1999).

Plant miRNAs differ from animal miRNAs in that many plant miRNAs have perfect homology to their target mRNAs and they act through the RNAi pathway to cause mRNA degradation (Rhoades et al. 2002). It is likely, however, that some plant miRNAs base-pair imperfectly with their miRNA target sites and act via a pathway similar to animal miRNAs (Figure 4). In plants and yeast there is also evidence that miRNAs are involved in repression of transcription by guiding chromatin methylation.

A key observation made by two laboratories is that mRNAs containing multiple, non-overlapping miRNA binding sites are more responsive to miRNA-induced translational repression than those containing a single miRNA binding site (Doench et al. 2003, Zeng et al. 2003). Furthermore, comparisons of repression by miRNAs bound to 2, 4 and 6 binding sites on a reporter construct indicate that translation decreases with each additional site (Zeng et al. 2003). This suggests that the expression of

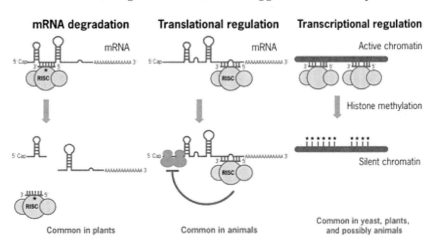

Figure 4: Mode of Action of miRNAs in Plants and Animals ((http://www.ambion. com/ techlib/resources/miRNA/mirna_pro.html)

(Color image of this figure appears in the color plate section at the end of the book.)

miRNA target genes can be fine-tuned in animals (and potentially plants) by altering the concentrations or identities of miRNAs within cells. This observation coupled with the predictions that many mRNAs have target sites for many different miRNAs suggests that gene expression in various tissues and cells can be greatly influenced by the miRNA populations in those cells. This could also explain why at least some miRNAs have such broad functionality and conversely why translational control of some genes is so complex. If miRNAs indeed regulate the translation of, but not the stability of target mRNAs, this might at least partially explain why gene expression profiles based on mRNA analysis do not always correlate with protein expression data (Kern et al. 2003).

FUNCTIONS OF miRNA

miRNA function as guide molecules in post-transcriptional genes silencing by base pairing with target mRNAs, which lead to mRNA cleavage or translational repression. With >200 members pre species in higher eukaryotes, miRNAs are one of the largest gene families, accounting for ~1% of the genome (He et al. 2004), recent studies have revealed that miRNAs have key roles in diverse regulatory pathways, including control of developmental timing, Haematopoietic cell differentiation, cell signaling, apoptosis, cancer and diseases and cell proliferation & organ development and stem cell (http://omics.org/index.php/MicroRNAome).

A. In Plants

In plants, miRNAs are implicated in diverse aspects of plant growth and development, including leaf morphology and polarity, lateral root formation, hormone signaling, transition from juvenile to adult vegetative phase and vegetative to flowering phase, flowering time, floral organ identity and reproduction (Mallory et al. 2006), Sunker et al. 2008). Several miRNAs are regulated in response to diverse stress conditions, which suggests that miRNA-directed post-transcriptional regulation of their respective target genes is important to cope with the stress (Chiou et al. 2006, Fahlgren et al. 2007, Fujii et al. 2005, Jones-Rhodes et al. 2004, Lu et al. 2005, Sunker et al. 2006, Sunker et al. 2007).

In plants, miRNAs have a propensity to pair to mRNAs with near-perfect complementarity, enabling convincing targets to be readily predicted for most known plant miRNAs (Bartel et al. 2003, Rhoades et al. 2002). Evolutionary conservation of the miRNA: mRNA pairing in *Arabidopsis* and rice, together with experimental evidence showing that miRNAs can direct cleavage of targeted mRNAs, supports the validity of these predictions (Kasschau et al. 2003, Llave et al. 2002a, Rhoades et al.

2002, Tang et al. 2003). The expression of several flowering plant miRNAs was detected in a gymnosperm and a fern, two (miR160 and miR390) were detected in a moss and one (miR160) was cloned from the moss *Polytrichum juniperinum* demonstrating that many plant miRNAs have remained essentially unchanged since before the emergence of flowering plants (Axtell et al. 2005). More recently, a microarray miRNA expression profile revealed the deep conservation of any plant miRNA families, with at least eight families conserved since before the emergence of seed plants (Axtell et al. 2005), Bartel 2004, Sunker et al. 2008, Tongwen et al. 2007).

B. In Animals

The phenotype of *mir-48; mir-84; mir-241* mutants in *C. elegan* is one of cellular overgrowth (Abbot et al. 2005). In *D. rerio*, the zygotic removal of Dicer results in a larval growth arrest (Wienhold et al. 2005). In additional, in *M. Musculus*, removal of Dicer in the limb mesoderm leads to a dramatic programmed cell death in the developing limb (Calin et al. 2004). *Drosophila* bantam encodes a 21 nucleotide regulatory miRNA that has an anti-apoptotic (Bartel et al. 2003) and also regulated growth (Abbot et al. 2005). Interest in the genes controlling developmental timing in *C. elegans* led to the cloning of the first miRNA, *line-4* and the identification of the first miRNA target, *lin-14* (Wightman et al. 1993). The developmental-timing, or heterochronic, pathway regulates stage-specific processes during *C. elegans* larval development. For example of the *C. elegans* life form, only at the adult stage, in line-4 mutant animals, the seam cells (the developmental fate of several stem cells in the lateral hypodermis) repeat the cell division pattern that characterized the first larval stage (L1) and fail to differentiate. This mutant phenotype has been interpreted as a heterochronic change with the developmental clock being stuck at the L1 stage. This result indicated that gain-of-function mutations in the *line-4* miRNA lead to an identifiable phenotype, whereas loss-of-function mutations in *line-14* result in an opposite where the seam cells skip the cell division of the first larval stage (Lau et al. 2001). Three miRNAs of the *let-7* family, *mir-48*, *mir-84* and *mir-241* are act redundantly to control the next developmental transition, from the second larval stage (L2) to the third larval stage (L3) (Abbot et al. 2005, Reinhart et al. 2000, D.P. Bartel 2004).

miRNA IN LEGUMES

Symbiotic association between leguminous plants and rhizobia bacteria results in specialized nitrogen-fixing structures called root nodules. The interaction between the symbiotic partners starts with the exchange of chemical signals. Legumes release specific flavonoids (a group of small phenolic compounds) as signal molecules into the soil through root exudates. Compatible rhizobia bacteria respond by producing specific lipochitooligosaccharide (LCO) bacterial signals which are in turn recognized by plants (S.R. Long 1989, G.B. Rolfe 1988, J.E. Cooper 2007) resulting in the attachment of bacterial cells to plant root hairs (Subramanian et al. 2008).

miRNA regulate nutritional balance in legumes. One of the earliest examples of a miRNA involved in nutrient sensing was miR395, whose expression increased with sulfate starvation (Jones-Rhoades et al. 2004). MiR395 was implicated in regulating sulfate homeostasis because it targets ATP sulfurylases and a low-affinity sulfate transporter (Allen et al. 2005, Jones-Rhoades et al. 2004). miRNA also responses to microbial challenge in leguminous plants. In the case of pathogen attack, the regulation of miRNAs leads to gene expression reprogramming as the sRNAs subsequently regulate the expression of genes involved in the defense response (H. Jin 2008). The Arabidopsis miR393 was the first miRNA shown to play an important role in plant immune responses. Based on miRNA target prediction, many of the targets were growth and disease-related transcripts, which suggests that the miRNA regulation created a tempered response that allowed for miRNA mediated repression of both positive and negative defense regulators to restrict organ development and pathogen growth (Stacey et al. 2009). At the time of nodulation, miRNA also plays an important role in leguminous plants. To identify potential regulators of the earliest stages of nodule development Subramanian et al. inoculated soybean roots with *Bradyrhizobium japonicum* and identified miRNAs 3 h post-inoculation. They found 20 conserved miRNA families and 35 novel miRNAs (Subramanian et al. 2008).

There is strong evidence that fluxes in hormone concentrations are critical for nodule formation and development (Ding et al. 2009). Plant hormones are regulators of development and growth and some miRNAs facilitate hormone-induced responses (Table I). One of the earliest reports that linked miRNAs and hormones was from a study in Arabidopsis describing a mutation in a double-stranded RNA-binding protein, HYL1, that caused decreased sensitivity to cytokinin, as well as an altered response to ABA and auxin (Lu et al. 2000). Some miRNAs implicated in regulation of auxin signaling target transcripts encoding ARFs. For example, miR167 in Arabidopsis and in cultured rice (*Oryza sativa*) cells targets ARFs, as does miR160 in Arabidopsis (Liu et al. 2009). Several miRNAs are regulated by more than one plant hormone and this may reflect a convergence of regulatory pathways or cross talk between pathways (Stacey et al. 2009).

Table 1: Interaction between miRNAs and hormones in plant development (Stacey et al. 2009)

miRNA	Hormone Signaling Involving a miRNA	Plant Developmental Process Mediated by miRNA
miR159	ABA, ethylene, GA	Germination, flower development
miR160	Auxin	Determination of leaf shape, hypocotyl elongation, flower development and reproduction (fertility)
miR164	Auxin	Root development
miR167	ABA and auxin	Reproduction (fertility)
miR172	Cytokinin	Flower development
miR319	ABA, cytokinin, GA, JA	Determination of leaf shape
miR393	ABA, auxin	Root development

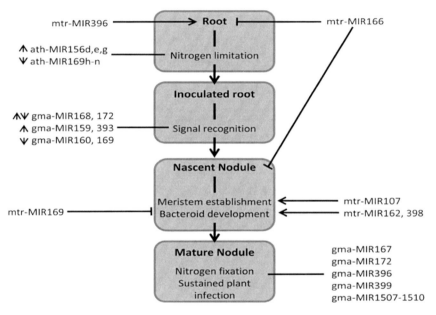

Figure 5: Model of miRNA interactions in symbiotic root nodule development. The interactions are inferred from the current knowledge of the distribution and abundance of miRNA in libraries of developing nodules. Lines indicate points of potential miRNA control of nodulation in soybean (gma) and *M. truncatula* (mtr) and the potential role of miRNAs in nitrogen homeostasis are extrapolated from Arabidopsis (ath) miRNAs; arrows indicate promotion of stage, blunted line indicate repression of stage; plain line indicates that the miRNA is present, but basic roles are uncharacterized. Small arrowheads indicate changes in levels of miRNA levels in response to specific events. Vertical arrows connecting boxed areas indicate the developmental progression from root to mature root nodule. (Stacey et al. 2009)

FUTURE PROSPECT

To date, there are thousands of miRNA which have been discovered and thousands of researches have also been performed but still there are some questions which have not been answered yet. So, by keeping above view in mind, we can unexplore the functions of miRNA in legume symbiosis, in plant-microbe interaction networks, anti-tumor activities in plants, animals as well as in humans.

ACKNOWLEDGEMENTS

The authors are thankful to Department of Biotechnology (DBT), India and Department of Science and Technology (DST), India for financial support for the present studies.

REFERENCES

Abbot, A. et al. 2005. The let-7 microRNA family member's mir-48, mir-84 and mir-241 function together to regulate developmental timing in *C. elegans*. *Dev. Cell*. **9**:403–414.

Allen, E., Z. Xie, A.M. Gustafson and J.C. Carrington. 2005. microRNA- directed phasing during trans-acting siRNA biogenesis in plants. *Cell*. **121**:207–221.

Ambros, V. 2001. microRNAs: Tiny regulators with great potential. *Cell*. **107(7)**:823–826.

Ambros, V. 2004. The functions of animal microRNAs. *Nature*. **431**:350–355.

Ambros, V., B. Bartel, D. P. Bartel, B.C. Burge, C.J.Carrington and X. Chen et al. 2003. A uniform system for microRNA annotation. *RNA*. **9**:277–279.

Aravin, A.A., M. Lagos-Quintana, A. Yalcin, M. Zavolan, D. Marks, B. Snyder, T. Gaasterland, J. Meyer and T. Tuschl. 2003. The small RNA profile during Drosophila melanogaster development. *Dev. Cell*. **5**:337–350.

Axtell, J.M. and D.P. Bartel. 2005. Antiquity of microRNAs and their targets in land plants. *Plant Cell*. **17**:1658–1673.

Bartel, B. and D.P. Bartel. 2003. MicroRNAs: At the root of plant development. *Plant Physiol*. **132**:709–717.

Bartel, D.P. 2004. MicroRNAs: genomics, biogenesis, mechanism and function. *Cell*. **116(2)**:281–297 (David P. Bartel *Cell*. Vol. 116, 281–297, January 23, 2004).

Brennecke, J., D.R. Hipfner, A. Stark, R.B. Russell and S.M. Cohen. 2003. Bantam encodes a developmentally regulated microRNA that controls cell proliferation and regulates the proapoptotic gene hid in Drosophila. *Cell*. **113**:25–36.

Cai, X., C.H. Hagedorn and B.R. Cullen. 2004. Human microRNAs are processed from capped, polyadenylated transcripts that can also function as mRNAs. *RNA*. Nov 3 (Epub ahead of print).

Calin, G.A. et al. 2004. Human microRNA genes are frequently located at gradile sites and genomic regions involved cancer. *Proc. Natl. Acad. Sci. USA.* **101**:2999–3004.

Carrington, J.C. and V. Ambros. 2003. Role of microRNAs in plant and animal development. *Science.* **301(5631)**:336–338.

Chen, X. 2003. A MicroRNA as a Translational Repressor of APETALA2 in Arabidopsis Flower Development. Science E-Pub.

Chiou, T.J., K. Aung, S.I. Lin, C.C. Wu, S.F. Chiang and C.L. Su. 2006. Regulation of phosphate homeostasis by microRNA in Arabidopsis. *Plant Cell.* **18**:412–421.

Cooper, J.E. 2007. Early interactions between legumes and rhizobia: disclosing complexity in a molecular dialogue. *J. Appl. Microbiol.* (Online Early Articles).

Cullen, B.R. 2006. Viruses and microRNAs. *Nature Genetics.* **38**:S25–S30.

Ding, Y. and G.E. Oldroyd. 2009. Positioning the nodule, the hormone dictum. *Plant Signal Behav.* **4**:89–93.

Doench, J.G., C.P. Petersen and P.A. Sharp. 2003. siRNAs can function as miRNAs. *Genes Dev.* **17**:438–42.

Dostie, J., Z. Mourelatos, M. Yang, A. Sharma and G. Dreyfuss. 2003. Numerous microRNPs in neuronal cells containing novel microRNAs. *RNA.* **9(2)**:180–6. *Erratum in: RNA.* **9(5)**:631–2.

Dugas, V.D. and B. Bartel. 2004. MicroRNA regulation of gene expression in plants. *Curr. Opin. Plant Biol.* **7**:512–520.

Fahlgren, N., M.D. Howell, K.D. Kasschau, E.J. Chapman, C.M. Sullivan, J.S. Cumbie, S.A. Givan, T.F. Law, S.R. Grant, J.L. Dang and J.C. Carrington. 2007. High-throughput sequencing of Arabidopsis microRNAs: Evidence for frequent birth and death of MIRNA Genes. *PLoS ONE.* **2**:e219.

Fujii, H., T.J. Chiou, S.I. Lin , K. Aung and J.K. Zhu. 2005. A miRNA involved in phosphate starvation response in Arabidopsis. *Current Biology.* **15**:2038–2043.

Griffiths-Jones, S. 2004. The microRNA registry. *Nucl. Acids Res.* **32**:D109–D111.

Hamilton, J.A. and C.D. Baulcombe. 1999. A species of small antisense RNA in ost transcriptional gene silencing in plants. *Science.* **286**:950–952.

He, L. and G.J. Hannon. 2004. MicroRNAs: small RNAs with a big role in gene regulation. *Nat. Rev. Genet.* **5**:522–531.

He, Z. and E.J. Sontheimer. 2004. SiRNAs and miRNAs: a meeting report on RNA silencing. *RNA.* **10**:1165–1173.

Hutvagner, G. and P.D. Zamore. 2002. A miRNA in a multiple-turnover RNAi enzyme complex. *Science.* **297**:2056–2060.

Jin, H. 2008. Endogenous small RNAs and antibacterial immunity in plants. *FEBS Lett.* **582**:2679–2684.

Jones-Rhoades, M.J. and D.P. Bartel. 2004. Computational identification of plant microRNAs and their targets, including a stressinduced miRNA. *Mol. Cell.* **14**:787–799.

Jones-Rhoades, M.W., D.P. Bartel and B. Bartel. 2006. MicroRNAs and their regulatory roles in plants. *Annual Review of Plant Biology.* **57**:19–53.

Kasschau, K.D., Z. Xie, E. Allen, C. Llave, E.J. Chapman, K.A. Krizan and J.C. Carrington. 2003. P1/HC-Pro, a viral suppressor of RNA silencing, interferes with Arabidopsis development and miRNA function. *Dev. Cell.* **4**:205–217.

Kern, W., A. Kohlmann, C. Wuchter, S. Schnittger, C. Schoch, S. Mergenthaler, R. Ratei, W.D. Ludwig, W. Hiddemann and T. Haferlach. 2003. Correlation of protein expression and gene expression in acute leukemia. *Cytometry.* **55B**:29–36.

Khvorova, A., A. Reynolds and S.D. Jayasena. 2003. Functional siRNAs and miRNAs exhibit strand bias. *Cell.* **115(2)**:209–16.

Kim, V.N. 2005. MicroRNA biogenesis: coordinated cropping and dicing. *Nat. Rev. Mol. Cell. Biol.* **6**:376–385.

Lagos-Quintana, M., R. Rauhut, A. Yalcin, J. Meyer, W. Lendeckel and T. Tuschl. 2002. Identification of tissue-specific microRNAs from mouse. *Curr. Biol.* **12**:735–739.

Lagos-Quintana, M., R. Rauhut, W. Lendeckel and T. Tuschl. 2001. Identification of novel genes coding for small expressed RNAs. *Science.* **294**:853–858.

Lai, E.C., P. Tomancak, R.W. Williams and G.M. Rubin. 2003. Computational identification of Drosophila microRNA genes. *Genome Biol.* **4(7)**:R42. Epub.

Lau, N.C., L.P. Lim, E.G. Weinstein and D.P. Bartel. 2001. An abundant class of tiny RNAs with probable regulatory roles in *Caenorhabditis elegans*. *Science.* **294**:858–862.

Lee, R.C., R.L. Feinbaum and V. Ambros. 1993. The C. elegans heterochronic gene lin-4 encodes small RNAs with antisense complementarity to lin-14. *Cell.* **75(5)**:843–854.

Lee, R.C. and V. Ambros. 2001. An extensive class of small RNAs in Caenorhabditis elegans. *Science.* **294**:862–864.

Lee, Y., K. Jeon, J.T. Lee, S. Kim and V.N. Kim. 2002. MiRNA maturation: stepwise processing and subcellular localization. *EMBO J.* **21**:4663–4670.

Lewis, B.P. et al. 2003. Prediction of mammalian microRNA targets. *Cell.* **115**:787–98.

Lim, L.P. et al. 2005. Microarray analysis shows that some microRNAs down regulated large numbers of target mRNA. *Nature.* **433**:767–773.

Lim, L.P., M.E. Glasner, S. Yekta, C.B. Burge and D.P. Bartel. 2003b. Vertebrate microRNA genes. *Science.* **299**:1540.

Lim, L.P., N.C. Lau, E.G. Weinstein, A. Abdelhakim, S. Yekta, M.W. Rhoades, C.B. Burge and D.P. Bartel. 2003a. The microRNAs of Caenorhabditis elegans. *Genes and Development.* **17**:991–1008.

Liu, Q. and Y.Q. Chen. 2009. Insights into the mechanism of plant development: interactions of miRNAs pathway with phytohormone response. *Biochem. Biophys. Res. Commun.* **384**:1–5.

Llave, C., K.D. Kasschau, M.A. Rector and J.C. Carrington. 2002a. Endogenous and silencing-associated small RNAs in plants. *Plant Cell.* **14**:1605–1619.

Long, S.R. 1989. Rhizobium-legume nodulation: life together in the under-ground. *Cell.* **56(2)**:203–214.

Lu, C. and N. Fedoroff. 2000. A mutation in the Arabidopsis HYL1 gene encoding a dsRNA binding protein affects responses to abscisic acid, auxin and cytokinin. *Plant Cell.* **12**:2351–2366.

Lu, S., Y-H. Sun, R. Shi, C. Clark, L. Li and V.L. Chiang. 2005. Novel and mechanical stress-responsive microRNAs in Populus trichocarpa that are absent from Arabidopsis. *Plant Cell.* **17:**2186-2203.

Lund, E., S. Guttinger, A. Calado, J.E. Dahlberg and U. Kutay. 2003. Nuclear Export of MicroRNA Precursors. *Science.* (Epub ahead of print).

Mallory, A. and H. Vaucheret. 2006. Functions of microRNAs and related small RNAs in plants. *Nat. Genet.* **38(suppl):**S31–36.

Olsen, P.H. and V. Ambros. 1999. The lin-4 regulatory RNA controls developmental timing in C. elegans by blocking LIN-14 protein synthesis after the initiation of translation. *Develop. Biol.* **216:**671-680

Palatnik, J.F., E. Allen, X. Wu, C. Schommer, R. Schwab, C. Carrington and D. Weigel. 2003. Control of leaf morphogenesis by microRNAs. *Nature.* **425:**257–263.

Park, W., J. Li, R. Song, J. Messing and X. Chen. 2002. CARPEL FACTORY, a Dicer homolog and HEN1, a novel protein, act in microRNA metabolism in Arabidopsis thaliana. *Curr. Biol.* **12:**1484–1495.

Pasquinelli, A.E., B.J. Reinhart, F. Slack, M.Q. Martindale, M. Kuroda, B. Maller,

A. Srinivasan, M. Fishman, D. Hayward, E. Ball and et al. 2000. Conservation across animal phylogeny of the sequence and temporal regulation of the 21 nucleotide let-7 heterochronic regulatory RNA. *Nature.* **408:**86–89.

Reinhart, B. et al. 2000. The 21-nucleotide let-7 RNA regulates developmental timing in *Caenohabditis elegans. Nature.* **403:**901–6.

Reinhart, J.B., G.E. Weinstein, W.M. Rhoades, B. Bartel and D.P. Bartel. 2002. MicroRNAs in plants. *Genes Dev.* **16:**1616–1626.

Rhoades, M.W., B.J. Reinhart, L.P. Lim, C.B. Burge, B. Bartel and D.P. Bartel. 2002. Prediction of plant microRNA targets. *Cell.* **110:**513–520.

Rolfe, G.B. 1988. Flavones and isoflavones as inducing substances of legume nodulation. *Biofactors.* **1(1):**3–10.

Schwarz, D.S., G. Hutvagner, T. Du, Z. Xu, N. Aronin and P.D. Zamore. 2003. Asymmetry in the assembly of the RNAi enzyme complex. *Cell.* **115(2):**199– 208.

Smalheiser, R.N. 2003. EST analyses predict the existence of a population of chimeric microRNA precursor-mRNA transcripts expressed in normal human and mouse tissues. *Genome Biol.* **4(7):**403. Epub.

Stacey, A.S., C.M. Blake and D. Sherrier. 2009. MicroRNAs in the Rhizobia Legume Symbiosis. *Plant Physiology.* **151:**1002–1008.

Subramanian, S., Y. Fu, R. Sunkar, W.B. Barbazuk, J.K. Zhu and O. Yu. 2008. Novel and nodulation-regulated microRNAs in soybean roots. *BMC Genomics.* **9:**160.

Sunkar, R., A. Kapoor and J.K. Zhu. 2006. Posttranscriptional induction of two Cu/Zn superoxide dismutase genes in Arabidopsis is mediated by downregulation of miR398 and important for oxidative stress tolerance. *Plant Cell.* **18:**2051–2065.

Sunkar, R., G. Jagadeeswaran. 2008. In silico identification of conserved microRNAs in large number of diverse plant species. *BMC Plant Biology.* **8:**37.

Sunkar, R., V. Chinnusamy, J. Zhu and J.K. Zhu. 2007. Small RNAs as big players in plant abiotic stress responses and nutrient deprivation. *Trends in Plant Sci.* **12(7)**:301–309.

Tang, G., B.J. Reinhart, D.P. Bartel and P.D. Zamore. 2003. A biochemical framework for RNA silencing in plants. *Genes Dev.* **17**:49–63.

Tongwen, Y., X. Lingui and A. Lizhe. 2007. Functional diversity of miRNA in plants. *Plant Science.* **172**:423–32.

Wienhold, E. et al. 2005. MicroRNA expression in zebrafish embryonic development. *Science.* **309**:310–1.

Wightman, B. et al. 1993. Posttranscriptional regulation of the heterochronic gene line-14 by line-4 mediates temporal pattern formation in *C. elegans. Cell.* **75**:855–62.

Xu, P., S.Y. Vernooy, M. Guo and B.A. Hay. 2003. The Drosophila microRNA Mir-14 suppresses cell death and is required for normal fat metabolism. *Curr. Biol.* **13(9)**:790–5.

Yi, R., Y. Qin, I.G. Macara and B.R. Cullen. 2003. Exportin-5 mediates the nuclear export of pre-microRNAs and short hairpin RNAs. *Genes Dev.* **17**:3011–3016.

Zeng, Y., R. Yi and B.R. Cullen. 2003. miRNAs and small interfering RNAs can inhibit mRNA expression by similar mechanisms. *Proc. Nat. Acad. Sc.* **100**:9779–9784.

Zhang, B.H., Q.L. Wang and X.P. Pan. 2007. MicroRNAs and their regulatory roles in animals and plants. *Journal of Cellular Physiology.* **210(2)**:279–289.

Zhang, B.H., X.P. Pan, G.P. Cobb and T.A. Anderson. 2006. Plant microRNA: A small regulatory molecule with big impact. *Developmental Biology.* **289(1)**:3–16.

CHAPTER 17

RNA Interference in Fungi

B.M. Bashyal* and Rashmi Aggarwal

ABSTRACT

RNAi has rapidly become a powerful reverse genetic tool, especially in organisms where gene targeting is inefficient and time-consuming. In filamentous fungi, RNAi is not currently used as widely as is gene targeting by homologous recombination that works with practical efficiencies in most model fungal species. The filamentous fungus *Neurospora crassa* is one of the first organisms used for RNAi studies. Quelling and meiotic silencing by unpaired DNA are two RNAi-related phenomena discovered in *Neurospora* and their characterizations have contributed significantly to our understanding of RNAi mechanisms in eukaryotes. The studies in *Cryphonectria*, *Mucor*, *Aspergillus* and other species indicate that RNAi is widely conserved in filamentous fungi and plays important roles in genome defense. This chapter summarizes different strategies and application of RNAi in filamentous fungi.

Keywords: Fungi, RNAi, functional genomics, gene silencing.

INTRODUCTION

Fungi have enormous impact on human affairs and the functioning of ecosystems. As a primary decomposer in the planet's ecosphere fungi have vital roles in nutrient recycling. Fungi can be directly used as food (mushrooms) they are important agents for the production of fermented foods (e.g., Cheese, bread, wine) and for the manufacture of enzymes and antibiotics. Fungi are also the most important group of plant pathogens causing serious losses in crop yield and marketability of produce worldwide. Cellular structure metabolic mechanisms and gene organization of fungi are similar to those of higher eukaryotes. Several species of fungi such as *Neurospora crassa*, *Saccharomyces cerevisiae*,

* Corresponding author *e-mail*: bishnumayabashyal@gmail.com

Aspergillus nidulans, have provided sophisticated genetic systems for basic science.

Only a limited number of post transcriptional gene silencing phenomena have been reported in fungi other than *Neurospora*. These included a co-suppression like phenomena in *Cladosporium fulvum* (Hamada and Spanu. 1998) and internuclear gene silencing in *Phytophthora infestans* (Van West et al. 1999). The story of RNAi in fungi began with a finding by Romano and Machino in 1992 in *Neurospora crassa,* whereby expression of the endogenous gene, *al-f* which is involved in carotenoid biosynthesis, was attenuated by a transformation with homologous *al-1* sequences, which was designated as quelling. However, studies on quelling deficient mutants of *N. crassa* (Cogoni and Macino 1997) revealed the involvement of RNA-dependent RNA polymerase (RdRP) and argonuate protein in the pathway, providing molecular evidence that quelling belongs to RNA interference. After the discovery of RNAi in 1998 (Fire et al. 1998) attempts were made to harness the technology for controlling gene expression in a variety of fungal species. Consequently, suppression of gene expression by a dsRNA expressing plasmid or related system has been shown in many fungal species including Ascomycota, Basidiomycota, Zygomycota and Oomycota.

RNAi is an important tool not only for elucidating the function of the many unknown genes but also for the identification of gene essential for growth and pathogenesis.

RNAi STRATEGIES IN FILAMENTOUS FUNGI

RNAi strategies common in filamentous fungi are:

1. RNA using a hairpin RNA (hpRNA) expressing plasmid
2. RNAi using an opposing dual promoter system
3. Direct delivery of siRNA/dsRNA into fungal cells

RNA using a hairpin RNA (hpRNA) expressing plasmid

Recently Goldoni et al. (2004) have shown that hpRNA-expressing constructs induce more efficient and stable silencing. The first example of fungal RNAi by an hpRNA-expressing plasmid was demonstrated by Liu et al. in the basidiomycetous yeast Cryptococcus neoformans (Liu et al. 2002). hpRNA or intron-containing hpRNA (ihpRNA) are the most prevalent and reliable platforms to induce RNAi in fungi (Kodatani et al. 2003). These types of vector have been successfully used to demonstrate RNAi using model genes and to explore gene function in a wide range of fungal species (Van West et al.1999) and Myxomycetes (slime molds) (Martens et al. 2002) (Table 1). pSilent-1 vector which facilitate the construction of an

Table 1. Various types of RNAi in fungi and fungus-like organisms.

Species	RNAi trigger	Transformation
Ascomycota		
Neurospora crassa	Homologous transgene	PEG-mediated method
Magnaporthe oryzae	IR	PEG-mediated method
Aspergillus nidulans	Synthetic siRNA	Uptake from culture medium
Basidiomycota		
Cryptococcus neoformans	IR	Electroporation
Coprinus cinereus	IR	Lithium acetate method
Schizophyllum commune	IR	PEG-mediated method
Zygomycota		
Mucor circinelloides	Homologous transgene	PEG-mediated method
Mortierella alpine	IR	Microparticle bombardment
Oomycota		
Phytophthora infestans	Homologous transgene	PEG-mediated method
Phytophthora infestans	Homologous transgene	Electroporation
Phytophthora infestans	dsRNA	Lipofectin*-mediated transfection

IR: hairpin RNA or inverted repeat RNA-expressing plasmid.
* Lipofectin was added to increase transformation efficiency.

ihpRNA-expressing plasmid by polymerase chain reaction (PCR)-based cloning was useful for ascomycetous fungi (Nakayashiki et al. 2005).

RNAi using an opposing-dual promoter system

ihpRNA-expressing application is generally limited to small or moderate scale analyses since the construction of such vectors normally requires two steps of orientated cloning. RNAi vectors with an opposing-dual promoter system, enables vector construction with a single, non orientated cloning step have been developed for *Histoplasma capsulatum, Magnaporthe oryzae* and *Cryptococcus. neoformans* (Rappleye et al. 2004). In these systems, sense and antisense RNA of the target gene, which is expected to form dsRNA in the cell, are transcribed independently under the control of the two opposing RNA polymerase II promoters. Recently, pSilent–Dual1 (pSD1), carrying opposing *trpC* and *gpd* promoters, has been constructed and used in *M. oryzae*. The efficacy of gene silencing by pSD1 was generally lower than that exhibited by ihpRNA-expressing vectors (Nguyen et al. 2008). Nevertheless, strong gene silencing (>80% reduction) was induced in a small fraction of the resulting transformants. In this screening system GFP fluorescence provides an effective indicator to select transformants in which interference was operating. This cosilencing-based screening has been successfully demonstrated in *Crpptococcus neoformans, Venturia inaequalis, Acremonium chrysogenum,* using GFP, DsRed or endogenous maker genes (Fitzgerald et al. 2004). Using the pSD1 system, a series of knockdown mutants of almost all known calcium signaling related genes in the genome of *M. oryzae* was generated and examined for phenotypic defects.

Direct delivery of siRNA/dsRNA into fungal cells

Even though the direct delivery of synthetic siRNA to cultured cells is a common method to introduce RNAi in mammalian systems, such applications have been very rarely reported in fungi. In *Aspergillus nidulans,* ornithine decarboxylase (ODC), a key polyamine biosynthesis gene, was specifically silenced by treating germinating spores with synthetic 23 nt siRNA duplex with 2 nt overhangs at the 3'-terminus. The treatment of fungal spores with 5–25 nM ODC siRNA duplex in culture medium caused a significant reduction in spore germination and germ tube growth together with silencing of the ODC gene (Whisson et al. 2005). Those results indicated that siRNA duplex could be taken up by germinating fungal spores from the culture medium, providing a rapid and convenient method to induce RNAi in fungal cells.

APPLICATION OF RNAi IN FILAMENTOUS FUNGI

Recent evidence indicates that noncoding small RNAs play critical roles in a greater variety of cellular processes than was once assumed. The proposed biological roles of small RNA-mediated gene silencing pathways (referred to hereafter as RNA silencing pathways) in eukaryotes include regulation of endogenous gene expression, resistance to viruses, transposon silencing, heterochromatin formation, programmed DNA elimination and maintenance of genome stability in reproductive cells. The RNA silencing pathways seem to have diversified during the evolution of eukaryotes and therefore some of the pathways are likely to be absent in certain eukaryotes, or present in only a limited class of organisms.

RNAi against transposons

siRNAs against transposons were detected in *N. crassa* and the transcript levels and copy number of the LINE1-like transposon *Tad* were significantly elevated in *qde-2* mutants. In addition, the *Tad* transcripts were also found to be up-regulated in the *dcl-1 dcl-2* double mutants (Chicas et al. 2005).

RNAi in heterochromation formation

RNAi has been shown to be involved in heterochromatin formation and/or DNA methylation in fission yeast (Moazed 2009). However, the known RNAi components in *Neurospora*, including three RdRPs (QDE-1, SAD-1 and RRP-3), two Argonaute proteins (QDE-2 and SMS-2), two dicer-like proteins (DCL-1 or SMS-3, DCL-2) and two RecQ helicases (QDE-3 and RecQ-2), are not required for the initiation or maintenance of heterochromatin formation and DNA methylation (Freitag et al. 2004). It was also demonstrated that the transgenic siRNA production is also not required for histone H3 Lys9 methylation (Nolan et al., 2005). Thus, the RNAi pathway does not appear to function in transcriptional gene silencing in *Neurospora*. However, the mutation of the histone Lys9H3 methyltransferase gene *dim-5* caused a low quelling efficiency and frequent reversion of the quelled transformants due to rapid loss of the integrated transgenic copies (Nolan et al. 2005). Thus, DIM-5 and histone methylation play an important role in stabilizing tandem copies of the transgene.

RNAi in antiviral defense mechanism

In *Cryphonectria parasitica* p29 is a papain-like protease encoded by the mycovirus *Cryphonectria* hypovirus 1 (CHV1), which is similar to the plant potyvirus-encoded suppressor of RNA silencing HC-Pro (Segers et al. 2006). Using the CHV1-EP713/*C. parasitica* system, Segers et al. demonstrated that p29 suppressed the hairpin RNA-induced silencing in *C. parasitica*. In addition, p29 also suppressed both the virus-induced and agroinfiltration-induced RNA silencing and systemic spread of silencing in GFP-expressing transgenic *Nicotiana benthamiana* (Segers et al. 2006). These results suggest that the antiviral defense mechanism of RNA silencing is conserved in both fungi and plants. On the other hand, infection of the *dcl-2* mutant by a hypovirus CHV1-EP713 mutant lacking the suppressor of RNA silencing p29 and the wild-type reovirus MyRV1-Cp9B21 exhibited elevated viral RNA levels compared to the wild-type. These results demonstrate that a fungal Dicer can function to regulate virus infection and RNAi plays an important role in antiviral defense in fungi (Segers et al. 2007).

RNAi as a viral defense mechanism

Aspergillus nidulans contains two genes encoding for RdRPs but they are not required for the inverted repeat triggered RNAi, suggesting that RdRPs are not involved in siRNA amplification (Hammond and Keller 2005). By stable infection of *A. nidulans* with three mycoviruses, Hammond and Keller demonstrated that the *Aspergillus* virus 1816 could suppress the inverted repeat transgene-induced RNA silencing. Although the mechanism for the suppression is not clear, this result suggests the existence of an RNA silencing suppressor encoded by this virus. On the other hand, the virus 341-derived siRNA was detected at a high level in an Argonaute mutant, indicating that this virus is targeted and processed into siRNA by the RNA silencing machinery. Together, these results suggest that there is an antagonistic relationship between mycoviruses and RNA silencing mechanism in *A. nidulans* and RNAi functions as a viral defense mechanism.

RNAi in functional genomics of fungi

One major advantage of RNA silencing is its applicability to down regulate gene expression without regard for gene targeting efficiency. Because of the compact and small-sized genome, there are a relatively large number of fungal species whose genomes have been completely sequenced. Those include *Aspergillus nidulans, Aspergillus oryzae, Fusarium graminearum, Magnaporthe oryzae, Ustilago maydis, Aspergillus fumigatus* and *Candida albicans*. To take full advantage of this wealth of genetic information for unraveling how the genes work, RNA silencing is one of the most powerful approaches. Since the operation of RNA silencing has been shown to exist in many fungi, RNA silencing is opening new avenues to explore the genomes of the fungi.

CONCLUSION

RNAi is useful for the studies of functional genomics in fungi, it can provide novel and rapid gene analysis applications that gene disruption methods cannot offer. The use of genetic modeling systems such as RNAi has been the key to understanding gene structure and function, the biology of cells and organisms; thus a combined approach of RNA silencing and gene knock-out technologies will without doubt greatly facilitate exploring gene functions in fungi.

REFERENCES

Chicas, A., E.C. Forrest, S. Sepich, C. Cogoni and G. Macino. 2005. Small interfering RNAs that trigger posttranscriptional gene silencing are not required for the histone H3 Lys9 methylation necessary for transgenic tandem repeat stabilization in *Neurospora crassa. Mol. Cell. Biol.* **25**:3793–3801.

Cogoni, C. and G. Macino. 1999. Gene silencing in *Neurospora crassa* requires a protein homologous to RNA-dependent RNA polymerase. *Nature.* **399**:166–169.

Fitzgerald, A., J.A. Van Kan and K.M. Plummer. 2004. Simultaneous silencing of multiple genes in the apple scab fungus, *Venturia inaequalis*, by expression of RNA with chimeric inverted repeats. *Fungal Genet. Biol.* **41**:963–971.

Fire, A., S. Xu, M.K. Montgomery, S.A. Kostas, S.E. Driver and C.C. Mello. 1998. Potent and specific genetic interference by double-stranded RNA in *Caenorhabditis elegans. Nature.* **391:**806–811.

Freitag, M., D.W. Lee, G.O.Kothe, R.J. Pratt, R. Aramayo and E.U. Selker. 2004. DNA methylation is independent of RNA interference in Neurospora. *Science.* **304:**1939.

Goldoni, M., G. Azzalin, G. Macino and C. Cogoni. 2004. Efficient gene silencing by expression of double stranded RNA in Neurospora crassa. *Fungal Genet Biol.* **41:**1016–1024.

Hamada, W., P.D. Spanu. 1988. Co-suppression of the hydr ophobin gene *HCf-1* is correlated with antisense RNA biosynthesis in *Cladosporium fulvum Mol. Gen. Genet.* **259:**630–638.

Hammond, T.M. and N.P. Keller. 2005 RNA silencing in *Aspergillus nidulans* is independent of RNA-dependent RNA polymerases. *Genetics.* **169:**607–617.

Kadotani, N., H. Nakayashiki, Y. Tosa Y and S. Mayama. 2003. RNA silencing in the phytopathogenic fungus *Magnaporthe oryzae. Mol. Plant Microbe. Interact.* **16:**769–776.

Liu, H., T.R. Cottrell, L.M. Pierini, W.E. Goldman and T.L. Doering. 2002. RNA interference in the pathogenic fungus *Cryptococcus neoformans. Genetics 2002.* **160:**463-470.

Martens, H., J. Novotny, J. Oberstrass, T.L. Steck, P. Postlethwait and W. Nellen. 2002. RNAi in *Dictyostelium:* the role of RNA-directed RNA polymerases and double-stranded RNase. *Mol. Biol. Cell.* **13:**445–453.

Moazed, D. 2009. Small RNAs in transcriptional gene silencing and genome defence. *Nature.* **457:**413–420.

Nakayashiki, H., S. Hanada, B.Q. Nguyen, N. Kadotani, Y. Tosa and S. Mayama. 2005. RNA silencing as a tool for exploring gene function in ascomycete fungi. *Fungal Genet. Biol.* **42:**275–283.

Nguyen, Q.B., N. Kadotani, S. Kasahara, Y. Tosa, S. Mayama and H. Nakayashiki. 2008. Systematic functional analysis of calciumsignalling proteins in the genome of the rice-blast fungus, *Magnaporthe oryzae,* using a high-throughput RNA-silencing system. *Mol. Microbiol.* **68:**1348–1365.

Nolan, T., L. Braccini, G. Azzalin, A. De Toni, G. Macino and C. Cogoni. 2005. The post-transcriptional gene silencing machinery functions independently of DNA methylation to repress a LINE1-like retrotransposon in *Neurospora crassa. Nucleic Acids Res.* **33:**1564–1573.

Rappleye, C.A., J.T. Engle and W.E. Goldman. 2004. RNA interference in *Histoplasma capsulatum* demonstrates a role for alpha-(1,3)-glucan in virulence. *Mol. Microbiol.* **53:**153–165.

Romano, N. and G. Macino. 1992. Quelling: transient inactivation of gene expression in Neurospora crassa by transformation with homologous sequences. *Mol. Microbiol.* **6:**3343–3353.

Segers, G.C., R. van Wezel, X. Zhang, Y. Hong and D.L. Nuss. 2006. Hypovirus papain-like protease p29 suppresses RNA silencing in the natural fungal host and in a heterologous plant system. *Eukaryot Cell.* **5:**896–904.

Segers, G.C., X. Zhang, F. Deng, Q. Sun and D.L. Nuss. 2007. Evidence that RNA silencing functions as an antiviral defense mechanism in fungi. *Proc. Natl. Acad. Sci. USA.* **104:**12902–12906.

van West, P., S. Kamoun, J.W. van't Klooster and F. Govers. 1999. Internuclear gene silencing in *Phytophthora infestans. Mol. Cell.* **3:**339–348.

Whisson, S.C. and A.O. Avrova, P.V. West and J.T. Jones. 2005. A method for double-stranded RNA-mediated transient gene silencing in *Phytophthora infestans. Mol. Plant Pathol.* **6:**153–163.

CHAPTER 18

Virulence of *Fusarium* spp. Causing *Wilt Disease* of Guava with Special Reference to RNAi

**V.K. Gupta,* P.K. Jain, A. Manimaran,
K.M. Turner, A. O'Donovan and M. Tuohy**

ABSTRACT

Wilt of *Psidium guajava* L., incited by *Fusarium oxysporum* f. sp. *psidii* and *Fusarium solani* is a serious soil borne disease of guava in India. The erratic spread and occurrence of guava wilt in different areas may be due to variable aggressiveness or virulence of different pathogenic isolates in the soil. Virulent gene related locus *viz.* Xyl, KHS1, PelA1, PG6/7, CHS1/2 and FMK1/MAPK1 were successfully expressed. This indicates that all the tested *Fusarium* sp. isolates of guava are having virulence factor in their genome. Since, the *Fusarium* spp. infecting guava plant were sequence for RNA genes and their expression studies was possibility an outcome of gene silencing techniques based on deletion/insertion of codan at ORF1 and ORF3 loci in the genome of *Fusarium* spp. responsible for causing wilt disease of guava. Therefore, in present review, an attempt has been made to discuss theses virulence factor in terms of RNAi.

Keywords: F. oxysporum f. sp. *psidii*, F. solani, Virulence genes, RNAi.

INTRODUCTION

Guava (*Psidium guajava* L.) is an important fruit crop of subtropical countries. In India, it is grown almost in all the states. Wilt is one of the most destructive diseases for guava plants in India and losses due to wilt disease are substantial. It can be said that wilt is the only disease of guava which

---

* Corresponding author *e-mail*: vijaifzd@gmail.com

has threatened guava cultivation in our country (Misra 2006). *Psidium guajava* wilt is prevalent in India, Latin America, Malaysia, Pakistan, South Africa, South Asia and Taiwan (Gupta et al. 2010a). *Fusarium*, in particular *F. oxysporum* f. sp. *psidii* is the predominant pathogen for wilt of guava. The other species *Fusarium solani* is also dominates in isolation (Misra and Gupta 2007).

Mortality induced in guava plants because of wilt, is a major loss for guava cultivars. Reports reveal wide variations in physiological characteristics and pathogenic potential of isolates of *F. oxysporum* and *F. solani* across the location. Pathogenicity tests is a very cumbersome and time consuming (50–300 days) process and is not sure as several factors which influence disease development and requires extensive facilities. Differential line tests of *F. oxysporum* can take over 40 days to complete.

RNAi (RNA interference) refers to the introduction of homologous double stranded RNA (dsRNA) to specifically target a gene's product, resulting in null or hypomorphic phenotypes. In this way, one can functionally delete from the genetic material of any gene that can be cloned.

This phenomenon was called co-suppression (by the plant biologists), quelling (by the fungal biologists) and RNA interference (by the animal developmental biologists). Often it is now called post-transcriptional gene silencing (PTGS). It appears that adding either the sense (in extremely large concentrations), antisense, or double-stranded gene into cells can cause the cells to destroy the mRNA homologous to that made by the introduced gene. This technique allows one to make organisms that are functionally "knockouts" (i.e., null mutants; loss-of-function mutants) for any gene that can be cloned. Whereas knockout technology in mice is laborious and takes a long time, RNAi can be triggered in nematodes by simply soaking the worms in double stranded DNA.

As a genetic tool, RNAi has several characteristics that contrast with those of the conventional gene disruption methods, which can make RNAi either advantageous or disadvantageous. First, RNAi causes only a partial reduction (knockdown) but not a complete loss (knockout) in gene expression. Incomplete gene suppression by RNAi is generally regarded as a drawback since it could result in phenotypic variations that sometimes make the interpretation of RNAi data difficult.

VIRULENCE FACTOR IN *FUSARIUM* spp.

Molecular studies of virulence in pathogenic fungi reveal a complex interaction between each fungal species and the plant/human host. Virulence in pathogenic fungi reveals a complex interaction between each fungus species and the human host. *Fusarium* species are normally a virulent in healthy people, but can be disseminated to deep tissues and cause fatal infections in patients with suppressed immune function. These infections present with non-specific symptoms, typically fever and are often difficult to diagnose as a result. This fungus straddles the boundary between 'true' and 'opportunistic' virulence, since it is known as a cause of community-acquired infection. The diversity of pathogenic potency, cellular form and route of invasion between fungal species makes it impossible to draw general conclusions about their molecular virulence attributes. In the case of opportunistic fungal pathogens, an understanding of host immune dysfunction is as important as an appreciation of which fungal genes may encode definable attributes of virulence.

The fungi that cause invasive disease differs considerably in their inherent pathogenicity, even though few, if any, approach the level of virulence of the best known fungal species like *Fusarium*. In plant pathology, the relationship between recognition phenomena and pathogenesis is somewhat contrary to that in cellular mycology. In the 'gene-for-gene' model established for plant-pathogen interactions, the loss or gain or mutation of an virulence genes in the fungus leads to host pathogen-recognition mechanisms and hence the onset of disease (Di Pietro et al. 2001). Studies of virulence at the molecular level have now revealed many factors that contribute to the overall pathogenicity of different fungal types (Odds et al. 2001). Di Pietro et al. (1998) studied on virulent factor related gene locus PG1 and PG2, exopolygalacturonase expressed by vascular wilt pathogen *F. oxysporum* f.sp. *lycopersici*, which is responsible for causing wilt in tomato.

Perception of the high incidence of fungal diseases, particularly those that threaten life, has led to considerable investment in research into fungal virulence, often based on cutting-edge molecular biological approaches. But the rapid progress made in identifying molecular virulence factors in several fungal pathogens has also raised some caveats and paradoxes that require resolution.

MOLECULAR BASIS OF FUNGAL VIRULENCE WITH SPECIAL CONCERN TO OF *FUSARIUM* WILT IN GUAVA

Recent advances in fungal molecular biology have begun to make it possible to circumvent these concerns by cloning the genes responsible for producing putative virulence factors and either introducing knockout mutations or expressing the factor in another organism. Specific gene products that might benefit from a molecular genetic approach, although sometimes it will only be possible to describe a virulence attribute. An intravenous model of infection would be a poor choice when studying a virulence factor that enables a fungus to survive in the pulmonary compartment after inhalation. Equally important is the readout of the model system. The end result chosen may define pathogenicity as lethality for the host, as alterations in tissue distribution of the organism in the host, as alterations in histopathological findings, or as alterations in the disease course from an acute disease to an indolent chronic infection. It is often argued that fungi have evolved for a saprophytic existence

Endo—1,4-xylanases are produced by a number of plant pathogenic fungi and it has been suggested that they may play a role during infection (Gupta et al. 2010b). The presence of xylanases in *F. oxysporum* and *F. solani* was previously reported (Ondøej et al. 2008, Gupta et al. 2009). Di Pietro et al. (2001) identified the gene fmk1 that encodes a MAPK in *F. oxysporum*. This gene is related to formation of infection hyphae, root attachment and invasive growth by *F. oxysporum* f. sp. *lycopersici* on tomato plants (Di Pietro et al. 2001). Chitin synthases (CHS) are the enzymes implicated in chitin synthesis and in fungi. In fact, chitin has been shown to play an important role in the pathotypic behavior towards vascular plants. Nothing is known about aspects of the pathogenicity of this organism. While substantial progress has been made in identifying virulence factors for some fungal pathogens, much work remains to be done for some of the more genetically intractable fungi, especially the dimorphic pathogens. Advances in molecular genetics for the dimorphic fungal pathogens have come only recently, Falkow has discussed the application of molecular Koch's postulates for the assessment of virulence factors and their genes (Falkow, 1988) from bacterial pathogens and these principles can be extended to studies of fungi. It was the goal of this review to assess the success of similar analyses of fungal pathogens when technology permitted and to illustrate candidate virulence factors in need of rigorous genetic testing. Rapid progress in molecular genetics will allow the development of alternative strategies for analysis of the virulence of these organisms and the development of better treatments for the diseases.

For virulence factor research of the opportunist fungus their Gene disruption in thisorganism is complicated by its state of permanent diploidy. Almost all investigators of its putative virulence factors have used disruption cassettes based on URA3 or HIS1 as selectable markers in the absence of other positive selectable markers (Fonzi and Irwin, 1993). Since 1991, at least 40 genes have been identified whose disruption leads to attenuated virulence of the mutant strain when injected intravenously into mice. These genes encode products which range from secreted hydrolytic enzymes through cell-wall polysaccharides and peptides to signal transduction pathway components. That expression of virulence by a fungus can depend on the particular host niche that is being affected. In plant pathology, the relationship between recognition phenomena and pathogenesis is somewhat contrary to that in medical mycology. In the 'gene-for-gene' model established for plant-pathogen interactions, the loss or mutation of a virulence genes in the fungus leads to escape of pathogen-recognition mechanisms and hence the onset of disease (De Wit 1992). Hence, mutants can be expected for which virulence is increased. So far, such mutations have not been reported in fungi pathogenic for humans, but we would do well to consider the gene-for gene model should hypervirulent mutations be revealed in future mutagenesis screens. Studies of virulence at the molecular level have now revealed many factors that contribute to the overall pathogenicity of different fungal types. The only common theme linking the virulence attributes of the four fungi discussed in this review is that no fungus depends upon any single molecule for its virulence. Fungal virulence is a polyvalent, complex process that requires the expression of multiple genes at different stages and different sites of infection. It is important, in future research, that investigators become increasingly self critical. They need to make a clear distinction between those molecules that serve a purely 'housekeeping' function that is equally essential for the fungus *in vitro* and *in vivo*, those that are expressed uniquely in infected tissues and interact directly with the host and those that regulate the expression of host-interactive virulence molecules. More widespread application of approaches such as IVET, transcript profiling of fungi *in vivo*, differential gene display and signature tagged mutagenesis are likely to facilitate understanding of fungal virulence at the molecular level.

Fusarium oxysporum and *Fusarium solani* are an asexual fungus that occurs in soils worldwide. As a species, they include morphologically indistinguishable pathogenic as well as non-pathogenic strains. The latter are defined as strains for which no host plants have been identified (yet). Pathogenic *F. oxysporum* and *Fusarium solani* strains can cause vascular wilt or root rot in over 100 plant species, among which are several economically important crops including banana, bulb flowers, cucumber, cutting flowers, date palm, melon, tomato and guava (Lievens et al. 2008, Gupta 2010a, b).

The interest in toxigenic *Fusarium* species is increasing world wide due to discovery of a growing number of naturally occurring *Fusarium* mycotoxin. The bean pathogen, *Fusarium solani* f.sp. *phaseoli* secretes a glycoenzyme, Kievitone hydratase, which catalyzes conversion of Kievitone to less toxic metabolites have been studied by Li et al. (1995) and revealed that Kievitone hydratase enzyme is virulence factor (KHS).

Therefore, pathogenic strains have been assigned to *formae speciales* based on host specificity and presently over 70 *formae speciales* have been described (Armstrong and Armstrong 1981). Most recently the research on virulence of *Fusarium* species causing wilt disease of guava in India was successfully done by Gupta et al. (2010a, b) both on field level pathogenicity study as well as *in vitro* amplication of virulent gene relation loci *viz.* Class V chitin synthase, FMK1 encoding mitogen activated protein kinase, Pectate lyase, Endo-polygalacturonase and Kievitone hydratase in the genome of *Fusarium* species causing wilt disease of guava in India (Gupta et al. 2010b).

APPLICATION OF RNAi IN EXPRESSION OF VIRULENCE IN *FUSARIUM* WILT PATHOGENS OF GUAVA

The best-studied and probably the most conserved RNA silencing pathway in fungi is the one thus far described in this review; here we call it, 'the RNAi pathway', which operates in vegetative (somatic) cells and uses siRNAs as specificity determinants. The RNAi pathway is known to be involved in transposon silencing in higher eukaryotes and that is also likely to be the case in fungi (Nolan et al. 2005, Murata et al. 2007). Another major RNA silencing pathway known in higher eukaryotes like in fungi is the miRNA pathway, in which miRNAs regulate the expression of endogenous mRNA post-transcriptionally via translational repression or mRNA degradation.

RNAi using a hairpin RNA (hpRNA)-expressing plasmid Although 'canonical' quelling is induced by the transformation with partial coding sequence that is homologous to an endogenous target (Romano and Macino 1992), more recently it has been shown that hpRNA-expressing constructs induce more efficient and stable silencing (Goldoni et al. 2004). The first example of fungal RNAi by an hpRNA-expressing plasmid was demonstrated by Liu et al. (2002). To date, plasmid constructs expressing hpRNA or intron-containing hpRNA (ihpRNA) are the most prevalent and reliable platforms to induce RNAi in fungi (Kadotani et al. 2003, Takeno et al. 2004, Namekawa et al. 2005, de Jong et al. 2006). These types of vector have been successfully used to demonstrate RNAi using model genes and to explore gene function in a wide range of fungal species and fungus-like organisms, such as Oomycetes, Basidiomycetes (Van West et al. 1999, Latijnhouwers et al. 2004) and Myxomycetes (Martens et al. 2002).

RNAi using an opposing-dual promoter system with an ihpRNA-expressing plasmid is useful for inducing RNAi in fungi, its applicability is generally limited to small or moderate scale analyses since the construction of such vectors normally requires two steps of orientated cloning. When the highly expressed gene inf1 was targeted for RNAi, a significant reduction in inf1 mRNA expression was usually detected only 12–15 days after transfection. In mammalian systems, synthetic siRNA-directed RNAi is typically activated within hours or one to two days and remains effective for several days after transfection. Since the original trigger dsRNA molecules may not be intact two weeks after transfection, some signal amplification mechanism is likely to be involved in the apparent late occurrence of gene silencing in *Fusarium* spp. causing wilt of guava (Gupta et al. 2011; *In press*). Advantages and disadvantages of RNAi as a genetic tool in fungal biology Knockout or knockdown?

Compared with conventional approaches for the analysis of such genes (such as the use of temperature-sensitive mutants), RNAi offers a more convenient and effective tool, especially by the combination with an inducible promoter that would allow gene expression to be diminished at specific stages during development (Bromley et al. 2006).

Recent evidence indicates that noncoding small RNAs play critical roles in a greater variety of cellular processes than was once assumed. RNA-mediated gene silencing in eukaryotes include regulation of endogenous gene expression, transposon silencing, heterochromatin formation, programmed DNA elimination and maintenance of genome stability in reproductive cells. Even within the kingdom fungi, RNA silencing pathways appear to have diversified signifi-cantly because the numbers of RNA silencing proteins differ considerably among fungal species (Jackson et al. 2006) (Figure 1). However, relatively little is known about the physiological roles of the fungal RNA silencing pathways.

Even though the molecular mechanisms leading to the developmental defects in the fungi are unknown, the fungal dicers appear to have a role in more than defense against viruses and transposable elements. The presence of dsRNAs has been reported in several *Fusarium* species, including *F. poae* and *F. solani* f. sp. *robiniae*, *Fusarium proliferatum* strain NFCCI 2053, *Fusarium chlamydosporum* isolate VKGFC01, *Fusarium solani* strain VKGFS1 & S2, *Fusarium oxysporum* f. sp. *psidii* strain VKGFO1 & FO2 (Compel et al. 1999, Fekete et al. 1995, Nogawa et al. 1993, 1996, Gupta et al. 2011). However, no morphological alterations or signs of degeneration have been observed to be caused by any of these dsRNA-containing isolates. Recently, we provided strong evidence that the dsRNA containing *Fusarium proliferatum* strain NFCCI 2053, *Fusarium chlamydosporum* isolate VKGFC01, *Fusarium solani* strain VKGFS1 & S2, *Fusarium oxysporum* f. sp. *psidii* strain VKGFO1 & FO2 displays alterations of several virulence-associated parameters, including reduced/high level mycelial growth, increased pigmentation and reduced/high level virulence on guava plants (Gupta et al. 2010a, b). The lack of sequence data impedes the

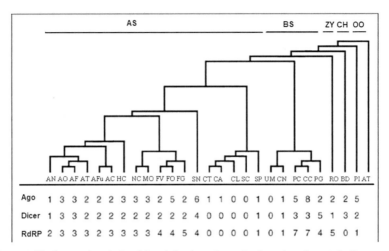

	AN	AO	AF	AT	AFu	AC	HC	NC	MO	FV	FO	FG	SN	CT	CA	CL	SC	SP	UM	CN	PC	CC	PG	RO	BD	PI	AT
Ago	1	3	3	2	2	2	3	3	3	2	5	2	6	1	1	0	0	1	0	1	5	8	2	2	2	5	
Dicer	1	3	3	2	2	2	2	2	2	2	2	4	0	0	0	0	1	0	1	3	3	5	1	3	2		
RdRP	2	3	3	3	2	3	3	3	3	4	4	5	4	0	0	0	0	1	0	1	7	7	4	5	0	1	

Figure 1. Phylogenetic relationship of the fungal species based on beta-tubulin sequences and the numbers of Argonaute (Ago)-, Dicer- and RNA-dependent RNA polymerase (RdRP)-like proteins predicted in the genome sequences. The amino acid alignment of the fungal beta-tubulin genes was generated by Clustal W (http://www. ddbj.nig.ac.jp/search/ clustalw-e.html) with the default parameter settings (Jackson et al. 2006, Nakayashiki and Nguyen 2008). AS, ascomycete; BS, basidiomycetes; ZY, zygomycete; OO, oomycete; AC, *Aspergillus clavatus*; AFl, *Aspergillus flavus*; AFu, *Aspergillus fumigatus*; AN, *Aspergillus nidulans*; AO, *Aspergillus oryzae*; AN, *Aspergillus terreus*; CA, *Candida albicans*; CC, *Coprinus cinereus*; CL, *Candida lusitaniae*; CN, *Cryptococcus neoformans*; CT, *Candida tropicalis*; FG, Fusarium graminearum; FO, Fusarium oxysporum; FV, *Fusarium verticillioides*; HC, *Histoplasma capsulatum*; MO, *Magnaporthe oryzae*; NC, *Neurospora crassa*; PC, *Phanerochaete chrysosporium*; PG, *Puccinia graminis*; PI, *Phytophthora infestans*; RO, *Rhizopus oryzae*; SC, *Saccharomyces cerevisiae*; SN, *Stagonospora nodorum*; SP, *Schizosaccharomyces pombe*; UM, *Ustilago maydis*; AT, *Arabidopsis thaliana*.

understanding of the molecular biology of *Fusarium* spp. strain NFCCI 2053, VKGFC01, VKGFS1 & S2, VKGFO1 & FO2 and its taxonomic relationships with other potentially related dsRNA *Fusarium* spp. Here, we report the partial nucleotide sequence and genome organization of *Fusarium proliferatum* strain NFCCI 2053, *Fusarium chlamydosporum* isolate VKGFC01, *Fusarium solani* strain VKGFS1 & S2, *Fusarium oxysporum* f. sp. *psidii* strain VKGFO1 & FO2. Examination of the these RNA sequence revealed conserved sequences of 20 nucleotides (nt) upstream of the open reading frame 1 (ORF1) and ORF3initiation codons.

Although the molecular mechanism by which *Fusarium* spp. strain NFCCI 2053, VKGFC01, VKGFS1 & S2, VKGFO1 & FO2 increases the virulence of *Fusarium* spp. is unknown, we have previously shown that the presence of NFCCI 2053, VKGFC01, VKGFS1 & S2, VKGFO1 & FO2 causes pronounced morphological changes, including reduced mycelial growth, increased pigmentation, increased wilting system and virulence towards decline physiology and metabolism of guava plant (Gupta et al. 2009,

2010a,b, 2011). The association between NFCCI 2053, VKGFC01, VKGFS1 & S2, VKGFO1 & FO2 infection and fungal virulence has been confirmed by hyphal fusion. The NFCCI 2053, VKGFC01, VKGFS1 & S2, VKGFO1 & FO2 dsRNA was able to be transferred to non-pathogenic strains and the recipient strain acquired the virulence associated phenotype of the donor strain, suggesting that the dsRNA of these *Fusarium* spp. encodes hyper virulence traits are related to successful amplification of virulence factor gene *viz.* . Class V chitin synthase, FMK1 encoding mitogen activated protein kinase, Pectate lyase, Endo-polygalacturonase and Kievitone hydratase in the genome of *Fusarium* species isolates of guava wilt. Therefore, it can be said that these disease related loci are present in the genome of Fop and Fs isolates are responsible for wilt producing character/expression in host plant (Gupta et al. 2010a, b, 2011).

CONCLUDING REMARKS

RNAi related sequence analyses of virulence locus demonstrated that they are unique to pathogenic forms of this vascular wilt pathogen. It implies that they play a role in pathogenesis. Differential expression of virulence genes upon infection of plant hosts, therefore, should in future be investigated. Such specific virulence proteins produced by pathogens can be bound for developing transgenic plants resistant to disease. For example, plant polygalacturonase inhibitor proteins (PGIPs) can bind fungal PGs. Inhibition specificities and kinetics, however, might vary within and among species.

   RNAi is a potentially powerful tool for a wide variety of gene silencing applications. RNAi is not limited to a convenient, although restricted, alternative to gene disruption. RNAi should prove useful for the studies of functional genomics in fungi in that it can provide novel and rapid gene analysis applications that gene disruption methods cannot offer. To make RNAi a better tool for gene function analysis in *Fusarium* spp., the further studies will be characterization of *Fusarium* spp. cells and to develop an inducible RNAi system with a combination of a strictly controlled promoter and a convenient inducer applicable to control the virulence related traits.

REFERENCES

Armstrong, G.M. and J.K. Armstrong. 1981. *Formae speciales* and races of *Fusarium oxysporum* causing wilt diseases, *In Fusarium: Disease, Biology and Taxonomy*, ed. by Nelson P.E., Toussoun T.A. and Cook R.J. Pennsylvania State University Press, University Park, PA, pp. 391–39.

Bromley, M., C. Gordon, N. Rovira-Graells and J. Oliver. 2006. The *Aspergillus fumigatus* cellobiohydrolase B (cbhB) promoter is tightly regulated and can be exploited for controlled protein expression and RNAi. *FEMS Microbiol. Lett.* **264**:246–254.

Compel, P., I. Papp, M. Bibo, C. Fekete and L. Hornok. 1999. Genetic interrelationships and genome organization of double-stranded RNA elements of *Fusarium poae*. *Virus Genes*. **18**:49–56.

de Jong, J.F., H.J. Deelstra, H.A. Wosten and L.G. Lugones. 2006. RNA-mediated gene silencing in monokaryons and dikaryons of *Schizophyllum commune*. *Appl. Environ. Microbiol.* **72**:1267–1269.

De Wit, P.M.G.M. 1992. Molecular characterization of gene-for-gene systems in plant-fungus interactions and the application of avirulence genes in control of plant pathogens. *Ann. Rev. Phytopathol.* **30**:391–418.

Di Pietro, A. and M.I.G. Roncero. 1998. Cloning, expression and role in pathogenicity of *pg1* encoding the major extracellular endopoly-galacturonase of the vascular wilt pathogen of *Fusarium oxysporum*. *Molecular Plant-Microbe Interat.* **11**:91–98.

Di Pietro, A., F.L. García-Maceira, E. Méglecz, J.F. Gutierrez-Corona and M.I.G. Roncero. 2001. A MAP kinase of the vascular wilt fungus *Fusarium oxysporum* is essential for root penetration and pathogenesis. *Molecular Microbiol.* **39**:1140–1152.

Falkow, S. 1988. Molecular Koch's postulates applied to microbial patho-genicity. *Rev. Infect. Dis.* **10**:S274–S276.

Fekete, C., G. Giczey, I. Papp, L. Szabo and L. Hornok. 1995. High-frequency occurrence of virus-like particles with double-stranded RNA genome in *Fusarium poae*. *FEMS Microbiol. Lett.* **131**:295–299.

Fonzi, W. and M. Irwin. 1993. Isogenic strain construction and gene mapping in *Candida albicans*. *Genetics*. **134**:717–728.

Goldoni, M., G. Azzalin, G. Macino and C. Cogoni. 2004. Efficient gene silencing by expression of double stranded RNA in *Neurospora crassa*. *Fungal Genet. Biol.* **41**:1016–1024.

Gupta, V.K., A. Manimaran and K.M. Turner. 2010b. Current Research Advancements on Wilt Disease of *Psidium guajava* L. LAP Lambert Academic Publishing, Germany. pp. 261

Gupta, V.K., R. Gaur, S.K. Yadava and N.S. Darmwal. 2009. Optimization of xylanase production by free and immobilized cells of *Fusarium solani* F7. *Bioresources*. **4**:932–945.

Gupta, V.K., A.K. Misra, R.K. Gaur, P.K. Jain, D. Gaur and S. Sharma. 2010a. Current Status of *Fusarium* Wilt Disease of Guava (*Psidium guajava* L.) in India. *Biotechnol.* **9**:176-192.

Gupta, V.K., J. vander Waals, A.K. Misra, P.K. Jain and R.K. Gaur. 2011. Virulence factor of *Fusarium* wilt pathogen of guava in relation to RNAi mediated ORF1 and ORF3 gene region. Molecular Plant pathology (Under Publication)

Jackson, A.L., J. Burchard, J. Schelter, B.N. Chau, M. Cleary, L. Lim and P.S. Linsley. 2006. Widespread siRNA "off-target" transcript silencing mediated by seed region sequence complementarity. *RNA*. **12**:1179–1187.

Kadotani, N., H. Nakayashiki, Y. Tosa and S. Mayama. 2003. RNA silencing in the phytopathogenic fungus *Magnaporthe oryzae*. *Mol. Plant. Microbe. Interact.* **16:**769–776.

Latijnhouwers, M., W. Ligterink, V.G.A.A. Vleeshouwers, P.V. West and F. Govers. 2004. A Galpha subunit controls zoospore motility and virulence in the potato late blight pathogen *Phytophthora infestans*. *Mol. Microbiol.* **51:**925–936.

Li, D., K.R. Chung, D.A. Smith and C.L. Schardl. 1995. The *Fusarium solani* gene encoding *kievitone hydratase*, a secreted enzyme that catalyzes detoxification of a bean phytoalexin. *Molecular Plant-Microbe Interaction.* **8:**388–97.

Lievens, B., R. Martijn and B.P.H.J. Thomma. 2008. Recent developments in the molecular discrimination of *formae speciales* of *Fusarium oxysporum*. *Pest Manag Sci.* **64:**781–788

Liu, H., T.R. Cottrell, L.M. Pierini, W.E. Goldman and T.L. Doering. 2002. RNA interference in the pathogenic fungus Cryptococcus neoformans. *Genetics.* **160:**463-470.

Martens, H., J. Novotny, J. Oberstrass, T.L. Steck, P. Postlethwait W. Nellen. 2002. RNAi in *Dictyostelium*: the role of RNA-directed RNA polymerases and double-stranded RNase. *Mol. Biol. Cell.* **13:**445–453.

Misra, A.K. and V.K. Gupta. 2007. Variability in *Fusarium solani*—a causal organism of wilt of Guava. *CISH Newsletter.* **8:**2.

Misra, A.K. 2006. Wilt of guava-a disease of national importance. *Indian Phytopathol.* **59:**269.

Murata, T., N. Kadotani, M. Yamaguchi, Y. Tosa, S. Mayama and H. Nakayashiki. 2007. siRNA-dependent and -independent posttranscriptional co-suppression of the LTR-retrotransposon MAGGY in the phytopathogenic fungus *Magnaporthe oryzae*. *Nucleic Acids Res.* **35:**5987–5994.

Nakayashiki, H. and Q.B. Nguyen. 2008 RNA interference: roles in fungal biology. *Current Opinion in Microbiology.* **11:**494–502.

Namekawa, S.H., K. Iwabata, H. Sugawara, F.N. Hamada, A. Koshiyama, H. Chiku, T. Kamada and K. Sakaguchi. 2005. Knockdown of LIM15/DMC1 in the mushroom *Coprinus cinereus* by doublestranded RNA-mediated gene silencing. *Microbiology.* **151:**3669–3678.

Nogawa, M., S.T. Kageyama, A. Nakatani, G. Taguchi and M. Shimosaka et al. 1996. Cloning and characterization of mycovirus double-stranded RNA from the plant pathogenic fungus, *Fusarium solani* f. sp. *robiniae*. *Biosci. Biotechnol. Biochem.* **60:**784–788.

Nogawa, M., S.T. Kageyama and M. Okazaki. 1993. A double-stranded RNA mycovirus from the plant pathogenic fungus, *Fusarium solani* f. sp. *robiniae*. *FEMS Microbiol. Lett.* **110:**153–158.

Nolan, T., L. Braccini, G. Azzalin, A. De Toni, G. Macino and C. Cogoni. 2005. The post-transcriptional gene silencing machinery functions independently of DNA methylation to repress a LINE1-like retrotransposon in *Neurospora crassa*. *Nucleic Acids Res.* **33:**1564–1573.

Odds, F.C., N.A.R. Gow and A.J.P. Brown. 2001. Fungal virulence studies come of age. *Genome Biology.* **2:**1009.1–1009.4.

Ondøej, M., R. Dostálová and R. Trojan. 2008. Evaluation of virulence of *Fusarium solani* isolates on pea. *Plant Protection Science.* **44**:9–18.

Romano, N. and G. Macino. 1992. Quelling: transient inactivation of gene expression in *Neurospora crassa* by transformation with homologous sequences. *Mol. Microbiol.* **6**:3343–3353.

Takeno, S., E. Sakuradani, S. Murata, M. Inohara-Ochiai, H. Kawashima, T. Ashikari and S. Shimizu. 2004. Establishment of an overall transformation system for an oil-producing filamentous fungus, *Mortierella alpina* 1S-4. *Appl. Microbiol. Biotechnol.* **65**:419–425.

Van West, P., S. Kamoun, J.W. van't Klooster and F. Govers. 1999. Internuclear gene silencing in *Phytophthora infestans. Mol. Cell.* **3**:339–348.

RNA Interference—Interactomics and Therapeutics

Sayak Ganguli and Abhijit Datta*

ABSTRACT

Evolutionarily, RNA interference (RNAi) is a highly conserved process of post-transcriptional gene silencing (PTGS) by which double stranded RNA (dsRNA), when introduced into a cell, causes sequence-specific degradation of specific targeted mRNA sequences. The main focus area of research over the past decade has been to identify miRNA and other small RNA genes in different organisms. Apart from basic experimental approaches computational algorithms have also been applied to screen the genes and store them in libraries and databases. However study of RNAi interactions has been a neglected area, though some basic sequence pattern identification and molecular modelling experiments have been done, basic study of RNA–Protein interactions are yet to be fully explored. Specific interacting residues of dicer and argonaute proteins with the cognate miRNAs were detected using molecular dynamic simulations. The presence of a residue bias indicates that sequence specificity and identity are essential for the proper functioning of the RNA Induced Silencing Complex (RISC). Recent years have also witnessed an increase in research activities using RNAi in therapeutics. However, there still remain certain weak areas which need significant exploration through both conventional as well as bioinformatics approaches.

Keywords: RNA interference, RNA induced silencing complex, interactomics.

INTRODUCTION

Gene silencing by the conserved RNAi-pathway as first described in *Caenorhabditis elegans* is initiated by the introduction of double-stranded RNAs (dsRNA), resulting in sequence-specific degradation of homologous endogenous mRNA (Almeida and Allshire 2005). Later, RNAi has been shown to function in a similar manner in every metazoan and has been

---

* Corresponding author *e-mail*: abhijit_datta21@yahoo.com

applied to study a wide variety of phenotypes *in vivo* and in cells. Experiments in *C. elegans* and *Drosophila* primarily make use of long dsRNAs, which are intracellularly diced into functional 21mer short interfering RNAs (siRNAs). However, long dsRNAs are not effective in most mammalian cells due to the induction of antiviral pathways that lead to host cell shutdown. This can be counteracted by transfecting synthetic or plasmid-encoded siRNA that in most cases do not elicit an interferon or other host cell response (Amarzguioui and Prydz 2004). These approaches have been effectively used to study many different biological pathways in loss-of-function analysis and several large-scale efforts have recently generated libraries that target every predicted gene in major model organisms and humans.

RNAi is a gene-silencing technique used in studying the absence of normal gene action by disrupting its activity *in vivo*. RNAi can help to determine gene function by blocking the expression of a specific messenger RNA (mRNA). By simply, effectively and specifically down-regulating the expression of genes in mammalian cells, researchers can develop new diagnostic and therapeutic strategies (Bartel 2004).

- **siRNA—Small interfering RNA (20–24 nt);**
- **miRNA—Micro RNA (20–24 nt);**
- **rasiRNA—Repeat-associated small interfering RNA (24–29 nt);**
- **piRNA—Piwi-interacting RNA (26–31 nt);**
- **tasiRNA—Trans-acting RNA (24–29 nt) in plant.**

RNAi LIBRARIES

Large-scale RNAi library construction involves the conglomeration of several strategies. Some of the approaches that have been pursued rely on the use of long dsRNA, chemically synthesized siRNAs or expression of short-hairpin RNAs (shRNAs). *C. elegans* and *Drosophila* RNAi libraries are mainly based on long dsRNAs, which are either synthesized *in vitro* or through experiments in *C. elegans* expressed in *Escherichia coli* that are fed to worms and elicit efficient gene silencing. Several libraries that target the majority of genes in *C. elegans* or *Drosophila* have become available and are in part distributed through public resource centres. In mammals, *in vitro* processing of long dsRNA by recombinant RNaseIII or Dicer can be used to generate pools of siRNA that provide effective target oriented silencing effects without eliciting an interferon response. Chemically synthesized siRNAs have higher efficiencies of transfection, but are relatively expensive and can result in non-specific off-target effects. In addition, vector-based expression of shRNAs has been used to silence gene expression from transiently or stably transfected vectors (Bertrand et al. 2002). This approach can also be adapted to lenti- or retroviral vectors

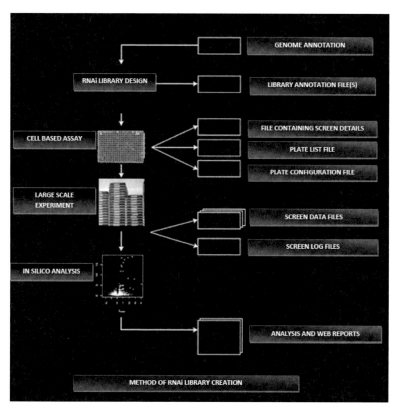

Figure 1: The method of RNAi library creation.

(Color image of this figure appears in the color plate section at the end of the book.)

to silence gene expression in cells that are otherwise difficult to transfect (Heale et al. 2005).

Both siRNA and shRNA libraries have been successfully used to screen human cells for various phenotypes. Several computational tools have been developed for the assessment of efficiency and specificity of RNAi probes, which provide means to evaluate or to rationally design long dsRNAs and siRNAs. As algorithms to predict efficient and specific siRNA sequences will be further improved, it is likely that currently available libraries will be modified and updated also to include changes in genome annotations.

DISCOVERING RNAi GENES AND THE ROLE OF BIOINFORMATICS

Cloning was the initial technique used for large scale microRNA discovery. However cloning can be useful for only those genes that are expressed in large amounts across a wide variety of tissue types. Those microRNAs that are expressed in low amounts and in response to a shift in environmental optima cannot be isolated with cloning. Bias based on sequences in cloning procedures can also result in some members being missed (Lewis et al. 2005).

Recent works with Dicer mRNAs have shown that they possess multiple catalytic motifs in their sequences indicative of a prebiotic origin (Ganguli et al. 2010). Molecular models of dicer generated using computational modelling methods reveals a basal helix loop-helix domain and as evolution has occurred the complexity of the structure has gradually increased with the addition of pleated sheets (Dhar et al. 2010).

Under such circumstances bioinformatics based approaches provide a suitable alternative to detect microRNA genes. The basic approach for bioinformatics based gene identification has been the detection of homologues, of known miRNAs, within the same genome and in the genome of other species (Amarzguioui and Prydz 2004). However, a more strenuous task is to identify miRNA genes which are novel, i.e. which are unrelated to previously known miRNAs. Micro RNAs of vertebrates, nematode and fly have been successfully detected using algorithms that detect sequence conservation and secondary structure conservation across a wide species range. However, the most successful approaches should involve machine learning techniques, which use the feature classifiers as filters to screen for the optimum match. For example, several experimental studies support that the first 8 nucleotides in the 5′ end of the microRNAs are particularly important for target site recognition. First 8 nucleotides sequences in the 5′ end of the microRNA sequences are converted to 'Compliment' RNA sequence. This motif can be used to search for target sites in the genes. First, all nucleotides in the motif must match, for it to be recognized as a target site, but some target sites do not require a complete match of all eight nucleotides (Marques and Williams 2005). The mutual target site can then be further aligned with each other (miRNA22 :: mRNA35) with the help of a suitable dynamic programming local alignment. However, it must be made sure that this alignment scores remain, based on sequence complementarities and not on sequence identity. In other words, the check looks for A:U (A:U=5) and G:C (G:C=5) matches and all other base pairs (mismatches) = –3 etc. The G:U (G:U=1) wobble pair is also permitted.

This miRNA::mRNA duplex can then be further classified with Support Vector Machine (SVM)-light for a post processing filtering. SVMs are a class of supervised learning algorithms and have been shown to produce superior results than other supervised learning methods in a wide range of applications. Given a set of labeled training feature vectors (in this case, the positive and the negative miRNA; target interaction pairs) an SVM learns to discriminate between the two classes. The result is a trained model that can be used to classify unlabeled inputs. This is a preferred behavior as it would be possible to rank the query samples according to their likelihood of being true positives.

Further approaches using bioinformatics techniques include the designing of antagomirs—antisense oligonucleotides which silence targeted siRNAs. These can be utilized to silence harmfull miRNAs or siRNAs which are implicated in cancer phenotypes. (Krutzfeldt et al. 2005).

INTERACTOMICS OF RNAi

A neglected area of research has been the study of RNA–RNA and RNA–Protein interactions or the study of RNA Induced Silencing Complex components and how they interact with each other. It is known that Argonautes interact with the target miRNA with the help of their PAZ domains; some however in lower groups do so with the help of their PIWI domains. Experimental methods have not been successful so far to identify specific residues which play essential role in identification and interaction of such complexes. Molecular dynamics simulation using flexible docking and fast Fourier transform can be utilized to identify the interacting sites and to specifically identify the residues which are essential for suitable interactions.

Using the workflow as shown above it was found that there was indeed residue conservation in both Argonaute and Dicer proteins, when they were involved in a complex with the miRNA. Below are certain interacting residues in different colours (Table 1: Color scheme used) Red indicates Dicer (RNASEIII) while blue is the miRNA.

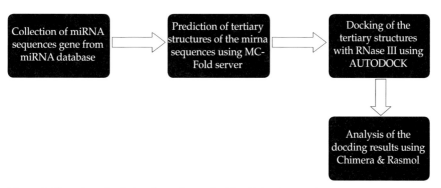

Figure 2: Flow chart depicting the pathway for identifying interacting partners.

Table 1: Color schemes used.

Amino Acids	Color
Arginine (Arg)	Pink
Aspargine (Asp)	Gray
Glutamic Acid (Glu)	Yellow
Histidine (His)	White
Methionine (Met)	Brown
Phenylalanine (Phe)	Purple
Proline (Pro)	Green
Tryptophan (Trp)	Orange

From the above work it is found that amino acids of the Rnase III, which have specific binding positions are—arginine (arg) at 708, aspargine (asn) at 721, glutamic acid (glu) at 726, histidine (his) at 802, methionine (met) at 722, phenylalanine (phe) at 182, proline (pro) at 180, tryptophan (trp) at 725, glutamine (gln) binds specifically at 186 and serine (ser) at 184.

Thus from the results we may conclude that RNASE III—miRNA interactions takes place at specific sites designated by a specific set of amino acids. This leads us to believe that if the pathway of RNA interference is required to be controlled or manipulated in future endeavours, these are the specific sites where SDM's could be induced and non specific amino acids could be incorporated, which might lead to an aberrant interactive pathway and thus create enough opportunities for manipulating the interference mechanisms.

A similar work was done using the Argonaute protein and its interactions with the target miRNAs. In this work also there has been an indication that sequence and residue specificity is an essential criterion in formation and maintenance of these complexes.

From all the observations it may be inferred that neutral but slightly polar amino acids like cysteine and tryptophan are completely absent from the interacting surface of Argonaute proteins, while asparagine which is polar and neutral is absent from the groove of the proteins. Neutral and non polar amino acids like proline, methionine and phenylalanine are the most abundant and conserved residues. Future endeavours using site directed mutagenesis (SDM's) etc. can be used at these specific residues to cripple and exert a level of control over the RNAi machinery of the cell.

RNAi AND THERAPEUTICS

Most bioactive drugs which rely on *in vivo* processing to exert their action face risk of saturating the intrinsic metabolic pathways, hence perturb the metabolome as a whole. RNAi triggers, which are externally

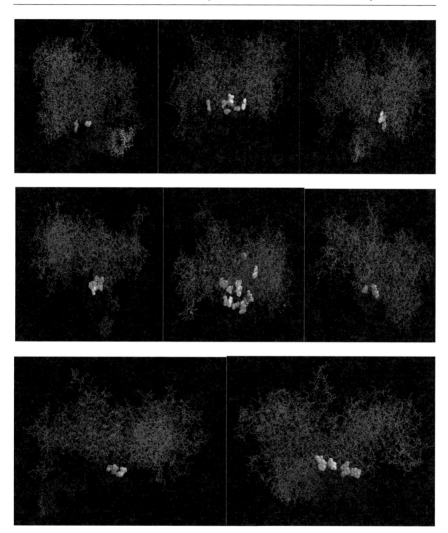

Figure 3: Different models of miRNA—DICER complexes.

(Color image of this figure appears in the color plate section at the end of the book.)

induced, behave in a similar manner since siRNA/shRNA relies on the endogenous microRNA machinery for identification and silencing of their targets (Sorensen et. al., 2003). Although our understanding of the natural role of RNAi in mammalian cells have expanded tremendously with the discovery of miRNA, other facets of this highly conserved biological system still remains to be discovered. Research with miRNA clearly indicates that they play important role for balancing gene activity.

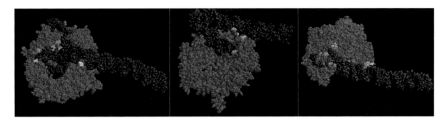

Figure 4: Argonaute (red) and miRNA (blue) interactions.

(Color image of this figure appears in the color plate section at the end of the book.)

Examples include the implication of miRNAs as oncogenes or as tumor suppressor genes. Comprehensive and conserved RNAi machinery is clearly essential for mammalian cells which have been established by the detection of lethality in early embryonic DICER knockouts.

STIMULATION OF INNATE IMMUNE RESPONSES

It was generally believed that double stranded RNA would induce innate responses in mammalians that would lead to shut down of cellular protein synthesis and eventually leads to cell death. However, mammalian cell could be transfected with dsRNA less than 30 nucleotides without inducing cellular toxicity. Long dsRNA has been known to rapidly induce interferon responses by binding to double-stranded-RNA-activated protein kinase (PKR), several Toll-like receptors (TLRs) or 22, 52-oligoadenylate synthetase-RNase L system. However there has been an evidence of certain siRNA sequence motifs invoked TLR7-dependent immune stimulation. A particular sequence motif (5′-GUCCUUCAA-3′) seems to be recognized by TLR7 in the endosomal compartment of plasmacytoid dendritic cells and this is responsible for activating immune responses. An interesting coinage—'Danger Motif', to describe GU-rich regions that has been shown to stimulate innate immune response leading to secretion of inflammatory cytokines in a particular cell-type in a sequence-specific manner have also been identified (Marques and Williams, 2005). While induction of the immune system could be beneficial in some clinical trials, these results have raised some doubts for the safe therapeutic use of RNAi. While knowledge of immune stimulatory properties requires research to proceed to animal models, *in vitro* use of human primary cells with a full repertoire for immune stimulation are needed. Despite these promising results great care and thorough testing are clearly needed before proceeding to clinical use. Notably, use of longer (27–29 bp) siRNAs (more potent Dicer substrates) should be carefully weighed with the risk of increased immune stimulation.

FUTURE IMPLICATIONS

The discovery of RNA interference in 1998 was a turning point in the history of molecular biology. Since its discovery it has become a powerful tool for studies of gene function in mammals. It is an important tool that provides the researcher with the ability to silence virtually any gene with artificial triggers of RNAi and utilizing the cellular machinery for efficient targeting of complementary transcripts. Both viral and non-viral delivery methods can provide selective and potent target gene suppression without any clear toxic effects as has been demonstrated in several clinical trials. Translational research using breakthroughs in basic RNAi research has taken place at warp speed and already there are several RNAi based human clinical trials in progress. There have also been a large number of animal model studies, including systemic delivery to non-human primates, but certain grey areas still remain which needs attention before establishing RNAi based therapeutics as the million dollar baby. These include off-target effects, triggering of type I interferon responses, competition with cellular RNAi components and effective delivery *in vivo*. Understanding of the basic mechanism is also vital to effective applications in human disease.

Bioinformatics and computational biology has immense role to play in the identification and characterization of novel RNAi genes in humans and its homologues. Gene identification and target site accessibility are the two most important areas on which the core bioinformatics research is now focused on. The use of machine learning techniques and the integration of biological information to analyze, screen and formulate new methods for such tasks are of paramount importance.

Given the pace of new findings and discoveries of applications, it can be anticipated that RNAi will be a major therapeutic alternative within the next several years. Hopefully RNAi will have the impetus to revolutionize the treatment of human disease and alter the threats to human existence.

REFERENCES

Almeida, R. and R.C. Allshire. 2005. RNA silencing and genome regulation. *Trends Cell. Biol.* **15**:251–258.

Amarzguioui M. and Prydz H. 2004. An algorithm for selection of functional siRNA sequences. *Biochem. Biophys. Res. Commun.* **316**:1050–1058.

Bartel D.P. 2004. MicroRNAs: genomics, biogenesis, mechanism and function. *Cell.* **116**:281–297.

Bertrand, J.R., M. Pottier, A. Vekris, P. Opolon, A. Maksimenko and C. Malvy. 2002. Comparison of antisense oligonucleotides and siRNAs in cell culture and *in vivo*. *Biochem. Biophys. Res. Commun.* **296**:1000–1004.

Dhar, P., S. Ganguli and A. Datta. 2010. Molecular modelling of Dicer and identification of Phosphorylation sites. *Bioinformation.* **4(9)**:412–416.

Ganguli, S.,S.K. Dey, P. Dhar, P. Basu, P. Roy and A. Datta. 2010. Catalytic RNA world relics in Dicer RNAs. *Intl. J. Genet.* **2(1)**:8–17.

Heale, B.S., H.S. Soifer, C. Bowers and J.J. Rossi, 2005. siRNA target site secondary structure predictions using local stable substructures. *Nucleic Acids Res.* **33**:e30.

Krutzfeldt, J., N. Rajewsky, R. Braich, K.G. Rajeev, T. Tuschl, M. Manoharan and M. Stoffel. 2005. Silencing of microRNAs *in vivo* with 'antagomirs' *Nature.* **438**:685–689.

Lewis, B.P., C.B. Burge and D.P. Bartel. 2005. Conserved seed pairing, often flanked by adenosines, indicates that thousands of human genes are microRNA targets. *Cell.* **120**:15–20.

Marques, J.T. and B.R. Williams. 2005. Activation of the mammalian immune system by siRNAs. *Nat. Biotechnol.* **23**:1399–1405.

Sorensen, D.R., M. Leirdal and M. Sioud. 2003. Gene silencing by systemic delivery of synthetic siRNAs in adult mice. *J. Mol. Biol.* **327**:761–766.

Index

Color Plate Section

Chapter 1

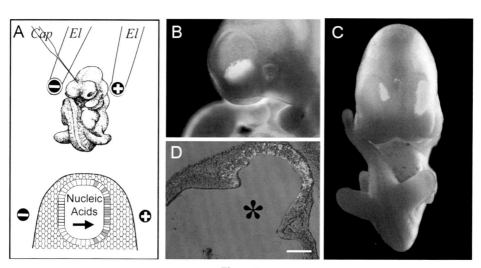

Figure 2

Chapter 2

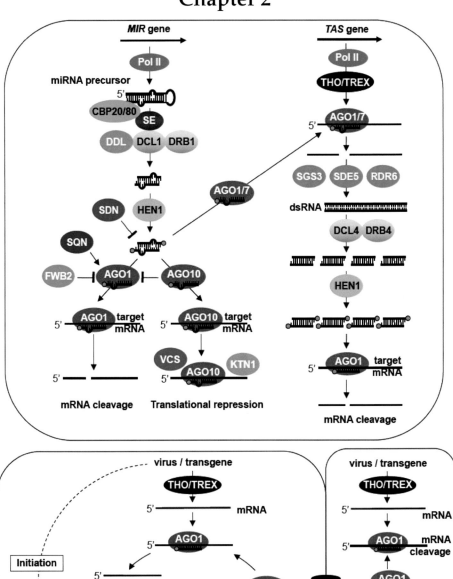

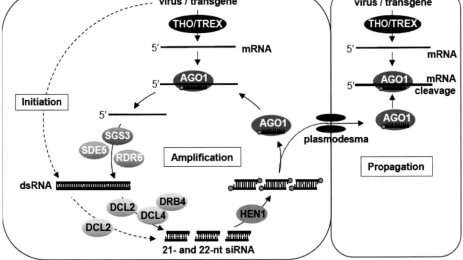

Figure 1

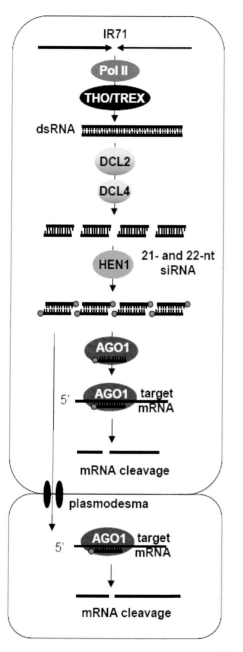
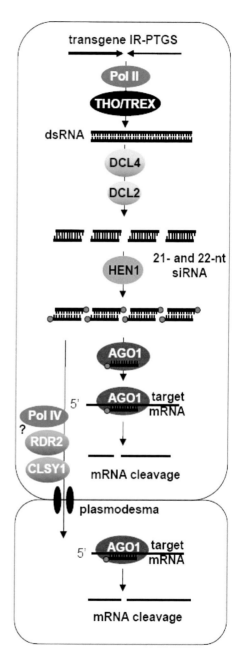

Figure 2

Chapter 4

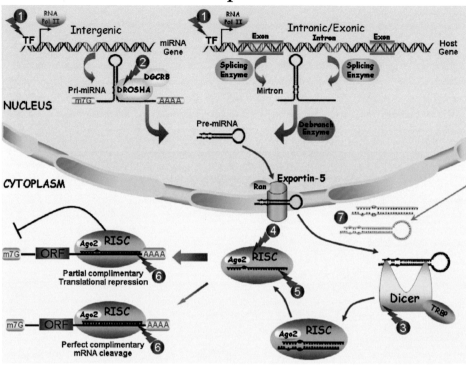

Figure 1

Chapter 5

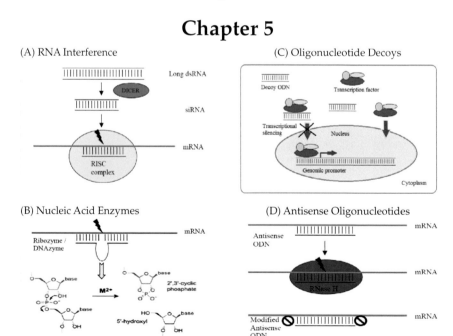

(A) RNA Interference

(C) Oligonucleotide Decoys

(B) Nucleic Acid Enzymes

(D) Antisense Oligonucleotides

Figure 1

Chapter 6

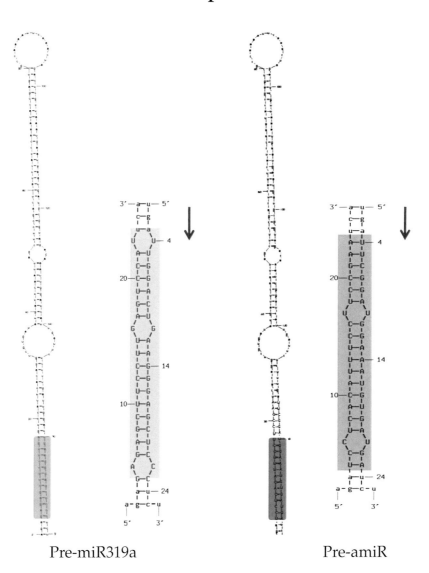

Pre-miR319a   Pre-amiR

Figure 1

Chapter 7

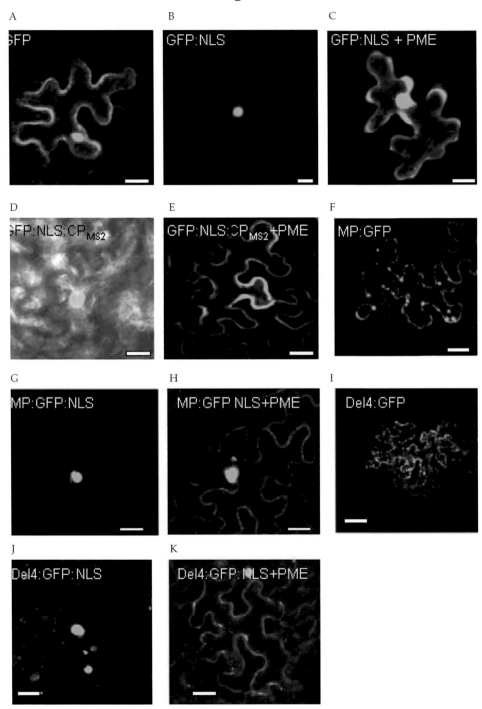

Figure 1

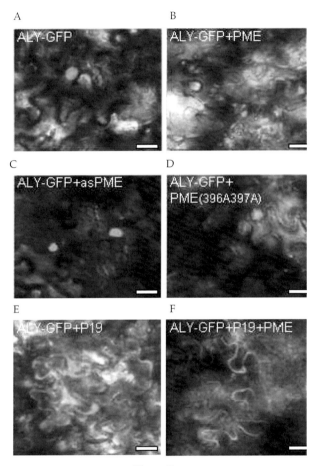

Figure 2

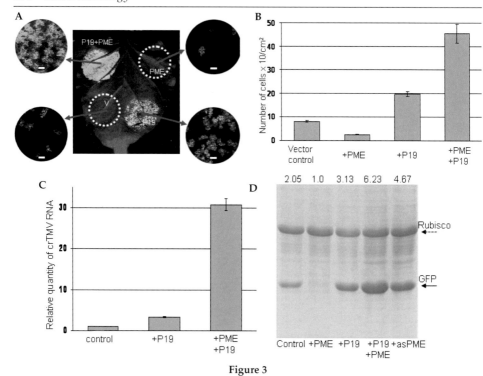

Figure 3

Chapter 8

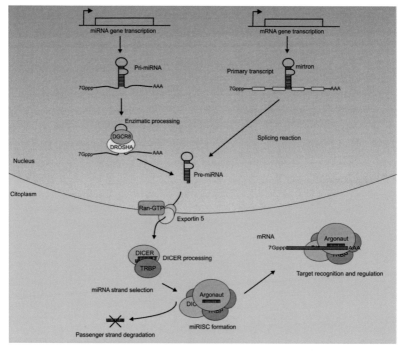

Figure 1

Chapter 9

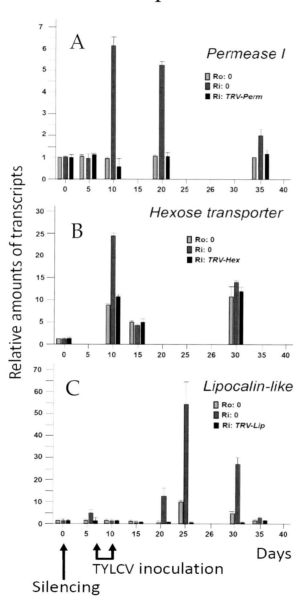

Figure 1

Chapter 10

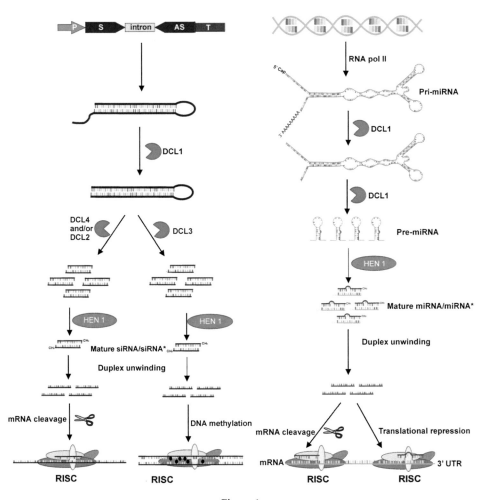

Figure 1

Figure 2

Chapter 15

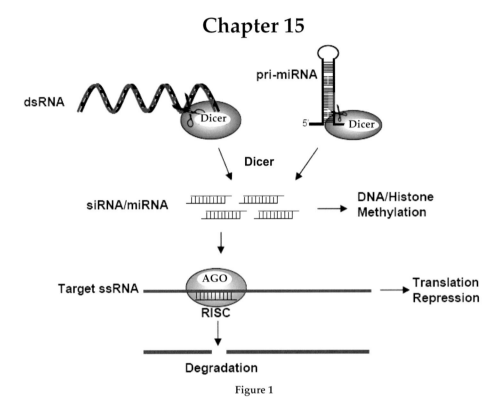

Figure 1

Chapter 16

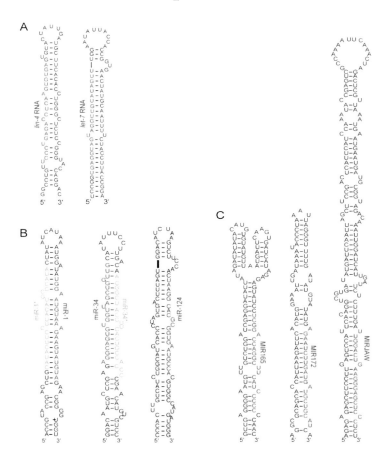

Figure 1

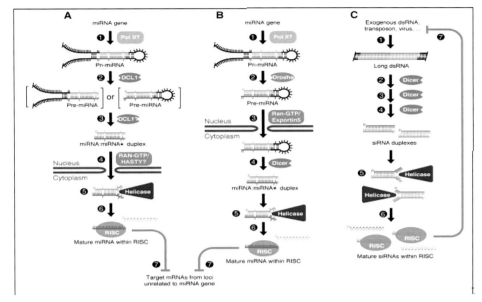

Figure 2

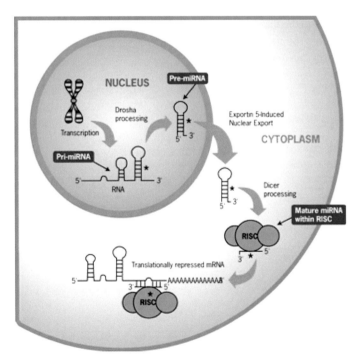

Figure 3

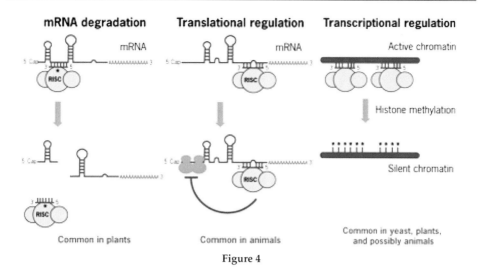

mRNA degradation **Translational regulation** **Transcriptional regulation**

Figure 4

Chapter 19

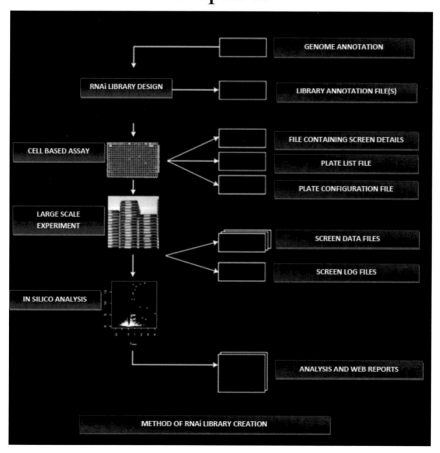

Figure 1

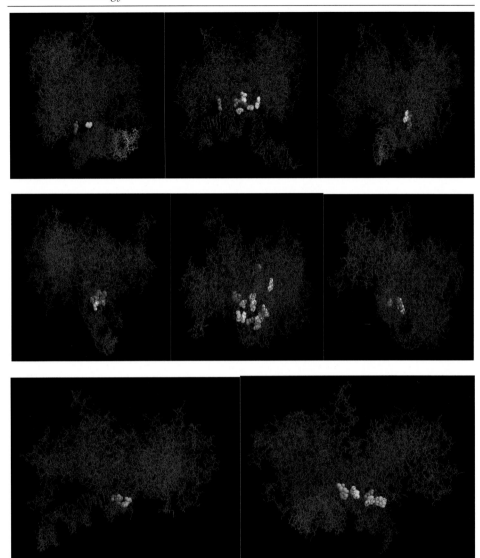

Figure 2

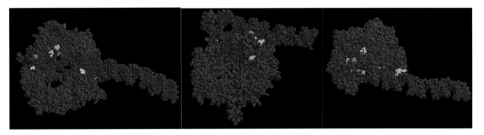

Figure 3

Printed and bound by CPI Group (UK) Ltd, Croydon, CR0 4YY

18/10/2024

01776270-0006